A FIRST COURSE IN PROBABILITY AND STATISTICS

A FIRST COURSE IN PROBABILITY AND STATISTICS

HENRICK J. MALIK
KENNETH MULLEN
University of Guelph

ADDISON-WESLEY PUBLISHING COMPANY
Reading, Massachusetts
Menlo Park, California · London · Amsterdam · Don Mills, Ontario · Sydney

Copyright © 1973 by Addison-Wesley Publishing Company, Inc. Philippines copyright 1973 by Addison-Wesley Publishing Company, Inc.

All rights reserved. No part of this publication may be reproduced, stored in a retrieval system, or transmitted, in any form or by any means, electronic, mechanical, photocopying, recording, or otherwise, without the prior written permission of the publisher. Printed in the United States of America. Published simultaneously in Canada. Library of Congress Catalog Card No. 72-1941.

ISBN 0-201-04413-7
ABCDEFGHIJ-MA-7987654

To Nancy and Toni

PREFACE

This book developed from the need for a text which emphasized equally, probability and statistics at the second-year level, and yet was flexible enough to serve separately, both as a probability text and as a statistics text.

The primary goals of the text are to acquaint students with the concepts of probability and ways of thinking needed to solve problems in probability and with the ideas and areas of applications of statistics. We have emphasized the use of examples, and these have covered a broad range of disciplines.

Emphasis has been placed on clarity and simplicity. Thus after a new concept has been discussed a formal definition is given. Theorems are similarly stated. Examples are profuse and many are worked in detail. At the end of each chapter there is a large number of exercises, emphasizing the concepts of the chapter. Many examples, begun in earlier chapters, are referred to later in order to illustrate new ideas. All definitions, theorems, examples, tables, and figures are numbered so that references can be made with ease.

No previous statistical background is necessary and no calculus-level mathematics is assumed, although a mature level of pre-calculus mathematics would be an asset.

The text contains sufficient material to allow for flexibility in the length of a course. There is undoubtedly more than enough material for a one-semester course. A semester course meeting three times a week should probably include Chapters 1, 2, and 3, Section 4.1, 4.2, and 4.3, Chapters 5, 6, 8, 9, 10, and 11.

At the University of Guelph, such a course is effectively taught with two or three hours of lectures per week (depending on the speed of the instructor) plus two hours of laboratory work.

We wish to thank all those who helped us in the preparation of this textbook. First of all, we express our deep gratitude to original thinkers of the past and the present, and the scholars who have written textbooks on probability and statistics. We gratefully acknowledge the contributions of Professors Byron Newton, of Oregon State University, Maurice Bryson, of Colorado State University, Evan Thweatt, of American River College, and Joyce Curry, of California

State Polytechnic College, who read the manuscript and provided comments, criticisms, and suggestions that have proved helpful in preparation of the final draft. The authors accept sole responsibility for any errors or omissions which may have been included. We are particularly thankful to Dr. T. D. Newton for his enthusiasm and encouragement, to Bovas Abraham for reading the final manuscript, to Mrs. Linda Selby for her careful and accurate typing, to Addison-Wesley Publishing company for its great help and cooperation, and to our wives and children for their patience and encouragement.

We are indebted to the Literary Executor of the late Sir Ronald Fisher, F.R.S., to Dr. Frank Yates, F.R.S., and to Oliver and Boyd, Edinburgh, for permission to reprint Table 9 from their book *Statistical Tables for Biological, Agricultural and Medical Research*; to Professor E. S. Pearson and the Biometrika Trustees for permission to use Tables 5, 6, 7, and 8 from *Biometrika Tables for Statisticians*, Vol. 1.

Guelph, Ontario, Canada H.J.M.
November 1972 K.M.

CONTENTS

Chapter 1 Introduction
  1.1 Statistics and probability 1
  1.2 The scope of the book 7
  1.3 The system of referencing 8

Chapter 2 Elementary Probability
  2.1 The notion of a set 9
  2.2 Operations on sets 11
  2.3 Sample spaces, sample points, and events 14
  2.4 Rules for counting sample points 17
  2.5 Probability of an event 23
  2.6 Some probability laws 26
  2.7 Conditional probability 28
  2.8 Independent events 30
  2.9 Bayes' theorem 31
  2.10 Summary 34

Chapter 3 Random Variables and Mathematical Expectation
  3.1 The notion of a random variable 40
  3.2 Distribution of a discrete random variable . . . 42
  3.3 Distribution of a continuous random variable . . 45
  3.4 The summation notation 48
  3.5 Two-dimensional random variables 51
  3.6 The expected value of a random variable 55
  3.7 Properties of expected value 58
  3.8 The variance of a random variable 63
  3.9 Properties of the variance of a random variable . 67
  3.10 Summary 69

Chapter 4 Some Discrete Probability Distributions
  4.1 Introduction 76
  4.2 The Bernoulli distribution 76
  4.3 The binomial distribution 77

4.4	The multinomial distribution	82
4.5	The geometric distribution	85
4.6	The negative binomial distribution	88
4.7	The hypergeometric distribution	91
4.8	The Poisson distribution	94
4.9	The uniform distribution	98
4.10	Summary	100

Chapter 5 The Normal Distribution

5.1	The normal distribution	106
5.2	The standard normal distribution	108
5.3	The normal approximation to the binomial distribution	116
5.4	Summary	119

Chapter 6 The Idea and Choice of a Sample

6.1	The concept of sampling	124
6.2	Selection of the sample	127
6.3	The use of random number tables	130
6.4	Stratified random sampling	132
6.5	Summary	132

Chapter 7 Organization and Analysis of Data

7.1	Introduction	135
7.2	Frequency tables	136
7.3	Graphical representation of the frequency table	141
7.4	Cumulative frequencies	143
7.5	Measures of location	144
7.6	Comparison of measures of location	150
7.7	Measures of dispersion	151
7.8	Summary	157

Chapter 8 Estimation

8.1	Introduction	162
8.2	Some properties of the sample mean and variance	163
8.3	Point estimation	166
8.4	Interval estimation of the population mean with known variance	168
8.5	Interval estimation of the population mean with unknown variance	172
8.6	Interval estimation of the population variance and standard deviation	175
8.7	Interval estimation of the difference of two normal means	178
8.8	Interval estimation of the ratio of two normal variances	183
8.9	Interval estimation of p in a binomial probability function	185
8.10	Interval estimation of the difference of two binomial parameters	187
8.11	Interval estimation of μ, the Poisson parameter	188
8.12	Summary	188

Chapter 9 Hypothesis Testing

- 9.1 Introduction 193
- 9.2 Testing simple hypotheses 194
- 9.3 Testing the mean of a normal population with known variance . 198
- 9.4 Testing the mean of a normal population with unknown variance 203
- 9.5 Comparison of confidence intervals and hypothesis testing . . 205
- 9.6 Testing the variance and standard deviation of a normal population 206
- 9.7 Testing the difference of two normal means 207
- 9.8 Testing the ratio of two normal variances 210
- 9.9 Testing the parameter of a binomial distribution 211
- 9.10 Testing the difference of two binomial parameters 212
- 9.11 Summary 213

Chapter 10 Tests of Significance Based on Chi-Square

- 10.1 Introduction 219
- 10.2 Goodness of fit test 223
- 10.3 Test of independence: contingency tables 226
- 10.4 Summary 231

Chapter 11 Regression and Correlation

- 11.1 Introduction 236
- 11.2 Linear regression 238
- 11.3 Partitioning the sum of squares and estimate of variance . . 244
- 11.4 Confidence intervals and tests of hypotheses 247
- 11.5 Multiple linear regression 252
- 11.6 Correlation 253
- 11.7 Significance of the coefficient of correlation 257
- 11.8 Summary 259

Chapter 12 The Analysis of Variance

- 12.1 Introduction 264
- 12.2 Notation 265
- 12.3 Testing equality of population means 266
- 12.4 Multiple comparisons 272
- 12.5 The analysis of variance with two factors 273
- 12.6 Fixed and random effects 277
- 12.7 Testing the equality of the variances 280
- 12.8 Summary 283

Chapter 13 Nonparametric Statistics

- 13.1 Introduction 289
- 13.2 Test of randomness: the runs test 290
- 13.3 Wilcoxon's test for paired data 292
- 13.4 Wilcoxon rank sum test 296
- 13.5 The Kruskall-Wallis test 298

13.6	Friedman's test for K related samples		299
13.7	Rank correlation		301
13.8	Summary		305

Appendix   Tables

Table 1	Squares and square roots	313
Table 2	Areas under the normal probability curve	314
Table 3	Random numbers	316
Table 4	Percentage points of the t-distribution	317
Table 5	Percentage points of the χ^2-distribution	318
Table 6a	95% significance points of the F-distribution	320
Table 6b	99% significant points of the F-distribution	322
Table 7a	Binomial confidence intervals with $1-\alpha=.95$	324
Table 7b	Binomial confidence intervals with $1-\alpha=.99$	325
Table 8	Poisson confidence intervals	326
Table 9	Values of Fisher's Z transformation: $Z=\tfrac{1}{2}\log_e (1+r)/(1-r)$	327
Table 10	Duncan's critical difference multipliers for $\alpha=.05$	328
Table 11	Transformation of percentage to arcsin $\sqrt{\text{percentage}}$	329
Table 12	Critical values for a 5% two-sided runs test	332
Table 13	Critical values for the Wilcoxon test for paired-data	333
Table 14	Critical values for the Wilcoxon rank sum test	334
Table 15a	Table of probabilities associated for obtaining values as large as observed values with Friedman's test. $k=3$	337
Table 15b	Table of probabilities associated for obtaining values as large as the observed values with Friedman's test. $k=4$	338

Answers to Exercises 341

Index 355

A FIRST COURSE IN PROBABILITY AND STATISTICS

Chapter 1
INTRODUCTION

1.1 STATISTICS AND PROBABILITY

Statistics is a discipline concerned with the application of mathematics to the study of data obtained from observation where the data may be qualitative or quantitative. Since observed data are the basis for the inductive sciences such as physics, chemistry, biology, sociology, economics, psychology, agriculture, engineering, and medicine, it is clear that the range of problems to which statistics can be applied is broad indeed.

The roots of statistics lie in the field of mathematics and in particular in the area of probability. In order therefore to understand the development of techniques used in statistics, it will be necessary to understand some of the concepts of probability. Thus the first few chapters of this book will be concerned with studying the concepts and ideas necessary for building the theory of probability, which will be used later to develop the theory of statistics.

Almost all observations (indeed some people would say all) are subject to chance variations; the study of the nature of these chance variations is the theory of probability. The concepts of chance and with it, luck, go back into the pre-history of man, but the mathematical study of them is only about three hundred years old, and was originally stimulated by games of chance, or gambling. These games involved tossing coins, drawing cards, and throwing dice, often with complicated sets of rules. The gamblers wished to know the best strategy needed in order to maximize their winnings. These early problems, and their solutions, are still of interest, both from an intellectual and a practical point of view; and in fact we shall initiate our formal study in Chapter 2 by considering such games. Although statistics is a natural outgrowth of probability, its development was somehow overlooked by early probabilists, and it was not until the late 19th century that statistics was considered as a discipline in its own right. Most prominent in its early development were Karl Pearson and R. A. Fisher. Much of the impetus for its development came from fields other than mathematics such as biology and agriculture.

Along with the outgrowth of the theory of statistics, called mathematical statistics, was a parallel growth of tools and techniques for use in applied statistics. In fact, some techniques were developed by non-mathematicians, used at first on an intuitive basis, and later had their use justified on the basis of the theory. It is appropriate to consider some examples of problems, arising in a wide variety of areas, whose whole or partial solution depends on the use of statistics or statistical analysis. Consider therefore the following examples.

Example 1.1.1 (Medicine) In comparing the effects of a new drug designed to reduce the blood pressure of hypertensive patients, an experimenter took two sets of ten hypertensive patients. To one set he gave the new drug, and to the other he gave a neutral substance. The blood pressure was measured before and after treatment yielding the following results.

TABLE 1.1.1

Blood pressure measured before and after treatment with a new drug and a neutral substance

Patient	New drug		Neutral substance	
	Before	After	Before	After
A	154	132	139	122
B	141	127	151	142
C	150	129	166	168
D	136	136	157	153
E	152	140	140	152
F	140	119	148	148
G	167	104	157	149
H	152	112	161	129
I	160	128	158	156
J	145	133	149	137

Since another experimenter performing the same experiment would use a different set of patients, it is clear that the patients themselves are not of prime interest in such an experiment. The results ought to be able to be generalized to, let us say, all hypertensive patients. Statistics is concerned with the problems of how to choose patients, how to allocate treatments, and how to analyze the results so that such generalizations can be made. It may happen for instance that the drug is not effective for a particular kind of patient, such as those with high potassium levels. Experiments can be designed to seek out answers to such questions.

Example 1.1.2 (Education) A typical problem in the field of education is that of calculating the degree of agreement between the scores obtained on two different tests. Consider a group of 30 children given a mathematics and an English test, and obtaining, on the scale 0–100, the following results.

TABLE 1.1.2

Showing English and mathematics results for 30 children

English	Mathematics	English	Mathematics
72	64	64	60
75	71	61	74
77	72	78	72
89	70	83	72
80	77	69	78
89	82	87	63
64	69	80	71
57	69	75	52
32	51	71	83
75	70	74	64
83	75	75	70
80	84	70	61
57	30	62	62
54	59	83	76
71	65	72	80

The two sets of results suggest that a relationship exists between them. Statistics provides a means for calculating this relationship and also the amount of confidence which can be placed in the result, allowing for the variability of the performance of the children.

Example 1.1.3 (Physics) In Table 1.1.3 is a set of 10 measurements made to obtain the specific gravity of mercury.

TABLE 1.1.3

Determinations of the specific gravity of mercury

13.696	13.695
13.699	13.697
13.683	13.688
13.692	13.690
13.705	13.707

It is assumed that all determinations were made under approximately the same experimental conditions. The results vary because these conditions were not exactly the same: i.e. there were conditions beyond the control of the measurer. In order to determine the true specific gravity of mercury it would have been necessary to continue the experiment indefinitely; however, for practical purposes it is necessary to stop after ten determinations, even though we still don't know the true specific gravity. An intuitively reasonable answer would be to calculate the average of the ten measurements, 13.695, and use it. Statistical theory allows us to calculate an interval, and to state, with a certain level of confidence, that the interval contains the true (but unknown) specific gravity. This same theory will show us that if we had taken more measurements, then the interval would have been smaller, so that our degree of ignorance would have been reduced.

Example 1.1.4 (Industry)   Iron used by a steel company can be obtained from two regions (A and B). Of particular importance is the melting point of the iron. Table 1.1.4 shows the melting points of 10 determinations from region A and 6 from region B.

TABLE 1.1.4

Melting points of iron from regions A and B

Region A	1493, 1519, 1518, 1512, 1512, 1514, 1489, 1508, 1508, 1494
Region B	1509, 1494, 1512, 1483, 1507, 1491

Statistics provides a means for saying whether the two regions differ in the melting points of their iron and also how reliable this statement is.

TABLE 1.1.5

A comparison of number of cattle deaths with the number immunized against tuberculosis

	Died or seriously affected	Unaffected or only slightly affected	Total
Inoculated	8	22	30
Not inoculated	10	4	14
Total	18	26	44

Example 1.1.5 (Agriculture) An experiment to measure the effects of immunization of cattle against tuberculosis was performed by deliberately infecting both those cattle which had been inoculated and those which had not. The results are presented in Table 1.1.5. Statistics provides an answer to the question of what is the degree of association between inoculation and exemption from the disease, and further it deals with the conditions under which the results on the above 44 animals can be generalized.

Example 1.1.6 (Genetics) The statistical aspects of genetics will be illustrated by considering an experiment first performed by the father of modern genetics, G. J. Mendel, in 1865, and comparing his results with those of later workers in this field. Pure bred sweet peas, one with yellow seeds and one with green seeds, were crossed. The offspring had yellow seeds. When these were pollinated with their own pollen approximately one quarter of the offspring had green seeds and bred true when self-fertilized. The remainder were yellow, which on being self-fertilized gave about $\frac{1}{3}$ true bred, while the other $\frac{2}{3}$ on being self-fertilized had yellow and green offspring in the ratio 3 : 1. The results of Mendel and later workers are given in Table 1.1.6.

TABLE 1.1.6

The proportions of yellow and green sweet peas, obtained by seven investigators in the offspring of yellow hybrids when self-fertilized

Investigator	Offspring when yellow hybrids are self-fertilized		
	Proportion yellow	Proportion green	Total
Mendel	.7505	.2495	8023
Correns	.7547	.2453	1847
Tschermak	.7505	.2495	4770
Hurst	.7464	.2536	1755
Bateson	.7530	.2470	15806
Lock	.7367	.2633	1952
Darbishire	.7509	.2491	145246

Reproduced from *A Course in Theoretical Statistics* by N. A. Rahman, by kind permission of C. Griffin and Co. Ltd., London.

One might ask if, on the basis of these data, the hypothesis that self-fertilized yellow hybrids give offspring in the ratio green : yellow : : 1 ; 3 is acceptable. The positive answer to that question is given by statistical analysis and agrees with a result obtained by the application of Mendel's theory of "particulate inheritance."

Example 1.1.7 (Pharmacology) A group of experimental mice are allocated at random to individual cages, and the cages are allocated in equal numbers of two treatments, a control A and a drug B. All animals are then infected with cancer cells. In the following days the mice are examined for tumors and Table 1.1.7 shows the day after treatment that the tumor was first noted.

TABLE 1.1.7

Number of days after treatment on which tumors were first noted

Control A	5, 6, 7, 7, 8, 9, 9, 9, 12, 13
Drug B	7, 8, 8, 8, 9, 9, 12, 14, 17, 20

With the help of statistical analysis one can decide whether the drug B was helpful in delaying the onset of the tumor in mice. As with earlier examples the drug has been tested on only a limited number of mice. One can use the ideas of statistics to ensure that even from this small sample, the results can be generalized to a larger group (or population) of mice.

Example 1.1.8 (Quality Control) Machine parts turned out by an automatic process are known to be, on the average, 1% defective. In order to judge each day's run, a quality control expert examines 200 similar parts each day. He rejects the whole day's run if three or more parts in the 200 are found to be defective. Since rejecting the whole run is an expensive procedure, one might reasonably ask whether he just happened to have bad luck in choosing three or more defective parts. In other words there are two reasons for obtaining such a result: (1) there are indeed too many defective parts in the run and it should be rejected, or (2) luck was against the quality control expert and he just happened to choose the wrong parts. The chance of such an event as (2) happening can be calculated from the theory of statistics.

We could go on giving similar examples from other areas; however, it should be clear by now that statistics can be used to solve or help solve a wide variety of problems. In each example the data were obtained by observation of a particular variable. So in Example 1.1.3, the variable measured is the specific gravity of mercury and in Example 1.1.7, it is the day on which the tumors develop. In Example 1.1.3, if there had been no uncontrollable effects or variation, each observation would have been the same. In Example 1.1.7, if there had been no difference between the ability of mice to use drug B, they all would have developed tumors at the same time. This variability between ostensibly the same experimental subjects is called experimental error. Note that the word error is used not in the usual sense meaning "blunder," but in a sense meaning variability. Even the best controlled experiments, for instance in physics, have sources of experimental error. In such a case the error would be small, but in sociological or biological experiments, where control is more difficult, the error can be very large indeed and the experiment must be designed to minimize it. If two observations, obtained under similar experimental conditions, are not the same, one says that the error is due to unassignable causes of variation. In some cases one changes the experiment deliberately to see the effect on the response; then one says that the difference in response is due to assignable causes of variation. For instance in Example 1.1.2, the different results for a particular child are due to the difference of the academic subjects, namely, English and mathematics, whereas the difference between the various children's scores in English are due to differences between children, and are uncontrollable in the experiment.

Hence the study of statistics must start with a study of the nature of variation, which, as we have previously stated, is the Theory of Probability. Games of chance are a convenient starting point for such a study since in general they are easy to describe, and there are usually no outside influences affecting their outcomes. One might reasonably ask how studying games of chance will help in answering some of the questions raised in Examples 1.1.1 through 1.1.8. Indeed the answer is not always obvious, but consider Example 1.1.8. We could consider that we had a hundred-sided coin or hundred-sided die, 99 of whose faces were marked "good" and one marked "defect." Thus the chance of "defect" showing would be one out of a hundred or 0.01. The drawing of 200 parts could be simulated by throwing the die 200 times and counting the number of times that "defect" appeared. If this simulation were repeated many times it would soon become apparent that obtaining three or more "defects" is not a very rare event. To the quality control expert, this would indicate that his policy of rejecting a day's run on the basis of three or more defectives is a rather severe policy and also economically wasteful. Hence the study of the nature of coin-tossing or dice throwing has an immediate analog in reality.

1.2 THE SCOPE OF THE BOOK

This book is intended to serve as a text for a first course in probability and statistics and is concerned with introducing the elementary concepts of both of these areas. It is introductory in the sense that no previous knowledge of statistics is required or presumed.

Chapters 2 through 5 are concerned with giving a logical structure to the study of probability. The reasoning in these chapters is purely deductive, so that the results can be derived on the basis of pure reason, starting from a set of axioms. A good knowledge of pre-calculus mathematics is assumed. Limiting ourselves to mathematics at this level means that some results will have to be given without proofs. Although the purpose is to acquaint the student with the underlying ideas of modern statistics, examples are given, so that he never loses touch with the underlying pragmatic development and impetus for so much of statistical thought.

The ideas of sampling from a population are introduced in Chapter 6, and Chapter 7 deals with methods for organizing and analyzing sample data. The remainder of the text is concerned with the problems of inference and hypothesis testing for many of the more important problems. Basically in total the object of this book is to introduce the ideas, terminology, and ways of reasoning in modern statistical thought, and to let the student see the types of problems which can be worked on. Although the contents deal with the theory of probability and statistics, it is part of the philosophy of its authors that "a little example goes a long way." In so far as is possible, ideas will be introduced with numerical examples, then the general cases will be discussed. Those who find formulas vague until an example or two are given should find this approach easier to follow. Those who desire, may skip over the examples and get to the general cases immediately. A generous number of problems are added at the end of each chapter, in the belief that no student in a mathematical subject can progress far without self-testing through problem solving.

1.3 THE SYSTEM OF REFERENCING

All chapters are divided into sections, and the numbers of the theorems, tables, examples, and definitions restart in each section. Thus Theorem 4.2.1 is Theorem 1 of Section 2 of Chapter 4. Equations are also numbered anew in each section and equation numbers are referred to in parentheses. Thus Equation (4.1.3) is Equation 3 of Section 1 of Chapter 4.

Chapter 2
ELEMENTARY PROBABILITY

2.1 THE NOTION OF A SET

The term set is regarded as one of the basic ideas of mathematics necessary for the study of probability and statistics. The idea of a set is common in everyday life. We speak of a set of books, a set of glasses, a set of people etc. A set is merely an aggregate or collection of objects of any sort. Anything that is a member of a set is called an element of the set. If a set has a finite number of elements, then it is called a finite set; otherwise it is called an infinite set.

Definition 2.1.1 A set is a well-defined collection of objects from a specified universe.

By well-defined we mean that for each particular object, we must be able to decide whether it does or does not belong to the set.

Example 2.1.1 The following are some examples of sets.
1. The set of citizens of the United States.
2. The set of stocks listed on the Toronto Stock Exchange.
3. The set of women who have run the mile in less than five minutes.
4. The set of positive integers.

It is customary to denote whole sets by capital letters such as A, B, C, and their elements by small letters such as a, b, c.

In common practice, there are two ways to describe a set. First, if the set has a finite number of elements, we may list its members, enclosing them in braces. Second, a set may be described by enclosing in braces a defining property that any object must meet in order to be a member of the set.

Example 2.1.2 The set A whose elements are the integers 1, 3, 5, and 7 is a finite set with four elements and may be written

$$A = \{1, 3, 5, 7\}.$$

Example 2.1.3 If $A = \{1, 3, 5, 7\}$ and B is the set of numbers that are squares of elements of A, specify the set B in two ways.

Solution B may be written by listing all its elements
$$B = \{1, 9, 25, 49\}$$
or in alternate notation
$$B = \{x^2 \mid x = 1, 3, 5, 7\}.$$

Definition 2.1.2 A set with no members is called an empty or null set. We denote this set by the symbol $\emptyset$.

The empty set in set theory plays a role similar to that of zero in the number system.

Example 2.1.4 The set of integer solutions of the equation $x^2 - 2 = 0$ is an empty set.

Definition 2.1.3 Two sets A and B are said to be equal or identical if and only if they have exactly the same elements.

If two sets A and B are equal, then every element that belongs to A also belongs to B, and every element that belongs to B also belongs to A. Note that the order in which we list the elements of a set is immaterial and that the sets $A = \{1, 2, 3\}$ and $B = \{2, 1, 3\}$ are equal.

In any particular discussion of sets, it is important to have the universe or the universal set clearly determined. We shall denote the universal set by U; we might then select special subsets.

Definition 2.1.4 A set A is a subset of a set B, if every element of a set A is also an element of a set B.

This is written as $A \subseteq B$ (A is contained in or is equal to B) or $B \supseteq A$ (B contains A or is equal to A). If every element of A is an element of B but some element of B is not an element of A, then A is a proper subset of B. This is written as $A \subset B$ or $B \supset A$. We note that the empty set $\emptyset$ is a subset of every set.

Example 2.1.5 Let the universal set $\mathsf{U} = \{a, b, c\}$. Then $\emptyset$, $\{a\}$, $\{b\}$, $\{c\}$, $\{a, b\}$, $\{a, c\}$, $\{b, c\}$, $\{a, b, c\}$ are all the subsets of the universal set $\mathsf{U} = \{a, b, c\}$.

This example illustrates that the number of subsets of a universal set containing three elements is $2^3 = 8$. In general, if U is a finite set with n elements, then there are 2^n different subsets of U, including U and $\emptyset$.

2.2 OPERATIONS ON SETS

A useful device when considering sets and operations on sets of a given universal set is a Venn diagram. The rectangle U in Fig. 2.2.1 represents the universal set **U**, and the elements of **U** are represented by the points inside and on the rectangle. A subset A of **U** is represented by the region within a circle (say) drawn inside the rectangle.

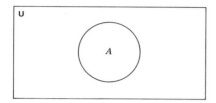

Fig. 2.2.1 Venn diagram.

Definition 2.2.1 The intersection of two sets A and B is the set of all elements of **U** that are members of both A and B.

We denote the intersection of A and B by $A \cap B$ (read "A intersection B"). In symbols

$$A \cap B = \{x \mid x \in A \quad \text{and} \quad x \in B\},$$

where the symbol $x \in A$ means "x is an element of A" and the symbol $x \in B$ means "x is an element of B." $A \cap B$, the intersection of the two sets A and B, is represented by the shaded region in Fig. 2.2.2.

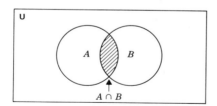

Fig. 2.2.2 A intersection B.

Example 2.2.1 Let $A = \{2, 4, 6\}$, $B = \{4, 6, 8\}$; then $A \cap B = \{4, 6\}$.

Example 2.2.2 Let $A = \{t \mid 10 \leq t \leq 50\}$, $B = \{t \mid 25 \leq t \leq 50\}$; then $A \cap B = \{t \mid 25 \leq t \leq 50\}$.

Definition 2.2.2 Two sets A and B are disjoint, or mutually exclusive, if they have no elements in common.

Symbolically, $A \cap B = \emptyset$. That is, A and B have no elements in common.

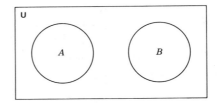

Fig. 2.2.3 Disjoint sets.

Example 2.2.3 Let $A = \{1, 2, 3, 4\}$, $B = \{5, 6, 7\}$; then $A \cap B = \emptyset$.

Definition 2.2.3 The union of two sets A and B is the set of all elements of **U** that belong either to A or to B or to both.

We denote the union of A and B by $A \cup B$ (read "A union B"). In symbols
$$A \cup B = \{x \mid x \in A \quad \text{or} \quad x \in B\}.$$

The union $A \cup B$ of two sets A and B is represented by the shaded area in Fig. 2.2.4.

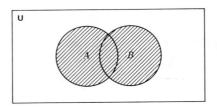

Fig. 2.2.4 A union B.

Example 2.2.4 Let **U** be the set of all students in a high school. Let M be the set of male students who plan to attend a college and F be the set of female students who plan to attend a college; the union $M \cup F$ is the set of all students in a high school who plan to attend a college.

Example 2.2.5 Let $A = \{1, 2, 3\}$, $B = \{2, 4, 6\}$; then $A \cup B = \{1, 2, 3, 4, 6\}$.

Example 2.2.6 Let $A = \{t \mid 10 \leq t \leq 50\}$, $B = \{t \mid 25 \leq t \leq 50\}$; then $A \cup B = \{t \mid 10 \leq t \leq 50\}$.

Definition 2.2.4 The complement of a set A is the set of all elements in **U** that are not contained in A.

We denote complement of A by A'. In symbols
$$A' = \{x \mid x \in \mathbf{U} \quad \text{and} \quad x \notin A\}$$
where the symbol $x \notin A$ means "x is not an element of A."

The complement of A is represented by the hatched region in Fig. 2.2.5.

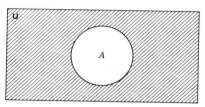

Fig. 2.2.5 Complement of A.

Example 2.2.7 Let the universal set be
$$\mathbf{U} = \{\text{Washington, Adams, Kennedy, Lincoln}\}.$$
The complement of the set A containing the names which begin with K is
$$A' = \{\text{Washington, Adams, Lincoln}\}.$$

Example 2.2.8 Let the universal set $\mathbf{U} = \{1, 2, 3, 4, 5, 6\}$ and $A = \{2, 4, 6\}$; then $A' = \{1, 3, 5\}$.

There are a number of other set operations involving union, intersection, and complementation. Some are listed in the following theorem. Their validity may be checked with the aid of a Venn diagram.

Theorem 2.2.1 Let A, B, C be subsets of a universal set **U**. Then the following relationships hold.

a) Identity laws:
$$A \cup \emptyset = A, \quad A \cap \emptyset = \emptyset,$$
$$A \cup \mathbf{U} = \mathbf{U}, \quad A \cap \mathbf{U} = A.$$

b) Complement laws:
$$A \cup A' = \mathbf{U}, \quad A \cap A' = \emptyset,$$
$$(A')' = A.$$

c) Commutative laws:
$$A \cup B = B \cup A, \quad A \cap B = B \cap A.$$

d) Idempotent laws:
$$A \cup A = A, \quad A \cap A = A.$$

14 Elementary probability

e) Associative laws:
$$(A \cup B) \cup C = A \cup (B \cup C),$$
$$(A \cap B) \cap C = A \cap (B \cap C).$$

f) Distributive laws:
$$A \cup (B \cap C) = (A \cup B) \cap (A \cup C),$$
$$A \cap (B \cup C) = (A \cap B) \cup (A \cap C).$$

g) De Morgan's laws:
$$(A \cup B)' = A' \cap B', \quad (A \cap B)' = A' \cup B'.$$

In our later work, we shall need to consider the union and the intersection of more than two sets, so we make the following definition.

Definition 2.2.5 Let n be any positive integer and let $A_1, A_2, \ldots, A_n$ be any given sets. Then the set of elements belonging to all the given sets is denoted by
$$A_1 \cap A_2 \cap \cdots \cap A_n,$$
and the set of elements belonging to at least one of the given sets is denoted by
$$A_1 \cup A_2 \cup \cdots \cup A_n.$$

The laws of union and intersection defined for just two sets in Theorem 2.2.1 can now be generalized to hold for unions and intersections of more than two sets. We list some of these laws in the following theorem.

Theorem 2.2.2 Let n be any positive integer and let $A, A_1, A_2, \ldots, A_n$ be subsets of a universal set **U**. Then the following relationships hold:

a) $(A_1 \cup A_2 \cup \cdots \cup A_n)' = A_1' \cap A_2' \cap \cdots \cap A_n'.$
b) $(A_1 \cap A_2 \cap \cdots \cap A_n)' = A_1' \cup A_2' \cup \cdots \cup A_n'.$
c) $A \cup (A_1 \cap A_2 \cap \cdots \cap A_n) = (A \cup A_1) \cap (A \cup A_2) \cap \cdots \cap (A \cup A_n).$
d) $A \cap (A_1 \cup A_2 \cup \cdots \cup A_n) = (A \cap A_1) \cup (A \cap A_2) \cup \cdots \cup (A \cap A_n).$

2.3 SAMPLE SPACES, SAMPLE POINTS, AND EVENTS

Suppose that we perform an experiment E and note all its possible outcomes. The word "experiment" is used in probability and statistics in a much broader sense than in everyday conversation. For example, tossing one or more coins or rolling a die will be considered as two separate experiments. Consider the following experiments.

1. A coin is tossed three times and the number of heads observed.
2. A die is rolled and a note is made of the number shown on top.

3. From an urn containing only five red and three black balls, a ball is drawn.
4. The political party which a voter chooses during a presidential campaign is observed.
5. A poll is taken across the nation asking whether a certain television program was seen on a particular evening.
6. Items are manufactured on a machine and the number of defective items during a specified period is counted.

Definition 2.3.1 A sample space S of an experiment E is the set of all possible outcomes. An element in a sample space is called a sample point.

Example 2.3.1 Consider the experiment of tossing three coins: a nickel, a dime, and a quarter. The sample space for this experiment is

$$S = \{HHH, HTH, THH, HHT, TTH, THT, HTT, TTT\},$$

where HHH means head for the nickel, head for the dime, and head for the quarter.

Example 2.3.2 An experiment consists of throwing two dice: a red and a black. Let r be the outcome for the red die and b be the outcome for the black die. Then the sample space S is the set of pairs (r, b) with r and b taking the values 1, 2, 3, 4, 5, 6.

$$S = \{(r, b) \mid 1 \leq r \leq 6, \quad 1 \leq b \leq 6\}.$$

It would serve our purpose just as well to throw a single die twice, the first throw corresponding to a red die, the second throw corresponding

TABLE 2.3.1

Sample space for two-dice experiment

Black die outcome

$r \backslash b$	1	2	3	4	5	6
1	(1, 1)	(1, 2)	(1, 3)	(1, 4)	(1, 5)	(1, 6)
2	(2, 1)	(2, 2)	(2, 3)	(2, 4)	(2, 5)	(2, 6)
3	(3, 1)	(3, 2)	(3, 3)	(3, 4)	(3, 5)	(3, 6)
4	(4, 1)	(4, 2)	(4, 3)	(4, 4)	(4, 5)	(4, 6)
5	(5, 1)	(5, 2)	(5, 3)	(5, 4)	(5, 5)	(5, 6)
6	(6, 1)	(6, 2)	(6, 3)	(6, 4)	(6, 5)	(6, 6)

Red die outcome

to a black die. Table 2.3.1 shows a sample space that lists the possible outcomes of the experiment. The entry (3, 4) in the third row and the fourth column represents the outcome "red die shows 3, black die shows 4."

A sample space S serves as the universal set for all questions related to an experiment. An event A with respect to a particular sample space S is simply a set of possible outcomes favorable to the event A. For example, consider the experiment of tossing a coin twice, the set

$$S = \{HH, HT, TH, TT\}$$

is the sample space for this experiment. For each outcome of the experiment we can determine whether a given event does or does not occur. We may be interested in the event "both tosses show the same face"; we find that this event occurs if the experiment results in an outcome corresponding to an element of the set

$$A = \{HH, TT\}.$$

We recognize A as a subset of the sample space S.

Definition 2.3.2 An "event" is a subset of a sample space S of an experiment.

Since an event is a subset of the sample space S, this means that S itself and the empty set $\emptyset$ are events. The event S is called the certain event, which always occurs, while $\emptyset$ is the impossible event, which can never occur.

Example 2.3.3 The following are some examples of events.
1. In the experiment of tossing a coin three times, "getting three heads" is an event; "getting three tails" is another event.
2. In the experiment of rolling a die, "an even number occurs" is an event. "An odd number occurs" is another event.

We can combine various events (sets) and obtain new events (sets).
1. Event "A or B." If A and B are events, $A \cup B$ is the event which occurs if and only if A or B or both occur.
2. Event "A and B." If A and B are events, $A \cap B$ is the event which occurs if and only if A and B occur.
3. Event "not A." If A is an event, A' is the event which occurs if and only if A does not occcur.
4. Event "at least one occurs." If $A_1, A_2, \ldots, A_n$ is any finite collection of events, then $A_1 \cup A_2 \cup \cdots \cup A_n$ is the event which occurs if and only if at least one of the events A_i occurs.

5. Event "all occur." If $A_1, A_2, \ldots, A_n$ is any finite collection of events, then $A_1 \cap A_2 \cap \cdots \cap A_n$ is the event which occurs if and only if all the events A_i occur.

Two events, A and B, are said to be mutually exclusive, if they cannot happen at the same time.

Definition 2.3.3 Two events, A and B, are mutually exclusive if $A \cap B = \varnothing$; that is, the intersection of A and B is the empty set.

Example 2.3.4 Consider the experiment of tossing a coin. The set
$$S = \{H, T\}$$
is the sample space for this experiment. Let A be the event "getting a head" and let B be the event "getting a tail." The intersection of the sets $A = \{H\}$ and $B = \{T\}$ is $A \cap B = \varnothing$, since they have no points in common.

The above definition of two mutually exclusive events may be extended to any finite number of events.

Definition 2.3.4 The events $A_1, A_2, \ldots, A_n$ are pairwise mutually exclusive events if $A_i \cap A_j = \varnothing$ $(i \neq j)$.

2.4 RULES OF COUNTING SAMPLE POINTS

We wish to count the set of all possible outcomes in a sample space with a finite number of elements. Up to now, we have considered experiments in which we were able to enumerate all possible outcomes. But if the experiment has a large number of outcomes it would be difficult and sometimes impossible to enumerate each one of them. We develop a few counting techniques in this section. Such techniques will enable us to determine the number of sample points in a sample space without having to list these sample points.

The Addition Principle. If two operations are mutually exclusive, and if the first operation can be performed in N_1 ways and the second operation can be performed in N_2 ways, then one or the other can be performed in $N_1 + N_2$ ways.

Example 2.4.1 A die is rolled once. In how many ways can an odd or even number occur?

Solution There are six possible ways in which the die can fall and of these, three show odd numbers, and three show even numbers. Therefore, $3 + 3 = 6$ is the number of ways in which an odd or even number appears.

The addition principle may be extended to any finite number of operations. The general case is stated in the following theorem.

Theorem 2.4.1 If there are K operations, and if the first operation can be performed in N_1 ways, and the second can be performed in N_2 ways, etc., then one of the K operations can be performed in $N_1 + N_2 + \cdots + N_K$ ways, assuming that no two operations are performed together.

Example 2.4.2 A bag contains four white, six black, and eight red balls. In how many ways can a white or a black or a red ball be drawn?

Solution Since a white ball can be drawn in four ways, a black ball in six ways, and a red ball in eight ways, the total number of ways in which a white or a black or a red ball can be drawn is $4 + 6 + 8 = 18$.

The Multiplication Principle. If there are N_1 ways of performing the first operation, and after it is performed in any one of these ways, a second can be performed in N_2 ways, then the two operations can be performed together in $N_1 \times N_2$ ways.

Example 2.4.3 A die is rolled twice. Determine the number of sample points in the sample space of this experiment.

Solution Since $N_1 = 6$, $N_2 = 6$, the number of sample points $= 6 \times 6 = 36$. The sample space consisting of 36 points is listed in Table 2.3.1.

The multiplication principle may be extended to any finite number of operations. The general case is stated in the following theorem.

Theorem 2.4.2 If there are K operations, and if the first can be performed in N_1 ways, and if, no matter how the first operation is performed, a second operation can be performed in N_2 ways, and if, no matter how operations one and two are performed, a third operation can be performed in N_3 ways, and so forth for K operations, then the K operations can be performed together in $N_1 \times N_2 \times \cdots \times N_k$ ways.

Example 2.4.4 A coin is tossed three times. Determine the number of sample points in the sample space of this experiment.

Solution Since $N_1 = 2$, $N_2 = 2$, $N_3 = 2$, the number of sample points is $2 \times 2 \times 2 = 8$.

Example 2.4.5 A coin is tossed, a die is rolled, and a card is drawn from a well-shuffled deck of cards. Determine the number of sample points in the sample space of this experiment.

Solution Since $N_1 = 2$, $N_2 = 6$, $N_3 = 52$, the number of sample points is $2 \times 6 \times 52 = 624$.

Another important counting problem deals with the different arrangements or orderings of a collection of objects. For example, we might want to know how many different batting orders are possible for a baseball team consisting of nine players, or we might ask in how many ways can five books be arranged on a shelf. The multiplication principle provides a general method for finding the number of arrangements of sets of such objects.

Definition 2.4.1 Arrangements, or orderings, of all or part of a set of objects are called permutations.

Example 2.4.6 In how many ways can three people form a line at a theatre box office?

Solution The first position can be filled in any one of three ways, the second in any one of two ways, and the last in exactly one way. Hence by the multiplication principle, the number of possible ways in which three people can form a line is $3 \times 2 \times 1 = 6$.

The six possible permutations or arrangements of three people A, B, C standing in a line is shown in Fig. 2.4.1.

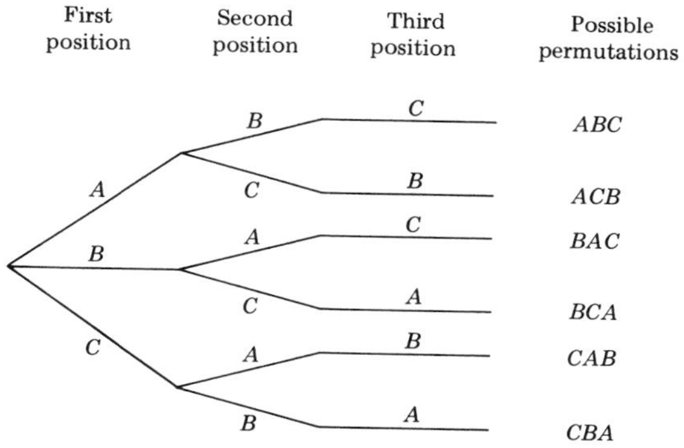

```
First       Second      Third       Possible
position    position    position    permutations

            B ───────── C ───────── ABC
    A
            C ───────── B ───────── ACB
    B       A ───────── C ───────── BAC
            C ───────── A ───────── BCA
    C       A ───────── B ───────── CAB
            B ───────── A ───────── CBA
```

Figure 2.4.1

In general, suppose we have n different objects, and we wish to arrange these objects in a line. The first place can be filled in n different ways. The second place must then be filled; this can be done in $n - 1$ ways. Then there are $n - 2$ choices for the third place, and so on. By the multiplication principle, we find that there are

$$n(n - 1)(n - 2) \cdots (2)(1)$$

different ways. This product is usually denoted by the symbol $n!$ and called "n factorial." In general, we define

$$n! = n(n - 1)(n - 2) \cdots (2)(1),$$

whenever n is a positive integer. We also define $1! = 1$ and $0! = 1$. Thus, the number of permutations of n different objects, taken altogether, is $n!$

Theorem 2.4.3 The number of permutations of a set of n different objects, taken all together, denoted by ^nP_n, is $n!$

Example 2.4.7 In how many ways can the supermarket manager display five brands of cereal on a shelf?

Solution The number of possible displays is

$$^5P_5 = 5! = 5 \times 4 \times 3 \times 2 \times 1 = 120.$$

Now consider permutations of n different objects in which some of the objects are used.

Example 2.4.8 In how many ways can the supermarket manager display five brands of cereal in three spaces on a shelf?

Solution The first space can be filled in five ways. After this has been done, the second space can be filled in four ways. Similarly, the third space can be filled in three ways. By the multiplication principle, the three spaces can be filled in $5 \times 4 \times 3$ ways.

In terms of factorial symbols, we have

$$5 \times 4 \times 3 = \frac{5 \times 4 \times 3 \times 2 \times 1}{2 \times 1} = \frac{5!}{2!}.$$

Let the symbol $^5P_3$ denote the number of permutations of five different objects, taken three at a time. Then

$$^5P_3 = \frac{5!}{2!} = \frac{5!}{(5 - 3)!}.$$

In general, suppose we have n different objects and r spaces to be filled. By the multiplication principle, the number of ways in which r spaces can be filled in is

$$n(n - 1)(n - 2) \cdots (n - r + 1).$$

Let ^nP_r denote the number of permutations of n different objects, taken r at a time. Then

$$^nP_r = n(n-1)(n-2)\cdots(n-r+1).$$

In terms of factorial symbols

$$^nP_r = \frac{n(n-1)(n-2)\cdots(n-r+1)(n-r)!}{(n-r)!}$$

$$= \frac{n!}{(n-r)!}.$$

Theorem 2.4.4 The number of permutations of n different objects taken r at a time, denoted by ^nP_r, is

$$^nP_r = \frac{n!}{(n-r)!}.$$

Example 2.4.9 In how many ways can an experimenter permute two treatments taken from a set of three different treatments?

Solution From Theorem 2.4.4, the number is

$$^3P_2 = \frac{3!}{(3-2)!} = \frac{3!}{1!}$$

$$= 3 \times 2 \times 1$$

$$= 6.$$

Now consider the example in which an experimenter is interested in selecting a set of two treatments from a set of three different treatments $\{T_1, T_2, T_3\}$. The number of possible sets is

$$\{T_1, T_2\}, \{T_1, T_3\}, \{T_2, T_3\}.$$

On the other hand, the number of permutations of three different treatments, taken two at a time, is six. The six possible permutations are listed in Table 2.4.1.

TABLE 2.4.1

T_1T_2	T_1T_3	T_2T_3
T_2T_1	T_3T_1	T_3T_2

The reader may have observed already that the two permutations listed in column one represent the same set of treatments $\{T_1, T_2\}$. Similarly, treatments given in column two represent the same set $\{T_1, T_3\}$, and the permutations given in column three represent the same set

$\{T_2, T_3\}$. Thus, the number of combinations of three different treatments, taken two at a time, is three, whereas the number of permutations of three different treatments, taken two at a time, is six. In other words, the number of sets or combinations of three different treatments taken two at a time is equal to the number of permutations of three different treatments taken two at a time divided by two (the number of permutations of treatments in the selection).

This example illustrates the difference between a permutation and a combination. That is, in a permutation, order counts; in a combination, order does not count.

In general, the number of combinations of n objects taken r at a time, denoted by $\binom{n}{r}$, is equal to the number of permutations of n objects taken r at a time divided by $r!$ That is,

$$\binom{n}{r} = {}^nP_r \div r!$$
$$= \frac{n!}{(n-r)!} \div r!$$
$$= \frac{n!}{r!(n-r)!}.$$

Theorem 2.4.5 The number of combinations of n different objects, taken r at a time, denoted by $\binom{n}{r}$, is

$$\binom{n}{r} = \frac{n!}{r!(n-r)!}.$$

Example 2.4.10 In how many ways can 13 cards be drawn from a standard deck of 52 different cards?

Solution From Theorem 2.4.5 the number of possible combinations is

$$\binom{52}{13} = \frac{52!}{13!(52-13)!}$$
$$= \frac{52!}{13!\,39!}.$$

Example 2.4.11 A bag contains three white and five black balls. In how many ways can three balls be drawn?

Solution Total number of balls is $3 + 5 = 8$. Three balls can be drawn in

$$\binom{8}{3} = \frac{8!}{3!\,5!} = 56 \text{ ways.}$$

The above theorem can now be extended to the general case, where the n objects consist of K kinds.

Theorem 2.4.6 The number of arrangements or permutations where the n objects consist of r_1 objects of one kind, r_2 objects of a second kind, r_K objects of the Kth kind, is

$$\binom{n}{r_1, r_2, \ldots, r_K} = \frac{n!}{r_1! \, r_2! \cdots r_K!}.$$

Example 2.4.12 In how many ways can the standard deck of 52 cards be dealt amongst four players?

Solution Since each player should have 13 cards, there are

$$\binom{52}{13, 13, 13, 13} = \frac{52!}{13! \, 13! \, 13! \, 13!}$$

different possible deals.

2.5 PROBABILITY OF AN EVENT

As mentioned in Chapter 1, the theory of probability has always been associated with games of chance such as dice, cards, etc. For example, suppose we have a die with six sides numbered 1 to 6 and that we want to know the chance of rolling a 3. Since there are six possible outcomes we regard each outcome as equally likely to occur and we say that the outcome 3 has a chance $\frac{1}{6}$ of occurring. Suppose we draw a card from a well-shuffled deck of 52 cards. Since each card has the same chance of being drawn as every other card, the chance of drawing a particular card is $\frac{1}{52}$.

When an experiment is performed, we set up a sample space S of all possible outcomes. In a sample space of N equally likely outcomes, we assign a chance or weight of $1/N$ to each sample point. We define probability on such a sample space as follows.

Definition 2.5.1 If an experiment can produce N different equally likely outcomes, and if exactly M of these outcomes are favorable to the event A, then the probability of event A, denoted by $P(A)$, is

$$P(A) = \frac{\text{Number of favorable outcomes}}{\text{Number of possible outcomes}} = \frac{M}{N}.$$

Example 2.5.1 A die is rolled. What is the probability that an even number occurs?

Solution The set $S = \{1, 2, 3, 4, 5, 6\}$ is a set of all possible outcomes. Let A be the event "even number occurs." Then

$$P(A) = \tfrac{3}{6} = \tfrac{1}{2}.$$

Example 2.5.2 A card is drawn from a standard deck of 52 cards. What is the probability that the card is
a) the ace of hearts?
b) a heart?
c) an ace?

Solution We define the following events.
A: The card is the ace of hearts.
B: The card is a heart.
C: The card is an ace.
a) Since there are 52 cards, and there is only one ace of hearts $P(A) = \frac{1}{52}$.
b) There are 13 heart cards. Therefore, $P(B) = \frac{13}{52} = \frac{1}{4}$.
c) There are four aces. Therefore $P(C) = \frac{4}{52} = \frac{1}{13}$.

Example 2.5.3 A bag contains four white and eight black balls. What is the probability of drawing a black ball?

Solution Let B be the event "a black ball is drawn." The total number of balls is 12 and any one is likely to be drawn. Therefore $P(B) = \frac{8}{12} = \frac{2}{3}$.

If there are no outcomes favorable to the event A, then $P(A) = 0$, and, if all possible outcomes are favorable to the event A, then $P(A) = 1$. Therefore $0 \leq P(A) \leq 1$.

Definition 2.5.2 Let E be an experiment. Let S be the sample space associated with E. Then the probability of the event A, denoted by $P(A)$, satisfies the following properties.
1) $0 \leq P(A) \leq 1$.
2) $P(\emptyset) = 0$.
3) $P(S) = 1$.

Example 2.5.4 A pair of dice is rolled. What is the probability of getting an 8?

Solution The sample space of this experiment consists of 36 points. (Refer to Table 2.3.1.) Let A be the event "getting an 8." Then the set

$$A = \{(2, 6), (3, 5), (4, 4), (5, 3), (6, 2)\}$$

is a set of favorable outcomes. The probability of getting an 8 is therefore

$$P(A) = \frac{5}{36}.$$

Example 2.5.5 A coin is tossed twice. What is the probability of getting
a) two heads?

b) two tails?
c) a tail and a head?

Solution The set
$$S = \{HH, HT, TH, TT\}$$
is the set of all possible outcomes.

a) There is only one sample point favorable to the event "two heads." Therefore $P(HH) = \frac{1}{4}$.
b) There is only one sample point favorable to the event "two tails." Therefore $P(TT) = \frac{1}{4}$.
c) There are two sample points favorable to the event "a head and a tail." Therefore $P(\text{head and tail}) = \frac{2}{4} = \frac{1}{2}$.

While Definition 2.5.1 is sufficient to define probability in games of chance or experiments with equally likely outcomes, it is an inadequate definition for experiments in which the outcomes are not equally likely. For example, what is the probability of a car accident on a particular road, or what is the probability that an electron tube is defective, or what is the probability that a person lives to age 70? Even in games of chance, Definition 2.5.1 is inadequate to calculate the probability of a head, using an unbalanced coin. Situations of the above type lead us to the following definition.

Definition 2.5.3 If an experiment is performed n times, and if an event A occurs f times, then the relative frequency of A is f/n. If we assume that as n increases, the relative frequency f/n approaches a limit, then this limit is called the probability of A.

In practice we estimate the probability of A by f/n for n large. For obvious reasons, Definition 2.5.3 is called the frequency definition of probability.

Example 2.5.6 A biased coin is tossed 1000 times yielding 640 heads. What is the probability of getting a head?

Solution Using the frequency definition of probability, the probability of a head is 0.64.

If no outcomes are favorable to the event A, then $f = 0$ and the probability of the event A is zero. If all outcomes are favorable, then f/n is equal to one and the probability of the event A is equal to 1.

Example 2.5.7 A die is rolled 100 times yielding the following results. What is the probability of getting a 1?

Number obtained	Frequency
1	21
2	18
3	14
4	17
5	10
6	20

Solution There are 100 rolls of a die of which 21 are favorable to getting a 1. Therefore probability of getting a 1 is

$$\frac{21}{100} = .21.$$

2.6 SOME PROBABILITY LAWS

Theorem 2.6.1 If A and B are any two events of a finite sample space S, then
$$P(A \cup B) = P(A) + P(B) - P(A \cap B).$$

Proof.

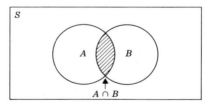

Fig. 2.6.1 Events in S.

Consider the Venn diagram in Fig. 2.6.1. The probability of $A \cup B$ is the sum of the probabilities of the points in $A \cup B$. Now $P(A) + P(B)$ is the sum of the probabilities of points in A plus the sum of the probabilities of points in B. Therefore we have added the probabilities of points in the intersection $A \cap B$ twice. If we subtract this probability $P(A \cap B)$ once, we have the sum of the probabilities of all points in $A \cup B$. Hence

$$P(A \cup B) = P(A) + P(B) - P(A \cap B).$$

Corollary 1 If A and B are mutually exclusive, then
$$P(A \cup B) = P(A) + P(B).$$

In words, the probability of A or B is the sum of their probabilities, provided the events are mutually exclusive.

The corollary follows from Theorem 2.6.1, since, if A and B are mutually exclusive, $A \cap B = \emptyset$, and $P(A \cap B) = P(\emptyset) = 0$.

The corollary 1 may be extended to any finite number of mutually exclusive events.

Corollary 2 Let $A_1, A_2, \ldots, A_n$ be mutually exclusive events in a finite sample space S, then

$$P(A_1 \cup A_2 \cup \cdots \cup A_n) = P(A_1) + \cdots + P(A_n).$$

In words, the probability of A_1 or A_2 or ... or A_n is the sum of their probabilities, provided the events are mutually exclusive.

Example 2.6.1 In a single throw with two dice, what is the probability of throwing the sum 7 or 10?

Solution (Refer to Table 2.3.1) Let A be the event that the sum is 7 and B be the event that the sum is 10. There are six sample points in favor of the event A and three for B. Since the corresponding sets do not overlap, the events A and B are mutually exclusive. Therefore

$$\begin{aligned} P(A \cup B) &= P(A) + P(B) \\ &= \tfrac{6}{36} + \tfrac{3}{36} \\ &= \tfrac{1}{4}. \end{aligned}$$

Example 2.6.2 Find the probability of drawing a diamond or an ace from a standard deck of 52 cards.

Solution Let A be the event "a diamond is drawn" and B be the event "an ace is drawn." There are 52 sample points in this sample space. Thus

$$P(A) = \tfrac{13}{52}, \qquad P(B) = \tfrac{4}{52}, \qquad P(A \cap B) = \tfrac{1}{52}.$$

Hence

$$\begin{aligned} P(A \cup B) &= P(A) + P(B) - P(A \cap B) \\ &= \tfrac{13}{52} + \tfrac{4}{52} - \tfrac{1}{52} \\ &= \tfrac{16}{52}. \end{aligned}$$

Theorem 2.6.2 If A and A' are complementary events in a finite sample space S, then $P(A') = 1 - P(A)$.

Proof. We may write

$$A \cup A' = S.$$

28 Elementary probability

By Theorem 2.6.1 we have
$$P(A) + P(A') - P(A \cap A') = P(S).$$
Since A and A' are mutually exclusive, it follows that
$$P(A) + P(A') = 1.$$
Therefore
$$P(A') = 1 - P(A).$$

Example 2.6.3 A coin is tossed three times. Find the probability that at least one head turns up.

Solution Let A be the event "getting a head at least once." The set
$$S = \{HHH, HTH, THH, HHT, TTH, THT, HTT, TTT\}$$
is the sample space for this experiment. There are seven sample points favorable to the event A. Therefore $P(A) = \frac{7}{8}$.

Alternative Solution Since we know that $P(A) + P(A') = 1$, we can just as well work with the complement of A. There is only one sample point favorable to A'. Therefore
$$\begin{aligned} P(A) &= 1 - P(A') \\ &= 1 - \tfrac{1}{8} \\ &= \tfrac{7}{8}. \end{aligned}$$

2.7 CONDITIONAL PROBABILITY

Suppose an experiment is performed and we are interested in the probability of an event A, given the additional information about the experiment that another event B has occurred. We would like to know how the probability of A is affected with this additional knowledge about B. The following example will help us to formulate a precise mathematical definition.

Example 2.7.1 Suppose a card is drawn at random from a standard deck of 52 cards. Let A denote the event "the card is a spade" and B denote the event "the card is black." We may then speak of the following probabilities.

1. $P(A)$ = probability that the card is a spade = $\frac{13}{52}$.
2. $P(B)$ = probability that the card is black = $\frac{26}{52}$.
3. $P(A \cap B)$ = probability that the card is both a spade and a black = $\frac{13}{52}$.

4. $P(A \mid B)$ = probability that the card is a spade knowing that it is black = $\frac{13}{26}$.

Knowing these probabilities, we notice that

$$\frac{P(A \cap B)}{P(B)} = \frac{13/52}{26/52} = \frac{13}{26},$$

which is exactly the same as $P(A \mid B)$. The symbol $P(A \mid B)$ is the conditional probability of an event A given B, so that

$$P(A) = \tfrac{1}{4} \quad \text{and} \quad P(A \mid B) = \tfrac{1}{2}.$$

Thus, in this example the probability of the event A is increased due to the added information that B has occurred. We now introduce a definition of conditional probability.

Definition 2.7.1 Let A and B be two events of a sample space S. The conditional probability of an event A, given B, denoted by $P(A \mid B)$, is defined by

$$P(A \mid B) = \frac{P(A \cap B)}{P(B)} \quad \text{if} \quad P(B) \neq 0. \tag{2.7.1}$$

Similarly

$$P(B \mid A) = \frac{P(A \cap B)}{P(A)} \quad \text{if} \quad P(A) \neq 0. \tag{2.7.2}$$

If we multiply both sides of Equation (2.7.1) by $P(B)$, we get

$$P(A \cap B) = P(B) P(A \mid B). \tag{2.7.3}$$

Since $A \cap B = B \cap A$, we also have

$$P(A \cap B) = P(A) P(B \mid A).$$

Example 2.7.2 (Industry) Consider the lot consisting of 10 defective and 40 nondefective items. If we choose two items at random, without replacement, what is the probability that both items are defective?

Solution Let the events A and B be defined as follows.
A: the first item is defective.
B: the second item is defective.
Then, $P(A) = \frac{10}{40} = \frac{1}{4}$ and $P(B \mid A) = \frac{9}{39}$. Therefore

$$\begin{aligned} P(A \cap B) &= P(A) P(B \mid A) \\ &= \tfrac{1}{4} \times \tfrac{9}{39} \\ &= \tfrac{9}{156}. \end{aligned}$$

Equation (2.7.3) is a particular case of the following theorem, usually known as the multiplication theorem.

Theorem 2.7.1 If $A_1, A_2, \ldots, A_n$ are events in the sample space S of an experiment, then

$$P(A_1 \cap A_2 \cap \cdots \cap A_n) = P(A_1)P(A_2 \mid A_1)P(A_3 \mid A_1 \cap A_2) \\ \cdots P(A_n \mid A_1 \cap A_2 \cap \cdots \cap A_{n-1}). \quad (2.7.4)$$

Example 2.7.3 Three cards are drawn without replacement from a standard deck of 52 cards. What is the probability that the three cards drawn are aces?

Solution We define the following events.
A_1: the first card is an ace.
A_2: the second card is an ace.
A_3: the third card is an ace.
Then $P(A_1) = \frac{4}{52}$, $P(A_2 \mid A_1) = \frac{3}{51}$, and $P(A_3 \mid A_1 \cap A_2) = \frac{2}{50}$.
According to Theorem 2.7.1, the probability of the event $A_1 \cap A_2 \cap A_3$ is equal to probability of obtaining an ace on the first draw multiplied by the probability of obtaining an ace on the second draw, given that an ace was obtained on the first draw, and multiplied by the probability of obtaining an ace on the third draw, given that an ace was obtained on the first and second draws. That is

$$\begin{aligned} P(A_1 \cap A_2 \cap A_3) &= P(A_1)P(A_2 \mid A_1)P(A_3 \mid A_1 \cap A_2) \\ &= \left(\tfrac{4}{52}\right)\left(\tfrac{3}{51}\right)\left(\tfrac{2}{50}\right) \\ &= \frac{24}{132{,}600}. \end{aligned}$$

2.8 INDEPENDENT EVENTS

The conditional probability, $P(A \mid B)$ of the event A given the event B is, in general, not the same as the unconditional probability $P(A)$. If $P(A \mid B) = P(A)$, we say that the event A is independent of the event B. Events A and B are said to be independent when the occurrence of one does not influence the probability of the occurrence of the other. From the definition of the conditional probability, we have

$$P(A \cap B) = P(B)P(A \mid B).$$

Since $P(A \mid B)$ is by assumption equal to $P(A)$, we find

$$P(A \cap B) = P(A)P(B)$$

and use this as our definition of independence.

Definition 2.8.1 Two events A and B are said to be independent if and only if

$$P(A \cap B) = P(A)P(B).$$

Example 2.8.1 A coin is tossed twice. What is the probability of obtaining two heads?

Solution Let A be the event getting a "head on the first toss" and B be the event "getting a head on the second toss." Since the two events are independent,
$$P(A \cap B) = P(A)P(B)$$
$$= \left(\tfrac{1}{2}\right)\left(\tfrac{1}{2}\right) = \tfrac{1}{4}.$$

Example 2.8.2 In a single throw with two dice, what is the probability of getting two aces?

Solution Let A be the event "getting an ace on the first die" and B be the event "getting an ace on the second die." Since the two events are independent,
$$P(A \cap B) = P(A)P(B)$$
$$= \left(\tfrac{1}{6}\right)\left(\tfrac{1}{6}\right) = \tfrac{1}{36}.$$

2.9 BAYES' THEOREM

Example 2.9.1 Urn I contains four black and six red balls. Urn II contains three black and two red balls. We select an urn at random and then draw one ball at random from that urn.
a) What is the probability of drawing a black ball?
b) What is the probability that urn I was chosen given that the black ball is drawn?

Solution We define the following events.
A: urn I is chosen.
B: urn II is chosen.
E: black ball is drawn.
Then
a) $P(\text{black}) = P(\text{urn I and black}) + P(\text{urn II and black})$. In terms of symbols
$$P(E) = P(A \cap E) + P(B \cap E)$$
or
$$P(E) = P(A)P(E \mid A) + P(B)P(E \mid B).$$
Using the probabilities given in the tree diagram (Fig. 2.9.1), we have
$$P(E) = \left(\tfrac{1}{2}\right)\left(\tfrac{4}{10}\right) + \left(\tfrac{1}{2}\right)\left(\tfrac{3}{5}\right)$$
$$= \tfrac{4}{20} + \tfrac{3}{10} = \tfrac{10}{20}.$$

32   Elementary probability

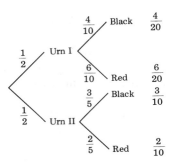

Figure 2.9.1

b) $P(\text{urn I} \mid \text{black}) = P(\text{urn I and black})/P(\text{black})$. In terms of symbols

$$P(A \mid E) = \frac{P(A \cap E)}{P(E)}.$$

Using the tree diagram

$$P(A \cap E) = \tfrac{4}{20}$$

and from Equation (2.9.1), we have

$$P(E) = \tfrac{10}{20}.$$

Therefore

$$P(A \mid E) = \frac{4/20}{10/20} = \frac{4}{10}.$$

A generalization of the above example is known as Bayes' Theorem.

Theorem 2.9.1 (Bayes' Theorem) Let $A_1, A_2, \ldots, A_n$ be mutually exclusive and exhaustive events of the sample space S, where $P(A_i) \neq 0$, $i = 1, 2, \ldots, n$. Let E be any event of S such that $P(E) \neq 0$. Then

$$P(A_1 \mid E) = \frac{P(A_1 \cap E)}{P(A_1 \cap E) + P(A_2 \cap E) + \cdots + P(A_n \cap E)},$$

$$P(A_2 \mid E) = \frac{P(A_2 \cap E)}{P(A_1 \cap E) + P(A_2 \cap E) + \cdots + P(A_n \cap E)}.$$

In general

$$P(A_i \mid E) = \frac{P(A_i \cap E)}{\sum_{i=1}^{n} P(A_i \cap E)}, \quad i = 1, 2, \ldots, n.$$

Proof. Consider the Venn diagram in Fig. 2.9.2.

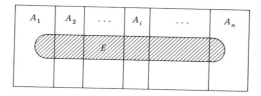

Figure 2.9.2

Since the events $A_1, A_2, \ldots, A_n$ are mutually exclusive and exhaustive, we have
$$S = A_1 \cup A_2 \cup \cdots \cup A_n$$
and
$$A_j \cap A_k = \emptyset: \quad j = 1, 2, \ldots, n; \quad k = 1, 2, \ldots, n; \quad j \neq k.$$
The event E is the union of mutually exclusive events
$$A_1 \cap E, A_2 \cap E, \ldots, A_n \cap E.$$
That is
$$E = (A_1 \cap E) \cup (A_2 \cap E) \cup \cdots \cup (A_n \cap E).$$
Since $A_1 \cap E, A_2 \cap E, \ldots, A_n \cap E$ are mutually exclusive, we have
$$P(E) = P(A_1 \cap E) + P(A_2 \cap E) + \cdots + P(A_n \cap E).$$
Applying the definition of conditional probability, we may write
$$P(A_i \mid E) = \frac{P(A_i \cap E)}{P(E)}$$
$$= \frac{P(A_i \cap E)}{P(A_1 \cap E) + P(A_2 \cap E) + \cdots + P(A_n \cap E)} \quad (2.9.2)$$
$$= \frac{P(A_i \cap E)}{\sum_{j=1}^{n} (A_j \cap E)}, \quad i = 1, 2, \ldots, n.$$

Equation (2.9.2) can also be written as
$$P(A_i \mid E)$$
$$= \frac{P(A_i)P(E \mid A_i)}{P(A_1)P(E \mid A_1) + P(A_2)P(E \mid A_2) + \cdots + P(A_n)P(E \mid A_n)} \quad (2.9.3)$$
$$= \frac{P(A_i)P(E \mid A_i)}{\sum_{j=1}^{n} P(A_j)P(E \mid A_j)}, \quad i = 1, 2, \ldots, n.$$

Example 2.9.2 (Industry) In a bolt factory machines A, B, C manufacture, respectively, 30, 30, and 40 percent of the total output. Of their output 1, 3, and 2 percent are defective bolts. A bolt is drawn from a day's output and is found defective. What is the probability it was manufactured by A? by B? by C?

Solution We define the following events.
E: defective bolt.
A: bolt manufactured by machine A.
B: bolt manufactured by machine B.
C: bolt manufactured by machine C.

Then $P(A) = .30$, $P(E \mid A) = .01$, $P(B) = .30$, $P(E \mid B) = .03$, and $P(C) = .40$, $P(E \mid C) = .02$. Now

$$P(A \cap E) = P(A)P(E \mid A) = (.30)(.01) = .003.$$
$$P(B \cap E) = P(B)P(E \mid B) = (.30)(.03) = .009.$$
$$P(C \cap E) = P(C)P(E \mid C) = (.40)(.02) = .008.$$

Using Bayes' theorem, we have

$$P(A \mid E) = \frac{P(A \cap E)}{P(A \cap E) + P(B \cap E) + P(C \cap E)}$$
$$= \frac{.003}{.003 + .009 + .008} = \frac{3}{20}.$$

Similarly

$$P(B \mid E) = \frac{.009}{.020} = \frac{9}{20}$$

and

$$P(C \mid E) = \frac{.008}{.020} = \frac{8}{20}.$$

2.10 SUMMARY

A sample space of an experiment is the set of all possible outcomes. An element in a sample space is called a sample point. An event is a subset of a sample space of an experiment. We can combine various events (sets) according to the operations of elementary set theory and obtain new events (sets). When an experiment is performed, we set up a sample space of all possible outcomes, and we assign a chance or weight, called the probability of that sample point. The probability of an event is a number which lies between 0 and 1 (inclusive).

Some probability laws about events are given in Section 2.6 and conditional probability of an event is discussed in Section 2.7 (two events are said to be independent when the occurrence of one does not influence the probability of the occurrence of the other). Definition of independence of two events is given in Section 2.8 and in Bayes' Theorem in Section 2.9.

EXERCISES

1. Denote each of the following sets in two ways employed in Section 1.
   a) The set of positive integers.
   b) The set of even positive integers.
   c) The set of outcomes when a coin is tossed.
   d) The set of outcomes when a die is rolled.

2. Let the universal set $\mathbf{U} = \{1, 2, 3, 4, 5, 6, 7, 8\}$, and let $A = \{1, 2, 3\}$, $B = \{1, 3, 5\}$, $C = \{2, 4, 6, 8\}$. Find the following sets:
   a) $A \cap B$
   b) $A \cup B$
   c) $A \cap C$
   d) $A \cup C$
   e) A'
   f) $(A \cup B)'$
   g) $A' \cap B'$
   h) $A \cap B \cap C$
   i) $A \cup B \cup C$

3. Let $A = \{a, b, c, d\}$. List the elements of each of the following sets:
   a) the subsets with exactly one element
   b) the subsets with exactly two elements
   c) the subsets with exactly three elements

4. Let the universal set $\mathbf{U} = \{t \mid t \geq 0\}$. Let $A = \{t \mid t > 50\}$, $B = \{t \mid 50 \leq t \leq 100\}$, and $C = \{t \mid t > 75\}$. Find the following sets:
   a) $A \cup B$
   b) $A \cap B$
   c) $A \cup C$
   d) $A \cap C$
   e) $A \cup B \cup C$
   f) A'

5. Use Venn diagrams to demonstrate relations listed in Theorem 2.2.1.

6. Use Venn diagrams to demonstrate relations listed in Theorem 2.2.2 for $n = 3$.

7. A set $\mathbf{U}$ consists of the subsets A, B, C and D. If from these construct the subsets $E = A \cup B$, $F = A \cup C$, $G = B \cup D$, then construct a Venn diagram showing A, B, C, D, E, F and G.

8. Given that a club has fifteen members.
   a) In how many ways can a committee of four be selected?
   b) In how many ways can a president and secretary be appointed?

9. Evaluate each of the following:
   a) $\binom{5}{3}$
   b) $\binom{15}{12}$
   c) $\binom{7}{0}$
   d) $\binom{9}{9}$
   e) $\binom{500}{1}$
   f) $\binom{50}{49}$

10. Evaluate each of the following:
    a) $\binom{n}{0}$
    b) $\binom{n}{n}$
    c) $\binom{n}{1}$
    d) $\binom{n}{2}$
    e) $\binom{n}{x}$
    f) $\binom{n}{x-1}$

11. Show that $\binom{n}{r} = \binom{n}{n-r}$.

12. In how many ways can a thirteen-card hand be drawn from a standard deck?

13. Six dice are tossed. How many ways are there for every possible number to appear?

14. Seven dice are tossed. How many ways are there for every possible number to appear?

15. Given that five airlines provide service between Toronto and New York, in how many distinct ways can a person select airlines for a trip from Toronto to New York and back if

    a) he must travel both ways by the same airline?
    b) he cannot travel both ways by the same airline?

16. If the license plates are to have two letters of the English alphabet followed by four digits, how many different license plates can be made?

17. In how many ways can five students be seated in a row?

18. In how many ways can five students be seated in a row of 10 seats?

19. How many different luncheons consisting of a soup, a meat dish, a vegetable, a dessert, and a beverage can be ordered from a menu which lists three soups, four meat dishes, five vegetables, six desserts, and three beverages?

20. A die is rolled once. What is the probability of getting

    a) an even number?
    b) a number less than 5?
    c) a number greater than 3?
    d) a number divisible by 3?

21. A ball is drawn at random from an urn containing three white and two black balls. What is the probability that it is white?

22. A five-digit number is formed of the digits 1, 2, 3, 4, 5 with no repetitions. What is the probability that

    a) the number is odd?
    b) the number is divisible by 5?

23. Two numbers are chosen at random without replacement from the set of numbers 1, 2, 3, ..., 17. Find the probability that their sum is 11.

24. If the letters of the word MATCHES be arranged at random, what is the probability that

    a) there are exactly two letters between A and E?
    b) A and E are together?
    c) the word will start with M?

d) the word will end with s?
   e) the word will start with M and end with s?
25. A pair of dice is rolled once. What is the probability of getting a
   (a) 9?   (b) 12?   (c) 9 or 12?
26. A card is drawn from an ordinary deck of 52 cards. What is the probability of getting a
   a) king?   b) queen?
   c) king or a queen? d) face card?
27. Three copies of a notice are to be sent to three people and three copies of another to three others. The six copies of the two notices are placed at random in the six envelopes addressed to those six persons. What is the probability that the notices have been correctly dispatched?
28. Using the results of Exercise 14, what is the probability of each number appearing at least once when seven dice are tossed?
29. An urn contains three red, four yellow and seven white balls. Three balls are drawn without replacement. What is the probability that two will be white, one yellow and one red?
30. What is the probability of getting a total of five points with three dice?
31. The probability that Harry will be late for work is $\frac{2}{3}$ if he walks, $\frac{1}{4}$ if he takes the bus, and $\frac{1}{6}$ if he drives the car. Assuming that one morning his choice is entirely at random, what is the probability of his being late for work?
32. Let A and B be the events in a finite sample space S, such that $P(A) = .6$, $P(B) = .5$, $P(A \text{ and } B) = .2$. Find the probabilities of the following events:
   a) A or B b) A' c) B'
   d) A' or B' e) A' and B' f) A' and B
33. The probability that Dick will be alive 35 years hence is $\frac{3}{8}$ and the probability that his brother will be alive 35 years hence is $\frac{4}{5}$. Assuming that these two events are independent, what is the probability that 35 years hence
   a) both will be alive?
   b) at least one of them will be alive?
   c) only Dick will be alive?
34. A couple hope to have two children. Assuming that the probability of a child being a girl is $\frac{1}{2}$, what is the probability that
   a) the first child will be a girl?
   b) both will be of the same sex?
   c) the children will be of opposite sex?
   d) both will be boys?

38 Elementary probability

35. A building has two elevators; one is waiting at floor 1 20% of the time and the other is waiting there 30% of the time. What is the probability that both are waiting at floor 1? What is the probability that neither is at floor 1? Assume that both elevators operate independently.

36. In a shooting test, the probabilities of hitting the target are $\frac{1}{2}$ for Tom, $\frac{1}{3}$ for Dick, and $\frac{3}{4}$ for Harry. If all of them fire independently at the same target, calculate the probability that

   a) the target will be hit.

   b) only one of them will hit the target.

37. In a sports meet the probabilities of A, B, C winning the 400-meter race are $\frac{2}{9}$, $\frac{1}{3}$, $\frac{1}{9}$, respectively. Assuming the events to be independent, find the probability that

   a) either A or B will win,

   b) neither B nor C will win,

   c) either A or B or C will win,

   d) any competitor other than A or B or C will win.

38. Fifty records are distributed among 10 persons who are seated in chairs numbered from 1 to 10. For each record a number between 1 and 10 is chosen at random and a record is given to a candidate corresponding to the number chosen. What is the probability that a particular person does not receive any record?

39. From six statisticians and five biologists a committee of seven is to be chosen. Find the probability that

   a) the committee includes exactly four statisticians,

   b) the committee includes at least four statisticians.

40. In a bus accident four out of twenty were reported injured. There were four tennis players in this bus. What is the probability that the four reported injured are the tennis players?

41. In a group of four people, what is the probability that there are two people with their birthdays falling on the same date?

42. A dealer of shirts has white, green, yellow, blue, and pink shirts for sale. The sales indicate that when a person buys a shirt the probability that he chooses a white, a green, a yellow, a blue, and a pink shirt are $\frac{1}{2}$, $\frac{1}{5}$, $\frac{1}{10}$, $\frac{3}{20}$, and $\frac{1}{20}$, respectively. Find the probability that he will choose

   a) either a blue or a green shirt,

   b) not a white shirt,

   c) neither a pink nor a yellow shirt.

43. The probability that a gloxinia seed germinates is $\frac{1}{25}$. The probability that it subsequently produces a flower is $\frac{5}{8}$. What is the probability that a gloxinia seed germinates and produces a flower?

44. Urn I contains five red and eight green balls and urn II contains six red and eleven green balls. A ball is drawn at random from one or the other of the two urns. Find the probability of drawing a green ball.

45. A club consists of 30% liberals and 70% conservatives. If 20% of the liberals and 40% of the conservatives smoke pipes, what is the probability that a pipe-smoking club member is a conservative?

46. Two purses contain one silver and two copper coins and two silver and one copper coins respectively. One coin is transferred from the first to the second purse, after which a coin is drawn from the second purse. What is the probability that it is a copper coin?

47. Let A and B be two events of a finite sample space S of an experiment. Let $P(A) = 0.4$, $P(B) = 0.6$, and $P(A \text{ and } B) = 0.2$. Find $P(A \mid B)$ and $P(B \mid A)$.

48. Let A and B be two events of a finite sample space S of an experiment. Let $P(A \mid B) = 0.3$ and $P(B \mid A) = 0.6$, and $P(A \text{ and } B) = 0.3$. Find $P(A)$ and $P(B)$.

49. Let A and B be two events of a finite sample space S of an experiment. Let $P(A \mid B) = .7$, $P(A \text{ and } B) = .2$, and $P(A') = .3$. Find $P(A \text{ or } B)$.

50. A pair of dice is tossed repeatedly. A win occurs if a 9 appears before a 10. What is the probability of winning?

51. A bridge player and his partner have five spades between them. What is the probability that they are split two-three?

52. If it is known that at least three heads appeared in six tosses of a coin, what is the probability that exactly five heads appeared?

53. An urn contains eight balls, numbered 1 through 8. Four balls are drawn, without replacement. What is the probability that the smallest number drawn was a 2?

54. Two dice are thrown. What is the probability that their sum is even if it is known that it is greater than 4?

Chapter 3

RANDOM VARIABLES AND MATHEMATICAL EXPECTATION

3.1 THE NOTION OF A RANDOM VARIABLE

In Chapter 2 we considered a number of random experiments such as throwing a die, tossing a coin, drawing a ball from an urn, drawing a card from a well-shuffled deck of cards. Basic concepts of probability were developed in terms of sample space and events. When an experiment is performed, the possible outcomes may be numerical quantities such as the outcomes in the case of rolling a die, or outcomes may be of a qualitative nature such as "heads or tails," "black or white," "defective or non-defective." Such quantities whose value is determined by the outcome of a random experiment are called random variables.

Definition 3.1.1 A variable whose value is determined by the outcome of a random experiment is called a random variable.

Example 3.1.1 A few examples of random variables are
1. the number of goals scored by an NHL team from game to game
2. the weekly or monthly number of cars sold by a car dealer
3. the number of children in a family
4. the number of different brands of cigarettes tested by a consumer until he finds one that is satisfactory
5. the life length of an electron tube
6. the percentage of alcohol in a certain compound

Definition 3.1.2 Let X be a random variable. If the number of possible values of X is finite or countably infinite (infinite number of values that are countable), then we call X a discrete random variable.

In Example 3.1.1, the first three random variables are discrete with a finite number of possible outcomes, and the fourth random variable is discrete with countably infinite number of possible outcomes.

Example 3.1.2 Consider the experiment of rolling a die. The sample space for this experiment is
$$S = \{1, 2, 3, 4, 5, 6\}.$$
This sample space has a finite number of elements, and a random variable X defined over this sample space is a discrete random variable.

Example 3.1.3 Consider the experiment of tossing a coin twice. The sample space for this experiment is
$$S = \{HH, HT, TH, TT\}.$$
Let the random variable X be "the number of heads obtained." Thus, possible values of X are 0, 1, 2 and X is a discrete random variable with a finite number of possible outcomes.

Example 3.1.4 Consider the experiment of tossing a coin until a head appears. Let the random variable N be the number of tosses required to get a first head.

The outcome of the experiment consists of the sequence of heads (H) and tails (T). The possible outcomes of the experiment are

$$H$$
$$T\ H$$
$$T\ T\ H$$
$$T\ T\ T\ H$$
$$.\ .\ .\ .\ .$$

Clearly, the number of possible outcomes is not finite. We could have an indefinitely long sequence of tails before getting a head. However, we can make a list or enumerate the set of all possible outcomes. The list might continue indefinitely, and hence the sample space of this experiment is countably infinite.

The countably infinite space has as many sample points as there are positive integers. That is, a one-to-one correspondence can be set up between the possible values of countably infinite sample points and the set of positive integers.

In many physical experiments the measurements can take every possible value in an interval. That is, the sample space is not countable and the random variable can take any real number to an outcome such as measurements of heights, weights, and time. Random variables of this type are called continuous.

Definition 3.1.3 Let X be a random variable. If the possible values of X form an interval or a collection of intervals, then we call X a continuous random variable.

In Example 3.1.1, the fifth and sixth random variables are continuous.

3.2 DISTRIBUTION OF A DISCRETE RANDOM VARIABLE

The probability function of a discrete random variable is the specification of its possible values together with their respective probabilities. We shall denote random variables by capital letters and their specific values by small letters.

Example 3.2.1 Consider the experiment of tossing a coin twice. Let the random variable X be the "number of heads obtained." The sample space for this experiment is
$$S = \{HH, HT, TH, TT\}.$$

The possible values of X are 0, 1, 2. Let $f(x)$ be the probability that the random variable X takes on the value x, then

$$f(x) = P(X = x).$$
Thus
$$f(0) = P(X = 0) = \tfrac{1}{4},$$
$$f(1) = P(X = 1) = \tfrac{1}{2},$$
$$f(2) = P(X = 2) = \tfrac{1}{4}.$$

We note that the values of X exhaust all possible values and the sum $f(0) + f(1) + f(2) = 1$.

The possible values of X, and their corresponding probabilities are shown in Table 3.2.1.

TABLE 3.2.1

$X = x$	0	1	2
$f(x) = P(X = x)$	$\tfrac{1}{4}$	$\tfrac{1}{2}$	$\tfrac{1}{4}$

Definition 3.2.1 Let X be a discrete random variable with finite number of values $x_1, x_2, \ldots, x_N$ and corresponding probabilities $f(x_i) = P(X = x_i)$, $i = 1, 2, \ldots, N$, then the function $f(x)$ is called the distribution of X or probability function of X if the following conditions are satisfied:

i) $f(x) \geq 0$ for all x.   (3.2.1)

ii) $\sum_{i=1}^{N} f(x_i) = 1.$ (3.2.2)

The probability function of the random variable X thus has the following form:

$X = x$	x_1	x_2	$\ldots$	x_N
$f(x) = P(X = x)$	$f(x_1)$	$f(x_2)$	$\ldots$	$f(x_N)$

We are often interested not only in the probability that a random variable X is equal to one of its possible values but also in the probability that X is less than or equal to one of its possible values. This accumulated probability is called the cumulative function or the distribution function of the random variable X, defined as follows.

Definition 3.2.2 The cumulative function or the distribution function of a random variable X is the probability $P(X \leq x)$ that X is less than or equal to x, and denoted by the symbol $F(x)$, then

$$F(x) = P(X \leq x) = \sum_{x_i \leq x} f(x_i). \qquad (3.2.3)$$

Example 3.2.2 A fair die is rolled once. Find the probability function of the number of points appearing on the top face.

Solution The sample space for this experiment is

$$S = \{1, 2, 3, 4, 5, 6\}.$$

Let the random variable X denote the number of points that appear on the top. Then X can be any integer from 1 to 6. There are six points in the sample space, each occurs with probability $\frac{1}{6}$. Thus, the probability function of the random variable X is

$X = x$	1	2	3	4	5	6
$f(x) = P(X = x)$	$\frac{1}{6}$	$\frac{1}{6}$	$\frac{1}{6}$	$\frac{1}{6}$	$\frac{1}{6}$	$\frac{1}{6}$

The above probability function can also be written as

$$f(x) = \tfrac{1}{6}, \quad x = 1, 2, 3, 4, 5, 6.$$

Example 3.2.3 A coin is tossed once. Find the probability function of the number of heads obtained.

Solution The sample space for this experiment is

$$S = \{H, T\}.$$

Let the random variable X denote the number of heads, then X takes on the values 0, 1. There are two sample points in the sample space; each occurs with probability $\frac{1}{2}$. Thus, the probability function of the random variable X is

$X = x$	0	1
$f(x) = P(X = x)$	$\frac{1}{2}$	$\frac{1}{2}$

Alternatively, the above probability function can be written as

$$f(x) = \tfrac{1}{2}, \quad x = 0, 1.$$

Definition 3.2.3 Let X be a discrete random variable with countably infinite number of values $x_1, x_2, \ldots, x_N, \ldots$ and corresponding probabilities $f(x_i) = P(X = x_i)$, $i = 1, 2, \ldots$, then the function $f(x)$ is called the probability distribution of X or the probability function of X if the following conditions are satisfied:

i) $f(x) \geq 0$, for all x. (3.2.4)

ii) $\sum\limits_{i=1}^{\infty} f(x_i) = 1.$ (3.2.5)

Thus the probability function of the random variable X has the following form:

$X = x$	x_1	x_2	$\ldots$
$f(x) = P(X = x)$	$f(x_1)$	$f(x_2)$	$\ldots$

Example 3.2.4 A coin is tossed until a head appears. Let N be the random variable representing the number of tosses required. Find the probability function of the random variable N.

Solution If H stands for head and T for tail, then the possible outcomes of this experiment and their corresponding probabilities are

	1st toss	H	$\frac{1}{2}$
	2nd toss	TH	$(\frac{1}{2})(\frac{1}{2})$
	3rd toss	TTH	$(\frac{1}{2})(\frac{1}{2})(\frac{1}{2})$
	.	.	.
	.	.	.
	.	.	.
	Nth toss	$TT \ldots TH$	$(\frac{1}{2})^{N-1}(\frac{1}{2})$

Thus the probability function of the random variable N is

$N = n$	1	2	...	n
$f(n) = P(N = n)$	$\frac{1}{2}$	$(\frac{1}{2})^2$	...	$(\frac{1}{2})^n$

Alternatively, the probability function of the random variable N can be written as

$$f(n) = (\tfrac{1}{2})^n, \quad n = 1, 2, \ldots.$$

To show that the sum of the probabilities is 1, we need to add the probabilities

$$\sum_{n=1}^{\infty} (\tfrac{1}{2})^n = \tfrac{1}{2} + (\tfrac{1}{2})^2 + (\tfrac{1}{2})^3 + \cdots.$$

This is an infinite geometric series and for such a series

$$a + ar + ar^2 + \cdots = \frac{a}{1-r}$$

where a is the first term and r, the common ratio between any two consecutive terms, is between -1 and $+1$.

In this case, since $a = \tfrac{1}{2}$, $r = \tfrac{1}{2}$, then

$$\sum_{n=1}^{\infty} f(n) = \frac{\tfrac{1}{2}}{1 - \tfrac{1}{2}} = 1.$$

3.3 DISTRIBUTION OF A CONTINUOUS RANDOM VARIABLE

Our discussion so far has been concerned with experiments with a finite or countably infinite number of outcomes. In many applications the random variable can take on any value within an interval or collection of intervals. For example, suppose an automatic machine is filling coffee jars with one pound of coffee. Due to some faults in the automatic process, the actual weight could vary from jar to jar. The weight of a jar is a random variable and the possible values of this random variable could be .90, .95, .99, 1.00, 1.01, 1.05 pounds and infinitely many more. Suppose we want to know the probability of selecting a jar at random with a weight lying between any two values, say .95 and 1.05 pounds. The possible values of this random variable lying between .95 and 1.05 pounds is not finite, and not even countably infinite. Therefore, the problem of assigning probabilities to continuous random variables cannot be treated as discrete random variables. However, the concept of area and its properties prove powerful in assigning probabilities to continuous random variables. The probability density function of a continuous random variable defined below plays a similar role to that of a probability function for a discrete random variable.

Definition 3.3.1 Let X be a continuous random variable defined over an interval (a, b) as shown in Fig. 3.3.1. Then the function $f(x)$ is called the probability density function of the random variable X if the following conditions are satisfied:

i) $f(x) \geq 0, \quad a < x < b$.
ii) The total area under the curve $f(x)$ above the x-axis and bounded by the ordinates $x = a$ and $x = b$ is equal to 1.

Further, if $f(x)$ satisfies (i) and (ii), then the probability that a continuous random variable X lies between c and d $(a < c < d < b)$, denoted by $P(c < X < d)$, is given by

$$P(c < X < d) = \text{Area under the curve } f(x) \text{ bounded by the } x\text{-axis and the two ordinates } x = c \text{ and } x = d.$$

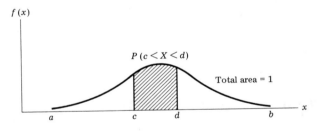

Figure 3.3.1

If the interval corresponding to the specific value of a continuous random variable X, say $X = x$, has no length, then $P(X = x) = 0$.

Example 3.3.1 Given the probability density function

$$f(x) = 1, \quad 0 < X < 1.$$

Find
a) $P(.25 < X < .75)$,
b) $P(X > .25)$.

Solution The graph of $f(x) = 1, 0 < X < 1$, is shown in Fig. 3.3.2.

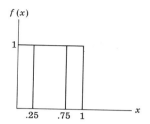

Figure 3.3.2

The region between the curve $f(x) = 1$ bounded by the x-axis and the ordinates $x = 0$ and $x = 1$ is a square, whose area is 1. Thus

a) $P(.25 < X < .75)$ = Area within the square bounded by the x-axis and the ordinates $x = .25$ and $x = .75$
= Area of a rectangle with length 1 and width .50
= $1 \times .50 = .50.$

Therefore $P(.25 < X < .75) = .50.$

b) $P(X > .25)$ = Area within the square and to the right of the ordinate $x = .25$
= Area of a rectangle with length 1 and width .75
= $1 \times .75 = .75.$

Example 3.3.2 Let the random variable X have the probability density function whose graph is shown in Fig. 3.3.3. Find
a) $P(0 < X < 1)$,
b) $P(1 < X < 2)$.

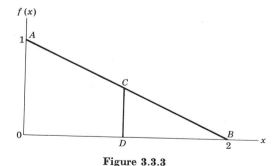

Figure 3.3.3

Solution The area of the triangle OAB (Fig. 3.3.3) is
$$\text{Area} = \tfrac{1}{2} \text{ base} \times \text{height}$$
$$= \tfrac{1}{2} \times 2 \times 1 = 1.$$

Thus we have

a) $P(0 < X < 1)$ = Area of trapezoid $OACD$
$= \frac{1}{2}(OA + CD) \times OD$
$= \frac{1}{2}(1.0 + .5) \times 1.0$
$= .75.$

b) $P(1 < X < 2)$ = Area of a triangle DCB
$= \frac{1}{2} DB \times CD$
$= \frac{1}{2} \times 1 \times .5$
$= .25.$

3.4 THE SUMMATION NOTATION

When an experiment is performed in probability and statistics, data consist of several outcomes, measurements, or observations of some characteristic of a number of items or individuals, such as the heights of male students at a certain university, annual incomes of a certain group of people etc. In order to differentiate between different outcomes, measurements, or observations, we let x_1 (read "x sub one") represent the first outcome, measurement, or observation; we let x_2 (read "x sub two") represent the second, and x_N (read "x sub N") represent the Nth outcome, measurement, or observation. To find the sums of large number of terms $x_1, x_2, \ldots, x_N$, a notational shortcut is needed to indicate summation.

Definition 3.4.1 Let the symbol x_i (read "x sub i") denote the N values $x_1, x_2, \ldots, x_N$. Then the symbol $\sum_{i=1}^{N} x_i$ (read "summation of x sub i" or "sigma x sub i") is used to denote the sum of all the x_i's from $i = 1$ to $i = N$, that is

$$\sum_{i=1}^{N} x_i = x_1 + x_2 + \cdots + x_N. \tag{3.4.1}$$

The Greek capital letter Σ denotes the sum and the summation index i assumes all integral values from 1 to N inclusive.

Example 3.4.1 Express each of the following summations in expanded form.

a) $\sum_{i=1}^{5} x_i$ b) $\sum_{i=1}^{4} x_i^2$ c) $\sum_{i=1}^{6} f_i x_i$

d) $\sum_{i=1}^{3} (x_i - 3)^2$ e) $\sum_{i=1}^{4} f_i(x_i - 3)^2$ f) $\left(\sum_{i=2}^{4} x_i\right)^2$

Solution By Definition 3.4.1 we have

a) $\sum_{i=1}^{5} x_i = x_1 + x_2 + x_3 + x_4 + x_5,$

b) $\sum_{i=1}^{4} x_i^2 = x_1^2 + x_2^2 + x_3^2 + x_4^2,$

c) $\sum_{i=1}^{6} f_i x_i = f_1 x_1 + f_2 x_2 + f_3 x_3 + f_4 x_4 + f_5 x_5 + f_6 x_6,$

d) $\sum_{i=1}^{3} (x_i - 3)^2 = (x_1 - 3)^2 + (x_2 - 3)^2 + (x_3 - 3)^2,$

e) $\sum_{i=1}^{4} f_i (x_i - 3)^2 = f_1 (x_1 - 3)^2 + f_2 (x_2 - 3)^2 + f_3 (x_3 - 3)^2 + f_4 (x_4 - 3)^2.$

f) $\left(\sum_{i=2}^{4} x_i \right)^2 = (x_2 + x_3 + x_4)^2$

Example 3.4.2 Given that $x_1 = 2$, $x_2 = 4$, $x_3 = 6$, $x_4 = 8$, find

a) $\sum_{i=1}^{4} x_i,$

b) $\left(\sum_{i=1}^{4} x_i \right)^2,$

c) $\sum_{i=1}^{4} x_i^2,$

d) $\sum_{i=1}^{4} (x_i - 5)^2.$

Solution

a) $\sum_{i=1}^{4} x_i = x_1 + x_2 + x_3 + x_4$
$= 2 + 4 + 6 + 8$
$= 20.$

b) $\left(\sum_{i=1}^{4} x_i \right)^2 = (x_1 + x_2 + x_3 + x_4)^2$
$= (2 + 4 + 6 + 8)^2$
$= 20^2 = 400.$

c) $\sum_{i=1}^{4} x_i^2 = x_1^2 + x_2^2 + x_3^2 + x_4^2$
$= 2^2 + 4^2 + 6^2 + 8^2$
$= 4 + 16 + 36 + 64$
$= 120.$

d) $\sum_{i=1}^{4} (x_i - 5)^2 = (x_1 - 5)^2 + (x_2 - 5)^2 + (x_3 - 5)^2 + (x_4 - 5)^2$
$= (2 - 5)^2 + (4 - 5)^2 + (6 - 5)^2 + (8 - 5)^2$
$= 9 + 1 + 1 + 9$
$= 20.$

Three properties which could be used to simplify operations using summation notation are given below.

Theorem 3.4.1 The summation of a constant multiplied by a variable is equal to the constant multiplied by the summation of the variable. Thus if c is a constant, then

$$\sum_{i=1}^{N} cx_i = c \sum_{i=1}^{N} x_i. \qquad (3.4.2)$$

Proof. Expanding the left-hand side of Equation (3.4.2) and rearranging terms, we get

$$\sum_{i=1}^{N} cx_i = cx_1 + cx_2 + \cdots + cx_N$$
$$= c(x_1 + x_2 + \cdots + x_N)$$
$$= c \sum_{i=1}^{N} x_i.$$

Example 3.4.3 Given that $x_1 = 2$, $x_2 = 3$, $x_3 = 4$, find

$$\sum_{i=1}^{3} 5x_i.$$

Solution By Theorem 3.4.1 we get

$$\sum_{i=1}^{3} 5x_i = 5 \sum_{i=1}^{3} x_i$$
$$= 5(x_1 + x_2 + x_3)$$
$$= 5(2 + 3 + 4)$$
$$= 45.$$

Theorem 3.4.2 The summation of a constant is equal to the number of terms in the summation multiplied by the constant. Thus if c is a constant, then

$$\sum_{i=1}^{N} c = Nc. \qquad (3.4.3)$$

Proof. Expanding the left-hand side of Equation (3.4.3), we get

$$\sum_{i=1}^{N} c = \underbrace{c + c + \cdots + c}_{N \text{ terms}} = Nc.$$

Example 3.4.4 Find $\sum_{i=1}^{5} 3$.

Solution By Theorem 3.4.2 we get

$$\sum_{i=1}^{5} 3 = (5)(3) = 15.$$

Theorem 3.4.3 The summation of the sum of two or more variables is equal to the sum of the summations of the individual variables. Thus

$$\sum_{i=1}^{N} (x_i + y_i + z_i) = \sum_{i=1}^{N} x_i + \sum_{i=1}^{N} y_i + \sum_{i=1}^{N} z_i. \quad (3.4.4)$$

Proof. Expanding the left-hand side of the Equation (3.4.3) and rearranging terms, we get

$$\sum_{i=1}^{N} (x_i + y_i + z_i) = (x_1 + y_1 + z_1) + (x_2 + y_2 + z_2) + \cdots + (x_N + y_N + z_N)$$
$$= (x_1 + x_2 + \cdots + x_N) + (y_1 + y_2 + \cdots + y_N)$$
$$+ (z_1 + z_2 + \cdots + z_N)$$
$$= \sum_{i=1}^{N} x_i + \sum_{i=1}^{N} y_i + \sum_{i=1}^{N} z_i.$$

Example 3.4.5 Given that $x_1 = 5$, $x_2 = -3$, $y_1 = 6$, $y_2 = -1$. Find

$$\sum_{i=1}^{2} (4x_i - 3y_i + 2).$$

Solution By Theorems 3.4.1, 3.4.2, and 3.4.3, we get

$$\sum_{i=1}^{2} (4x_i - 3y_i + 2) = \sum_{i=1}^{2} 4x_i - \sum_{i=1}^{2} 3y_i + \sum_{i=1}^{2} 2$$
$$= 4\sum_{i=1}^{2} x_i - 3\sum_{i=1}^{2} y_i + (2)(2)$$
$$= 4(x_1 + x_2) - 3(y_1 + y_2) + 4$$
$$= 4(5 - 3) - 3(6 - 1) + 4$$
$$= -3.$$

3.5 TWO-DIMENSIONAL RANDOM VARIABLES

When an experiment is performed, it is frequently necessary to consider the simultaneous behavior of several random variables associated with the same experiment. For example, we might study the height and the weight of some person drawn at random from a given population. In a sample survey we might consider the joint distribution of income and age of householders. Quite often we take observations on two or more characteristics of an object or an individual and then study their joint behavior.

Example 3.5.1 In three tosses of a fair coin let X be the total number of heads and let Y be the number of heads on the first two tosses. Then the sample space S and the corresponding values of X and Y are

Element of S	Value of X	Value of Y
HHH	3	2
HHT	2	2
HTH	2	1
THH	2	1
HTT	1	1
THT	1	1
TTH	1	0
TTT	0	0

We want to determine the probability of each pair of values of X and Y. For example, the event that $X = 3$ and $Y = 2$ simultaneously is the event $\{HHH\}$ which has probability $\frac{1}{8}$ and we write

$$P(X = 3, Y = 2) = \tfrac{1}{8}$$

using a comma in place of $\cap$ to denote the intersection of the two events $X = 3$ and $Y = 2$. We similarly find

$$P(X = 0, Y = 1) = P(\varnothing) = 0, \text{ etc.}$$

The probabilities of all possible values of X and Y are given in Table 3.5.1.

TABLE 3.5.1

X \ Y	0	1	2	Row totals
0	$\frac{1}{8}$	0	0	$\frac{1}{8}$
1	$\frac{1}{8}$	$\frac{1}{4}$	0	$\frac{3}{8}$
2	0	$\frac{1}{4}$	$\frac{1}{8}$	$\frac{3}{8}$
3	0	0	$\frac{1}{8}$	$\frac{1}{8}$
Column totals	$\frac{1}{4}$	$\frac{1}{2}$	$\frac{1}{4}$	1

Probability function for X (row totals)

Probability function for Y (column totals)

We call this table the joint probability table of X and Y. In general, the entry in row i, where $i = 0, 1, 2, 3$ and column j, where $j = 0, 1, 2$ is the probability that $X = i$, $Y = j$. We denote this probability by $P(X = i, Y = j)$.

The probability function of the random variable X is found by summing the elements in each row. Similarly, the probability function of the random variable Y is found by summing the elements in each column. For example, the probability of the event $X = 1$ is obtained by adding the probabilities

$$P(X = 1, Y = 0) + P(X = 1, Y = 1) + P(X = 1, Y = 2)$$
$$= \tfrac{1}{8} + \tfrac{1}{4} + 0 = \tfrac{3}{8}.$$

Definition 3.5.1 Let (X, Y) be a two-dimensional discrete random variable defined on the sample space S. Let (x_i, y_j), $i = 1, 2, \ldots, M$; $j = 1, 2, \ldots, N$ be the possible outcomes of (X, Y). Then the function

$$f(x_i, y_j) = P(X = x_i, Y = y_j) \tag{3.5.1}$$

is called the joint probability distribution or the joint probability function of the random variables X and Y if the following conditions are satisfied:

i) $f(x_i, y_j) \geq 0$,   for all (x, y). $\tag{3.5.2}$

ii) $\sum_{i=1}^{M} \sum_{j=1}^{N} f(x_i, y_j) = 1.$ $\tag{3.5.3}$

The joint probability function of the random variables X and Y thus has the form shown in Table 3.5.2.

TABLE 3.5.2

X \ Y	y_1	y_2	$\ldots$	y_j	$\ldots$	y_N	$P(X = x)$
x_1	$f(x_1, y_1)$	$f(x_1, y_2)$	$\ldots$	$f(x_1, y_j)$	$\ldots$	$f(x_1, y_N)$	$f(x_1)$
x_2	$f(x_2, y_1)$	$f(x_2, y_2)$	$\ldots$	$f(x_2, y_j)$	$\ldots$	$f(x_2, y_N)$	$f(x_2)$
.	.	.		.		.	.
.	.	.		.		.	.
x_i	$f(x_i, y_1)$	$f(x_i, y_2)$	$\ldots$	$f(x_i, y_j)$	$\ldots$	$f(x_i, y_N)$	$f(x_i)$
.	.	.		.		.	.
.	.	.		.		.	.
x_M	$f(x_M, y_1)$	$f(x_M, y_2)$	$\ldots$	$f(x_M, y_j)$	$\ldots$	$f(x_M, y_N)$	$f(x_M)$
$P(Y = y)$	$f(y_1)$	$f(y_2)$	$\ldots$	$f(y_j)$	$\ldots$	$f(y_N)$	1

The probabilities listed in the last column of this table provide the probability distribution for X; it is called the marginal probability function for X. The marginal probability function for X has the following properties:

i) $f(x_i) = \sum_{j=1}^{N} f(x_i, y_j), \qquad i = 1, 2, \ldots, M.$ (3.5.4)

ii) $\sum_{i=1}^{M} f(x_i) = 1.$ (3.5.5)

Similarly, the probabilities listed in the last row provide the probability distribution for Y; it is called the marginal probability function for Y. The marginal probability function for Y has the following properties:

i) $f(y_j) = \sum_{i=1}^{M} f(x_i, y_j), \qquad j = 1, 2, \ldots, N.$ (3.5.6)

ii) $\sum_{j=1}^{N} f(y_j) = 1.$ (3.5.7)

Example 3.5.2 An urn contains two white and four black balls. A random sample of two balls is drawn without replacement. Find the joint probability distribution of the two balls drawn.

Solution We define the following random variables.
  i) $X = 0$ if the first ball is white.
  ii) $X = 1$ if the first ball is black.
  iii) $Y = 0$ if the second ball is white.
  iv) $Y = 1$ if the second ball is black.

The joint probability function of the random variables X and Y is shown in Table 3.5.3.

TABLE 3.5.3

X \ Y	0	1	$P(X = x)$
0	$\frac{2}{6} \cdot \frac{1}{5}$	$\frac{2}{6} \cdot \frac{4}{5}$	$\frac{10}{30}$
1	$\frac{4}{6} \cdot \frac{2}{5}$	$\frac{4}{6} \cdot \frac{3}{5}$	$\frac{20}{30}$
$P(Y = y)$	$\frac{10}{30}$	$\frac{20}{30}$	1

You will recall that we have earlier defined the independence of two or more events. Two events A and B are independent if

$$P(A \cap B) = P(A)P(B). \tag{3.5.8}$$

Suppose in the above example, we let A be the event $X = 0$ and B be the event $Y = 0$. Then from Table 3.5.3 we have

$$P(A \cap B) = \tfrac{2}{30}, \qquad P(A) = \tfrac{1}{3}, \qquad P(B) = \tfrac{1}{3}.$$

Equation (3.5.8) is not satisfied. Thus, the two events A and B are not independent. We now introduce the idea of independent random variables.

Definition 3.5.2 Let (X, Y) be a two-dimensional discrete random variable defined on the sample space S. Let (x_i, y_j), $i = 1, 2, \ldots, M$; $j = 1, 2, \ldots, N$ be the possible outcomes of X, Y. Then the random variables X and Y are independent if and only if

$$P(X = x_i, Y = y_j) = P(X = x_i)P(Y = y_j) \tag{3.5.9}$$

for all pairs of (x_i, y_j).

Random variables that are not independent are said to be dependent. In Example 3.5.2, the random variables "the first ball is white" and "the second ball is black" are dependent.

3.6 THE EXPECTED VALUE OF A RANDOM VARIABLE

Suppose that a player tosses a die, and if the even number appears, he will receive 6 cents and if the odd number appears he loses 4 cents. That is, for single toss of a die, he would either receive 6 cents or lose 4 cents. Thus, the player's pay-offs are 6, −4 and their respective probabilities are $\tfrac{1}{2}, \tfrac{1}{2}$. The average amount the player would make if he plays this game over and over is

$$6(\tfrac{1}{2}) - 4(\tfrac{1}{2}) = 1 \text{ cent}.$$

In an infinite number of trials we expect him to make an average of 1 cent per toss. This long-run average of 1 cent per toss is called the mathematical expectation.

Definition 3.6.1 Let X be a discrete random variable with the following probability function:

$X = x$	x_1	x_2	$\ldots$	(x_N)
$f(x) = P(X = x)$	$f(x_1)$	$f(x_2)$	$\ldots$	$f(x_N)$

Then the mathematical expectation or the expected value of X, denoted by $E(X)$, is defined as

$$E(X) = x_1 f(x_1) + x_2 f(x_2) + \cdots + x_N f(x_N) = \sum_{i=1}^{N} x_i f(x_i). \quad (3.6.1)$$

If the random variable X has a countably infinite number of outcomes $x_1, x_2, \ldots,$ then

$$E(X) = x_1 f(x_1) + x_2 f(x_2) + \cdots = \sum_{i=1}^{\infty} x_i f(x_i). \quad (3.6.2)$$

This number $E(X)$ is also called the mean of the random variable X or the population mean. The Greek letter μ (mu) is often used to denote the expected value or the mean. When several random variables X, Y, ... are being studied we may use subscripts on μ to indicate the means. Thus we would write

$$\mu_X = E(X) \quad \text{and} \quad \mu_Y = E(Y).$$

When there is only one random variable under consideration, the subscript is usually omitted. Hence

$$E(X) = \mu_X = \mu.$$

Example 3.6.1 A fair die is thrown. What is the mathematical expectation of the number of points that appear on the top face?

Solution Let X denote the number of points obtained on the top face of the die. The possible values of X are $1, 2, \ldots, 6$, and each occurs with probability $\frac{1}{6}$. Hence by Equation (3.6.1)

$$E(X) = 1(\tfrac{1}{6}) + 2(\tfrac{1}{6}) + 3(\tfrac{1}{6}) + 4(\tfrac{1}{6}) + 5(\tfrac{1}{6}) + 6(\tfrac{1}{6}) = \tfrac{21}{6} = \tfrac{7}{2}.$$

Example 3.6.2 A fair coin is tossed three times. What is the mathematical expectation of the number of heads obtained?

TABLE 3.6.1

Sample point	Number of heads	Probability
HHH	3	$\tfrac{1}{8}$
HHT	2	$\tfrac{1}{8}$
HTH	2	$\tfrac{1}{8}$
THH	2	$\tfrac{1}{8}$
HTT	1	$\tfrac{1}{8}$
THT	1	$\tfrac{1}{8}$
TTH	1	$\tfrac{1}{8}$
TTT	0	$\tfrac{1}{8}$

Solution The sample space for this experiment and the corresponding probabilities are shown in Table 3.6.1.

Let X denote the number of heads obtained when the coin is tossed three times. The probability function of X is given in Table 3.6.2.

TABLE 3.6.2

$X = x$	0	1	2	3
$f(x) = P(X = x)$	$\frac{1}{8}$	$\frac{3}{8}$	$\frac{3}{8}$	$\frac{1}{8}$

Hence by Equation (3.6.1)

$$E(X) = \sum_{i=1}^{3} x_i f(x_i)$$
$$= 0(\tfrac{1}{8}) + 1(\tfrac{3}{8}) + 2(\tfrac{3}{8}) + 3(\tfrac{1}{8})$$
$$= \tfrac{12}{8} = \tfrac{3}{2}.$$

The graph of the probability function given in Table 3.6.2 is shown in Fig. 3.6.1.

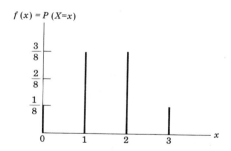

Figure 3.6.1

Example 3.6.3 From an urn containing three white and two black balls, a man draws two balls at random with replacement. If he gets $100 for every white ball he draws and loses $50 for every black ball, what is the mathematical expectation of this game?

Solution Let the random variable X be the number of dollars won. Table 3.6.3 shows a sample space of possible outcomes, corresponding probabilities, and the profits in dollars.

TABLE 3.6.3

		Second ball	
		W	B
First ball	W	$\frac{9}{25}$: $200	$\frac{6}{25}$: $50
	B	$\frac{6}{25}$: $50	$\frac{4}{25}$: $-$100

The expected value of the random variable X, the profit on one play, is

$$E(X) = 200(\tfrac{9}{25}) + 50(\tfrac{6}{25}) + 50(\tfrac{6}{25}) - 100(\tfrac{4}{25}) = \$80.$$

3.7 PROPERTIES OF EXPECTED VALUE

In this section we shall develop some basic properties of expected value. These basic properties will simplify our computations of expectation of functions of random variables. For example, if X and Y are two random variables, we might be interested in such functions as $X + Y$, XY, or X/Y. The results listed below are valid for both discrete and continuous random variables. To avoid calculus, proofs will be given only for discrete random variables.

Definition 3.7.1 Let X be a discrete random variable with the following probability function:

$X = x$	x_1	x_2	$\cdots$	x_N
$f(x) = P(X = x)$	$f(x_1)$	$f(x_2)$	$\cdots$	$f(x_N)$

Let Y be a function of X. Then the mean, or expected value, of the new random variable $Y = g(X)$ is given by

$$E[g(X)] = g(x_1)f(x_1) + g(x_2)f(x_2) + \cdots + g(x_N)f(x_N)$$

or

$$E(Y) = E[g(X)] = \sum_{i=1}^{N} g(x_i)f(x_i). \qquad (3.7.1)$$

Example 3.7.1 The random variable X has the following probability function:

$X = x$	1	2	3	4	5	6
$f(x) = P(X = x)$	$\tfrac{1}{6}$	$\tfrac{1}{6}$	$\tfrac{1}{6}$	$\tfrac{1}{6}$	$\tfrac{1}{6}$	$\tfrac{1}{6}$

Compute

a) $E(X^2)$, b) $E(2X)$, c) $E(2X - 1)$.

Solution (a) The possible values of X^2 and their corresponding probabilities are

X^2	1	4	9	16	25	36
$f(x) = P(X = x)$	$\frac{1}{6}$	$\frac{1}{6}$	$\frac{1}{6}$	$\frac{1}{6}$	$\frac{1}{6}$	$\frac{1}{6}$

Then from Equation (3.6.1) we find

$$E(X^2) = 1(\tfrac{1}{6}) + 4(\tfrac{1}{6}) + 9(\tfrac{1}{6}) + 16(\tfrac{1}{6}) + 25(\tfrac{1}{6}) + 36(\tfrac{1}{6}) = \tfrac{91}{6}.$$

Note that $E(X^2) \neq [E(X)]^2$; that is, the mean of the square does not equal the square of its mean.

b) The possible values of $2X$, and their corresponding probabilities are

$2X$	2	4	6	8	10	12
$f(x) = P(X = x)$	$\frac{1}{6}$	$\frac{1}{6}$	$\frac{1}{6}$	$\frac{1}{6}$	$\frac{1}{6}$	$\frac{1}{6}$

Then from Equation (3.6.1) we find

$$\mu_{2X} = E(2X) = 2(\tfrac{1}{6}) + 4(\tfrac{1}{6}) + 6(\tfrac{1}{6}) + 8(\tfrac{1}{6}) + 10(\tfrac{1}{6}) + 12(\tfrac{1}{6}) = \tfrac{42}{6} = 7.$$

c) The possible values of $2X - 1$, and their corresponding probabilities are

$2X - 1$	1	3	5	7	9	11
$f(x) = P(X = x)$	$\frac{1}{6}$	$\frac{1}{6}$	$\frac{1}{6}$	$\frac{1}{6}$	$\frac{1}{6}$	$\frac{1}{6}$

Then from Equation (3.6.1) we find

$$\mu_{2X-1} = E(2X - 1) = 1(\tfrac{1}{6}) + 3(\tfrac{1}{6}) + 5(\tfrac{1}{6}) + 7(\tfrac{1}{6}) + 9(\tfrac{1}{6}) + 11(\tfrac{1}{6}) = \tfrac{36}{6} = 6.$$

Theorem 3.7.1 If a and b are constants and X is a random variable, then

$$E(aX + b) = aE(X) + b.$$

Proof. Let $g(X) = aX + b$, then by Definition 3.7.1

$$E[g(X)] = E[aX + b] = \sum_{i=1}^{N} (ax_i + b)f(x_i)$$

$$= (ax_1 + b)f(x_1) + (ax_2 + b)f(x_2) + \cdots + (ax_N + b)f(x_N)$$
$$= ax_1 f(x_1) + ax_2 f(x_2) + \cdots + ax_N f(x_N)$$
$$+ bf(x_1) + bf(x_2) + \cdots + bf(x_N)$$
$$= a \sum_{i=1}^{N} x_i f(x_i) + b \sum_{i=1}^{N} f(x_i).$$

The first sum on the right is $E(X) = \sum_{i=1}^{N} x_i f(x_i)$, the second sum is one since $\sum_{i=1}^{N} f(x_i) = 1$.

Therefore we have
$$E(aX + b) = aE(X) + b.$$

Corollary 1 If $a = 0$, then $E(b) = b$.

Corollary 2 If $a = 1$, then $E(X + b) = E(X) + b$. That is, adding a fixed amount to every value of a random variable changes the mean of the random variable by the same amount.

Corollary 3 If $b = 0$, then $E(aX) = aE(X)$. That is, multiplying every value of a random variable by the same factor multiplies the mean by that factor.

Corollary 4 If $a = 1$, $b = -E(X) = -\mu$, then $E[X - E(X)] = E[X - \mu] = 0$. That is, the mean deviation of X from its mean is zero.

Definition 3.7.2 Let $f(x_i, y_j)$; $i = 1, 2, \ldots, M$, $j = 1, 2, \ldots, N$ be the joint probability density function of two random variables X and Y. Then the mathematical expectation or expected value of $g(X, Y)$, denoted by $E[g(X, Y)]$, is defined as

$$E[g(X, Y)] = \sum_{i=1}^{M} \sum_{j=1}^{N} g(x, y) f(x_i, y_j). \tag{3.7.2}$$

Theorem 3.7.2 Let $f(x_i, y_j)$; $i = 1, 2, \ldots, M$, $j = 1, 2, \ldots, N$ be the joint probability density function of two random variables X and Y with means $E(X)$ and $E(Y)$. Then

$$E(X + Y) = E(X) + E(Y). \tag{3.7.3}$$

In words, the mean of the sum of two random variables is equal to the sum of their means.

Proof. Let $g(X, Y) = X + Y$. Then by Definition 3.7.2

$$E[g(X, Y)] = E(X + Y) = \sum_{i=1}^{M} \sum_{j=1}^{N} (x_i + y_j) f(x_i, y_j)$$

or

$$E(X + Y) = \sum_{i=1}^{M} \sum_{j=1}^{N} x_i f(x_i, y_j) + \sum_{i=1}^{M} \sum_{j=1}^{N} y_j f(x_i, y_j). \quad (3.7.4)$$

Recalling that

$$\sum_{j=1}^{N} f(x_i, y_j) = f(x_i), \quad i = 1, 2, \ldots, M \quad (3.7.5)$$

and

$$\sum_{i=1}^{M} f(x_i, y_j) = f(y_j), \quad j = 1, 2, \ldots, N, \quad (3.7.6)$$

equation (3.7.4) can be written as

$$E(X + Y) = \sum_{i=1}^{M} x_i f(x_i) + \sum_{j=1}^{N} y_j f(y_j) = E(X) + E(Y).$$

The above result can easily be extended by induction to any finite number of random variables.

Theorem 3.7.3 Let $X_1, X_2, \ldots, X_N$ be N independent random variables with means $E(X_1), E(X_2), \ldots, E(X_N)$, respectively, then

$$E(X_1 + X_2 + \cdots + X_N) = E(X_1) + E(X_2) + \cdots + E(X_N). \quad (3.7.7)$$

In words, the mean of the sum of N random variables is equal to the sum of the N respective means.

Example 3.7.2 N dice are rolled. What is the mathematical expectation of the sum of points that appear on the top?

Solution Let $X_i, i = 1, 2, \ldots, N$ be the number of points that appear on the ith die, then the sum of points on N dice is

$$S = X_1 + X_2 + \cdots + X_N.$$

Therefore

$$E(S) = E(X_1 + X_2 + \cdots + X_N)$$
$$= E(X_1) + E(X_2) + \cdots + E(X_N).$$

But, from Example 3.6.1, $E(X_i) = \frac{7}{2}$, $i = 1, 2, \ldots, N$. Hence $E(X) = 7N/2$.

Theorem 3.7.4 Let $f(x_i, y_j); i = 1, 2, \ldots, M, j = 1, 2, \ldots, N$ be the joint probability function of two random variables X and Y with means $E(X)$ and $E(Y)$. Let X and Y be independent. Then

$$E(XY) = E(X)\,E(Y). \tag{3.7.8}$$

In words, the mean of a product of two independent random variables is equal to the product of their means.

Proof. Let $g(X, Y) = XY$. Then

$$E[g(X, Y)] = E(XY) = \sum_{i=1}^{M} \sum_{j=1}^{N} x_i y_j f(x_i, y_j).$$

From Definition 3.5.2, the independence of X and Y means that

$$f(x_i, y_j) = f(x_i)f(y_j), \quad \text{for all } i \text{ and } j.$$

Hence
$$E(XY) = \sum_{i=1}^{M} \sum_{j=1}^{N} x_i y_j f(x_i) f(y_j)$$

$$= \sum_{i=1}^{M} x_i f(x_i) \sum_{j=1}^{N} y_j f(y_j)$$

$$= E(X)E(Y).$$

Example 3.7.3 Consider again the Example 3.5.1, of tossing a fair coin three times. Let the random variable X be the total number of heads and let the random variable Y be the number of heads on the first two tosses. Find

a) $E(X)$,
b) $E(Y)$,
c) $E(XY)$.
d) Are the random variables X and Y independent?

Solution The joint probability function of the random variables X and Y is shown in Table 3.7.1.

TABLE 3.7.1

$X = x$ \ $Y = y$	0	1	2	$P(X = x)$
0	$\frac{1}{8}$	0	0	$\frac{1}{8}$
1	$\frac{1}{8}$	$\frac{1}{4}$	0	$\frac{3}{8}$
2	0	$\frac{1}{4}$	$\frac{1}{8}$	$\frac{3}{8}$
3	0	0	$\frac{1}{8}$	$\frac{1}{8}$
$P(Y = y)$	$\frac{1}{4}$	$\frac{1}{2}$	$\frac{1}{4}$	1

a) $E(X) = \sum\limits_{x=0}^{3} xf(x)$

$= 0(\frac{1}{8}) + 1(\frac{3}{8}) + 2(\frac{3}{8}) + 3(\frac{1}{8})$
$= \frac{3}{2}.$

b) $E(Y) = \sum\limits_{y=0}^{2} yf(y)$

$= 0(\frac{1}{4}) + 1(\frac{1}{2}) + 2(\frac{1}{4})$
$= 1.$

c) $E(XY) = \sum\limits_{x=0}^{3} \sum\limits_{y=0}^{2} xy f(x, y)$

$= (0)(0)(\frac{1}{8}) + (0)(1)(0) + (0)(2)(0) + (1)(0)(\frac{1}{8})$
$\;\; + (1)(1)(\frac{1}{4}) + (1)(2)(0) + (2)(0)(0) + (2)(1)(\frac{1}{4})$
$\;\; + (2)(2)(\frac{1}{8}) + (3)(0)(0) + (3)(1)(0) + (3)(2)(\frac{1}{8})$
$= \frac{1}{4} + \frac{2}{4} + \frac{4}{8} + \frac{6}{8}$
$= 2.$

d) We found that $E(XY) = 2$, and $E(X) = \frac{3}{2}$, $E(Y) = 1$. Thus $E(XY) \neq E(X)E(Y)$. Hence the random variables X and Y are not independent.

3.8 THE VARIANCE OF A RANDOM VARIABLE

The expected value or the mean of a random variable tells us about the center of the probability function. It gives us quick information about the long run average when an experiment is performed over and over. But it does not give any information about the "spread" or "dispersion" of the values of a random variable from one experiment to another. We now introduce the most commonly used measure of dispersion called the variance.

Definition 3.8.1 Let X be a discrete random variable with the following probability function:

$X = x$	x_1	x_2	$\ldots$	x_N
$f(x) = P(X = x)$	$f(x_1)$	$f(x_2)$	$\ldots$	$f(x_N)$

Let $E(X) = \mu$ be the mean of X. The variance of X, denoted by $\mathrm{Var}(X)$ or σ_X^2 (read: "sigma sub-X square"), is defined by

$$\sigma_X^2 = \mathrm{Var}(X) = E[(X - \mu)^2] \tag{3.8.1}$$

or equivalently, by Definition 3.7.1

$$\sigma_X^2 = \text{Var}(X) = \sum_{j=1}^{N} (x_i - \mu)^2 f(x_i). \quad (3.8.2)$$

In words, the variance of X is the mean squared deviation of X from its mean.

We note that the units of variance of X are squares of the units of X and so variance does not measure dispersion in the same units as the values of X. In order to have a measure of dispersion in the same units as the values of X, we define the standard deviation as the positive square root of the variance.

Definition 3.8.2 Let X be a discrete random variable with mean μ. The standard deviation of X, denoted by σ_X (read "sigma sub-X"), is the positive square root of the variance, and is given by

$$\sigma_X = \sqrt{\text{Var}(X)} = \sqrt{E(X - \mu)^2}. \quad (3.8.3)$$

Example 3.8.1 A coin is tossed three times. Let X be the number of heads that appear. Table 3.8.1 gives the probability function of X, together with the computation of σ_X^2 and σ_X.

TABLE 3.8.1

$X = x$	$f(x)$	$xf(x)$	$[x - E(X)]^2$	$[x - E(X)]^2 f(x)$
0	$\frac{1}{8}$	0	$(0 - \frac{3}{2})^2 = \frac{9}{4}$	$\frac{9}{32}$
1	$\frac{3}{8}$	$\frac{3}{8}$	$(1 - \frac{3}{2})^2 = \frac{1}{4}$	$\frac{3}{32}$
2	$\frac{3}{8}$	$\frac{6}{8}$	$(2 - \frac{3}{2})^2 = \frac{1}{4}$	$\frac{3}{32}$
3	$\frac{1}{8}$	$\frac{3}{8}$	$(3 - \frac{3}{2})^2 = \frac{9}{4}$	$\frac{9}{32}$
		$E(X) = \frac{12}{8} = \frac{3}{2}$		$\sigma_X^2 = \frac{24}{32} = \frac{3}{4}$

Thus $\sigma_X^2 = \frac{3}{4}$ and $\sigma_X = \sqrt{\frac{3}{4}} = \frac{\sqrt{3}}{2}$.

Theorem 3.8.1 Let X be a random variable with $E(X) = \mu$ and variance $V(X) = \sigma^2$. Then

$$\sigma^2 = E(X^2) - [E(X)]^2 = E(X^2) - \mu^2. \quad (3.8.4)$$

Proof. By Definition 3.8.1

$$V(X) = \sigma^2 = E\{[X - E(X)]^2\}$$
$$= E\{X^2 - 2XE(X) + [E(X)]^2\}.$$

By Theorem 3.7.2 we have

$$V(X) = \sigma^2 = E(X^2) - 2E(X)E(X) + [E(X)]^2$$
$$= E(X^2) - [E(X)]^2.$$

Example 3.8.2 Consider Example 3.6.2 again, of tossing a coin three times. Table 3.8.2 gives the computation of σ_X^2 using Theorem 3.8.1.

TABLE 3.8.2

$X = x$	$f(x) = P(X = x)$	$xf(x)$	$x^2 f(x)$
0	$\frac{1}{8}$	0	$(0)^2 \frac{1}{8} = 0$
1	$\frac{3}{8}$	$\frac{3}{8}$	$(1)^2 \frac{3}{8} = \frac{3}{8}$
2	$\frac{3}{8}$	$\frac{6}{8}$	$(2)^2 \frac{3}{8} = \frac{12}{8}$
3	$\frac{1}{8}$	$\frac{3}{8}$	$(3)^2 \frac{1}{8} = \frac{9}{8}$
		$E(X) = \frac{3}{2}$	$E(X^2) = \frac{24}{8} = 3$

Thus
$$\text{Var}(X) = \sigma^2 = E(X^2) - [E(X)]^2 = 3 - (\tfrac{3}{2})^2 = 3 - \tfrac{9}{4} = \tfrac{3}{4}.$$

Example 3.8.3 A fair die is rolled. Let the random variable X be the number that appears on the top. Find the variance of X.

TABLE 3.8.3

$X = x$	$f(x) = P(X = x)$	$xf(x)$	$x^2 f(x)$
1	$\frac{1}{6}$	$\frac{1}{6}$	$(1)^2 (\frac{1}{6}) = \frac{1}{6}$
2	$\frac{1}{6}$	$\frac{2}{6}$	$(2)^2 (\frac{1}{6}) = \frac{4}{6}$
3	$\frac{1}{6}$	$\frac{3}{6}$	$(3)^2 (\frac{1}{6}) = \frac{9}{6}$
4	$\frac{1}{6}$	$\frac{4}{6}$	$(4)^2 (\frac{1}{6}) = \frac{16}{6}$
5	$\frac{1}{6}$	$\frac{5}{6}$	$(5)^2 (\frac{1}{6}) = \frac{25}{6}$
6	$\frac{1}{6}$	$\frac{6}{6}$	$(6)^2 (\frac{1}{6}) = \frac{36}{6}$
		$E(X) = \frac{21}{6} = \frac{7}{2}$	$E(X^2) = \frac{91}{6}$

Therefore $\sigma_X^2 = E(X^2) - [E(X)]^2 = \frac{91}{6} - (\frac{7}{2})^2 = \frac{91}{6} - \frac{49}{4} = \frac{35}{12}$.

Example 3.8.4 A ball is drawn at random from an urn containing N balls numbered $1, 2, \ldots, N$. Let the random variable X be the number on the ball. Find

a) $E(X)$,
b) $E(X^2)$,
c) σ_X^2.

Solution Table 3.8.4 gives the probability distribution of the random variable X, together with the computation of $E(X)$, $E(X^2)$ and σ_X^2.

TABLE 3.8.4

$X = x$	$f(x) = P(X = x)$	$xf(x)$	$x^2 f(x)$
1	$\frac{1}{N}$	$\frac{1}{N}$	$\frac{1^2}{N}$
2	$\frac{1}{N}$	$\frac{2}{N}$	$\frac{2^2}{N}$
.	.	.	.
.	.	.	.
N	$\frac{1}{N}$	$\frac{N}{N}$	$\frac{N^2}{N}$

Thus

a) $E(X) = \sum_{x=1}^{N} xf(x) = \frac{1}{N}(1 + 2 + \cdots + N)$.

Using the fact that $1 + 2 + \cdots + N = N(N + 1)/2$

$$E(X) = \frac{1}{N} \frac{N(N + 1)}{2} = \frac{N + 1}{2}.$$

b) $E(X^2) = \sum_{x=1}^{N} x^2 f(x) = \frac{1}{N}(1^2 + 2^2 + \cdots + N^2)$.

Using the fact that $1^2 + 2^2 + \cdots + N^2 = N(N + 1)(2N + 1)/6$

$$E(X^2) = \frac{1}{N} \frac{N(N + 1)(2N + 1)}{6} = \frac{(N + 1)(2N + 1)}{6}.$$

c) $\sigma_X^2 = E(X^2) - [E(X)]^2$

$$= \frac{(N + 1)(2N + 1)}{6} - \left[\frac{N + 1}{2}\right]^2$$

$$= \frac{N^2 - 1}{12}.$$

Note that Example 3.8.3 is a special case of Example 3.8.4.

3.9 PROPERTIES OF THE VARIANCE OF A RANDOM VARIABLE

There are various important properties of the variance of a random variable. These properties will simplify our computations of variance of functions of random variables.

Theorem 3.9.1 Let a be a constant and X be a random variable. Then
$$\text{Var}(aX) = a^2 \text{Var}(X) \qquad (3.9.1)$$
or
$$\sigma_{aX}^2 = a^2 \sigma_X^2. \qquad (3.9.2)$$

In words, multiplying each random variable by the same factor a multiplies the variance by a^2.

Proof. By the definition of the variance,
$$\begin{aligned}\text{Var}(aX) &= E\{[aX - E(aX)]^2\} \\ &= E\{a^2[X - E(X)]^2\} \\ &= a^2 E[X - E(X)]^2,\end{aligned}$$
which from Definition 3.8.1 gives
$$\text{Var}(aX) = a^2 \text{Var}(X), \qquad \text{or} \qquad \sigma_{aX}^2 = a^2 \sigma_X^2.$$

Example 3.9.1 Let the variance of a random variable X be .50; what is the variance of the random variable

a) $2X$?

b) $\dfrac{X}{2}$?

a) By Equation (3.9.1) we have
$$\text{Var}(2X) = \sigma_{2X}^2 = 4\sigma_X^2 = 4(.50) = 2.$$

b) By Equation (3.9.1) we have
$$\text{Var}(\tfrac{X}{2}) = \sigma_{X/2}^2 = \tfrac{1}{4} \text{Var}(X) = \tfrac{1}{4}(.50) = .125.$$

Theorem 3.9.2 Let b be a constant and X be a random variable. Then
$$\text{Var}(X + b) = \text{Var}(X) \qquad (3.9.3)$$
or
$$\sigma_{X+b}^2 = \sigma_X^2. \qquad (3.9.4)$$

In words, adding a fixed amount to every value of a random variable has no effect on the variance.

Proof. By the definition of the variance

$$\begin{aligned}\operatorname{Var}(X+b) &= E\{[X+b-E(X+b)]^2\} \\ &= E\{[X+b-E(X)-b]^2\} \\ &= E[X-E(X)]^2 \\ &= \operatorname{Var}(X)\end{aligned}$$

or

$$\sigma^2_{X+b} = \sigma^2_X.$$

Example 3.9.2 Let the variance of the random variable X be 5. What is the variance of the random variable

a) $X+3$?
b) $X-6$?

Solution

a) By Equation (3.9.3) we have

$$\sigma^2_{X+3} = \operatorname{Var}(X+3) = \operatorname{Var}(X) = 5.$$

b) By Equation (3.9.3) we have

$$\sigma^2_{X-6} = \operatorname{Var}(X-6) = \operatorname{Var}(X) = 5.$$

Theorem 3.9.3 Let X and Y be independent random variables with means $E(X) = \mu_X$, $E(Y) = \mu_Y$, and variances $\operatorname{Var}(X) = \sigma^2_X$, $\operatorname{Var}(Y) = \sigma^2_Y$, respectively, then

$$\operatorname{Var}(X+Y) = \operatorname{Var}(X) + \operatorname{Var}(Y) \qquad (3.9.5)$$

or

$$\sigma^2_{X+Y} = \sigma^2_X + \sigma^2_Y. \qquad (3.9.6)$$

In words, the variance of the sum of two independent random variables is equal to the sum of their variances.

Proof.
$$\begin{aligned}\operatorname{Var}(X+Y) &= E[\{(X+Y)-E(X+Y)\}^2] \\ &= E([\{X-E(X)\}+\{Y-E(Y)\}]^2) \\ &= E([X-E(X)]^2) + E([Y-E(Y)]^2) \\ &\quad + 2E([X-E(X)][Y-E(Y)]).\end{aligned}$$

The first two terms on the right are $\operatorname{Var}(X)$ and $\operatorname{Var}(Y)$. We must therefore show that the last term is zero. Thus

$$\begin{aligned}E([X-E(X)][Y-E(Y)]) &= E[XY - XE(Y) - E(X)Y + E(X)E(Y)] \\ &= E(XY) - E(X)E(Y) - E(X)E(Y) \\ &\quad + E(X)E(Y) \\ &= E(XY) - E(X)E(Y) \\ &= 0\end{aligned}$$

since $E(XY) = E(X)E(Y)$ for independent random variables. Therefore $\text{Var}(X + Y) = \text{Var}(X) + \text{Var}(Y)$ or

$$\sigma^2_{X+Y} = \sigma^2_X + \sigma^2_Y.$$

The above theorem can be generalized to any finite number of independent random variables.

Theorem 3.9.4 Let $X_1, X_2, \ldots, X_N$ be N independent random variables with means $E(X_i) = \mu_i$, $i = 1, 2, \ldots, N$ and variances $\text{Var}(X_i) = \sigma_i^2$, $i = 1, 2, \ldots, N$, then

$$\text{Var}(X_1 + X_2 + \cdots + X_N) = \text{Var}(X_1) + \text{Var}(X_2) + \cdots + \text{Var}(X_N) \tag{3.9.7}$$

or

$$\sigma^2_{X_1 + X_2 + \cdots + X_N} = \sigma^2_{X_1} + \sigma^2_{X_2} + \cdots + \sigma^2_{X_N}. \tag{3.9.8}$$

Example 3.9.3 N dice are rolled. What is the variance of the sum of points that appear?

Solution Let X_i, $i = 1, 2, \ldots, N$ be the number of points that appear on the ith die. Then the sum of points on N dice is

$$S = X_1 + X_2 + \cdots + X_N.$$

Therefore
$$\begin{aligned}\text{Var}(S) &= \text{Var}(X_1 + X_2 + \cdots + X_N) \\ &= \text{Var}(X_1) + \text{Var}(X_2) + \cdots + \text{Var}(X_N).\end{aligned}$$

But from Example 3.8.3 $\text{Var}(X_i) = \frac{35}{12}$, $i = 1, 2, \ldots, N$. Hence $\text{Var}(S) = 35N/12$.

3.10 SUMMARY

A random variable is a number whose value is determined by the outcome of a random experiment. If the number of possible values of a random variable X is a finite or infinite number of values that are countable, then X is called a discrete random variable. If the possible values of a random variable X lie in an interval or collection of intervals, then X is called a continuous random variable. The probability function of a discrete random variable is the specification of its possible values together with their respective probabilities. The sum of these probabilities is equal to 1.

The joint probability function of two random variables and their marginal distributions are discussed in Section 3.5.

The concept of expected value of a random variable is introduced in

Section 3.6 and some basic properties of expected value are given in Section 3.7. In Section 3.8 the most commonly used measure of dispersion called variance is defined, and some important properties of the variance of a random variable are given in Section 3.9.

EXERCISES

1. List some random variables in the following experiments.

   a) Balls are drawn with replacement, from an urn containing black, red and green balls.
   b) One thousand families are asked if among their children there are any twins.
   c) Experimental animals are fed a new diet and their weight gain is measured.
   d) Two pennies are tossed until both show heads simultaneously.
   e) The number of straights and flushes in an evening's poker are noted.

2. Classify the following random variables as discrete or continuous.

   a) The number of customers in a bank during a lunch hour on a particular day.
   b) The weekly or monthly volume of gasoline sold by a certain gas dealer.
   c) The number of items of a particular stock sold daily at a certain store.
   d) The weekly consumption of milk by a family.
   e) The length of time spent daily by one student in studying statistics.

3. An urn contains four white, five red and three black balls. Two balls are drawn with replacement.

   a) Describe the sample space.
   b) How many points are there in the space?
   c) Attach a probability to each point in the sample space.
   d) If all the black balls are removed, do parts (a), (b) and (c) again.

4. Four coins are tossed. Describe the sample space of this experiment. If the random variable X is the number of heads showing, make a table showing the possible values of X, and their corresponding probabilities.

5. Describe two phenomena with which you are familiar, one with a continuous and one with a discrete random variable.

6. Let the probability function of the random variable X be of the following form, where c is some constant:

$$f(x) = c\binom{5}{x}, \quad x = 0, 1, 2, \ldots, 5.$$

   Determine the value of c.

7. Let the probability function of the random variable X be of the following form where c is some constant

$$f(x) = cx, \quad x = 3, 4, 5, 6.$$

   a) Determine the value of c.
   b) Find $E(X)$.
   c) Find Var (X).

8. Let the random variable X represent the number of accidents occurring on a given day in the city of New York. Let the probability function of the random variable X be of the following form where c is some constant:

$$f(x) = \begin{cases} c & \text{if} & x = 0, \\ 2c & \text{if} & x = 1, \\ 3c & \text{if} & x = 2, \\ 4c & \text{if} & x = 3, \\ 1.5c & \text{if} & x = 4, \\ 0.5c & \text{if} & x = 5. \end{cases}$$

   a) Determine the value of c.
   b) Find

   i) $P(X < 3)$,
   ii) $P(0 < X \leq 4)$,
   iii) $P(0 < X < 2)$.

   c) Determine the distribution function of X.

9. Four cards are dealt from a standard deck. (a) What is the distribution of the number of hearts? (b) What is the distribution of the number of twos?

10. In Exercise 9 show that (a) $f(x) \geq 0$, (b) $\sum_{\text{all } x} f(x) = 1$.

11. A penny is thrown until a head appears. If X, the random variable, is the number of throws until a head appears, what is the probability function of X?

12. Two pennies are tossed until two heads appear simultaneously. What is the distribution of the number of tosses required?

13. From one dozen eggs of which four are rotten, three are selected at random without replacement. Let the random variable X be the number of rotten eggs in the sample. Find the probability function of the random variable X.

14. In the two-dice experiment let the random variable X denote the difference on the top face of the two dice. What are the possible values of the random variable X? How many points are there in this sample space?

15. One student is to be selected from a group of 10 students numbered from 1 to 10. Let the random variable X denote the number of the student. Find the probability function of the random variable X.

16. Let the random variable X represent the number of boys in families with five children. Assuming equal probability for boys and girls, obtain the probability function for X.

17. An urn contains five white and three green balls. Two balls are drawn at random. Let the random variable X be the number of green balls drawn.
    a) Find the probability function of X.
    b) Find the distribution function of X.

18. Let X be a random variable with density function
$$f(x) = \tfrac{1}{2}, \qquad 0 < X < 2.$$
    Find
    a) $P[.5 < X < 1.5]$,
    b) $P[X > .25]$,
    c) $P[X < .75]$,
    d) $P[X > 3]$.

19. Let $f(x) = \dfrac{(2-x)}{2}$, $\quad 0 < X < 2$.
    Find
    a) $P[.5 < X < 1]$,
    b) $P[X > 1.5]$,
    c) $P[X < .3]$,
    d) $P[0 < X < 2]$.

20. Let
$$f(x) = \begin{cases} x, & 0 < X < 1, \\ 2-x, & 1 < X < 2. \end{cases}$$
    Find
    a) $P[0 < X < 1]$,
    b) $P[X > 1.5]$,
    c) $P[.5 < X < 1.5]$.

21. Express each of the following summations in expanded form.
    a) $\sum_{i=1}^{10} f_i x_i^2$
    b) $\sum_{i=6}^{9} x_i^2$
    c) $[\sum_{i=1}^{4} y_i(y_i - 2)]^2$

22. Write each of the following expressions, using a summation sign.
    a) $x_1 + x_2 + x_3 + x_4 + x_5$
    b) $f_1 x_1 + f_2 x_2 + \cdots + f_N x_N$

c) $x_2^2 + x_3^2 + x_4^2 + x_5^2$
d) $(x_1 - 2)^2 + (x_2 - 2)^2 + (x_3 - 2)^2 + (x_4 - 2)^2$
e) $f_1(x_1 - 4)^2 + f_2(x_2 - 4)^2 + \cdots + f_{10}(x_{10} - 4)^2$

23. Let $x_1 = 5$, $x_2 = 2$, $x_3 = -2$, $x_4 = 3$. Find

  a) $\sum_{i=1}^{4} x_i$,

  b) $\sum_{i=1}^{4} (x_i - 2)$,

  c) $\sum_{i=1}^{4} (x_i - 2)^2$.

24. Find $E(X)$ in Exercise 4.

25. Find $E(X)$ in Exercise 9.

26. Find $E(X)$ in Exercise 13.

27. Find the probability function and the expectation of the random variable X, where X represents the sum of the top numbers when two dice are thrown together.

28. For the experiment of Exercise 13, find (a) $E(X^2)$, (b) $E(2X)$.

29. A man purchases a lottery ticket. He can win a first prize of $4000 or a second prize of $3000 with probabilities .005 and .008, respectively. What should be a fair price to pay for the ticket?

30. Let $Y = 3X - 5$ and $E(X) = 4$, $\text{Var}(X) = 2$, what are the mean and variance of Y?

31. Given that a random variable X has mean 10 and variance 6, find $E(X^2 + 3X)$.

32. Let $\mu = E(X)$ be finite, prove
  a) $E(X - \mu) = 0$,
  b) $E(X - c)^2 = E(X - \mu)^2 + (\mu - c)^2$, where c is some constant.

33. The joint probability function of the random variables X and Y is of the following form, where c is some constant

$$f(x, y) = c(x^2 + y^2), \quad \begin{cases} x = 0, 1, 2, 3, \\ y = 0, 1. \end{cases}$$

  a) Determine the value of c.
  b) Find the marginal probability function for X.
  c) Find the marginal probability function for Y.

34. The joint probability function of X and Y is given by

$$f(x, y) = xy/36, \quad \begin{cases} x = 1, 2, 3, \\ y = 1, 2, 3. \end{cases}$$

a) Find the marginal probability function for X.
b) Find the marginal probability function for Y.

35. Let the joint probability function of the random variables X and Y be given by
$$f(x, y) = (\tfrac{2}{3})^{x+y} (\tfrac{1}{3})^{2-x-y}, \qquad \begin{cases} x = 0, 1, \\ y = 0, 1. \end{cases}$$

a) Find the marginal probability function for X.
b) Find the marginal probability function for Y.

36. A five card hand is dealt from a standard deck. If X is the number of spades and Y the number of diamonds in the hand, what is the joint distribution of X and Y?

37. Consider an experiment where a fair coin is tossed four times. Let X be the total number of heads and Y the number of heads on the second toss.

a) Write out the whole sample space.
b) Write the joint probability function of X and Y in the form of a two-way table.
c) Verify that
  i) $f(x_i, y_j) \geq 0$,
  ii) $\sum\limits_{\text{all } x} \sum\limits_{\text{all } y} f(x_i, y_j) = 1$.

38. Using Exercise 37, find the marginal distributions of X and of Y.

39. A population is classified concerning its preference for tea and coffee. It is found that 20% like neither tea nor coffee, 50% like both tea and coffee and 10% like coffee but not tea. What proportion like (a) tea? (b) coffee?

40. Three separate experiments are performed. In one, a single coin is tossed twice, in the second two coins are tossed twice, and in the third three coins are tossed twice. Let X be the total number of heads obtained in the three experiments. Find $E(X)$, using Theorem 3.7.3.

41. Given that X is the number of errors on a page of typing and $f(0) = .9$, $f(1) = .05$, $f(2) = .03$, $f(3) = .02$. Find the expected number of errors in 300 pages.

42. Typing errors are of type A or B. X is the number of type A errors on a page and Y is the number of type B on a page. Given $f(x, y)$ is as follows

Y \ X	0	1	2	3
0	.70	.10	.05	.05
1	.04	.00	.01	.00
2	.01	.01	.01	.00
3	.01	.01	.00	.00

a) Find the distribution of X.
b) Find the distribution of Y.
c) Find $E(X)$, $E(Y)$, $E(X+Y)$, $E(XY)$.
d) Are X and Y independent?

43. Find the variance of X in Exercise 4.

44. Find the variance of X in Exercise 9.

45. Find the variance of X in Exercise 13.

46. In Exercise 42 find (a) $\text{Var}(X)$, (b) $\text{Var}(Y)$.

47. Two people independently flip pairs of coins four times. Let X be the number of heads for the first person, and Y the number of heads for the second person. Find (a) $E(X+Y)$, (b) $\text{Var}(X+Y)$.

48. Let X and Y be independent random variables with means μ_X, μ_Y and variances σ_X^2, σ_Y^2, respectively. Find (a) μ_{X-Y}, (b) σ_{X-Y}^2.

Chapter 4

SOME DISCRETE PROBABILITY DISTRIBUTIONS

4.1 INTRODUCTION

In this chapter we shall study some discrete random variables that occur frequently in many applied problems. We shall perform many experiments under different sets of assumptions. Because of their importance and wide areas of relevant application, these experiments and their outcomes will be given special names. We shall derive probability distributions of random variables associated with these experiments and study some of their general properties.

4.2 THE BERNOULLI DISTRIBUTION

We have considered numerous experiments in Chapters 2 and 3 involving one or more trials of tosses of a coin, rolls of a die, and draws of a card from a deck of cards. We are often interested in the outcome of an individual trial. Each trial may have two or more possible outcomes. In many applications we are interested in a trial which has only two possible outcomes generally called success or failure. Such trials are called Bernoulli trials, named after James Bernoulli, 1654–1705.

Definition 4.2.1. A random variable X is called a Bernoulli random variable if it has two possible outcomes.
    The two possible outcomes of an experiment are usually coded with the values 0 and 1. The value 1 is assigned to the success of a certain event and the value 0 to its failure.

Example 4.2.1 The following experiments involve Bernoulli random variables.
1. Tossing a coin.
2. Drawing a ball from an urn containing M black and N white balls

3. Picking an item from a box containing defective and non-defective items

Definition 4.2.2 Let X be a random variable with values 0 and 1. Then the probability function of the random variable X, called the Bernoulli distribution, is

$$P(X = 1) = p,$$
$$P(X = 0) = q = 1 - p \quad (4.2.1)$$

or alternatively by

$$f(x) = P(X = x) = p^x(1 - p)^{1-x}, \quad x = 0, 1. \quad (4.2.2)$$

Theorem 4.2.1 Let the random variable X have the Bernoulli distribution

$$f(x) = p^x(1 - p)^{1-x}, \quad x = 0, 1.$$

Then the mean μ and the variance σ^2 of the Bernoulli distribution are

$$\mu = E(X) = p \quad (4.2.3)$$

and

$$\sigma^2 = E(X)^2 - [E(X)]^2 = pq. \quad (4.2.4)$$

Proof. By Definition 3.6.1 the mean is

$$\mu = E(X) = \sum_{x=0}^{1} x f(x)$$

$$= \sum_{x=0}^{1} x p^x(1 - p)^{1-x}$$

$$= p.$$

The variance is given by

$$\sigma^2 = E(X^2) - [E(X)]^2. \quad (4.2.5)$$

By Definition 3.7.1 we have

$$E(X^2) = \sum_{x=0}^{1} x^2 p^x(1 - p)^{1-x}$$

$$= p.$$

Therefore from Equation (4.2.5)

$$\sigma^2 = p - p^2 = p(1 - p) = pq.$$

4.3 THE BINOMIAL DISTRIBUTION

Consider the experiment in which a sequence of Bernoulli trials is performed. An experiment of this type is called a binomial experiment. The binomial random variable is defined as follows.

Definition 4.3.1 Let the random variable X be the total number of successes in n independent trials. Let p be the probability of success and $q = 1 - p$ be the probability of failure on a single trial, then X is called the binomial random variable, if the following conditions are satisfied;
 i) Each trial has only two outcomes: either S (success) or F (failure).
 ii) The probability p of a success is the same for each trial.
 iii) The trials are independent.

Example 4.3.1 The following experiments involve binomial random variables.
1. Toss a coin 10 times. Then the random variable X is the number of heads observed.
2. Draw three balls with replacement from an urn containing eight black and four white balls. Then the random variable X is the number of black balls drawn.
3. Pick four items with replacement from a box containing three defective and seven nondefective items. Then the random variable X is the number of defective items picked.

Theorem 4.3.1 If X is binomial random variable in n independent trials with probability p of success and $q = 1 - p$ of failure on a single trial, then the probability function of the random variable X, called the binomial distribution, is

$$f(x) = \binom{n}{x} p^x q^{n-x} \qquad x = 0, 1, 2, \ldots, n. \qquad (4.3.1)$$

Proof. The number of successes X in n independent trials may be 0, 1, 2, ..., x, ..., n. Let us consider the sequence

$$\underbrace{SS \ldots S}_{x} \underbrace{FF \ldots F}_{n-x}$$

where S denotes a success and F a failure. By the multiplication theorem, the probability of the above sequence, that is, the probability that the first x trials are successes and the remaining $n - x$ are failures, is

$$p^x q^{n-x}.$$

Since the trials are independent, the probability of any other sequence of x successes and $n - x$ failures is also $p^x q^{n-x}$. The number of distinct sequences of n outcomes into two groups with x in one group and $n - x$ in the other is $\binom{n}{x}$. That is, the number of distinct events in which x successes and $n - x$ failures can occur, is $\binom{n}{x}$. These events are mutually exclusive, since only one event can occur at a time. Therefore, by the addition theorem, the probability function of the random variable X, the number of successes in n trials, is

$$f(x) = \binom{n}{x} p^x q^{n-x} \qquad x = 0, 1, 2, \ldots, n. \qquad (4.3.2)$$

The probability function (4.3.2) is called the binomial distribution. The probabilities of $0, 1, 2, \ldots, n$ successes correspond to the successive terms in the binomial expansion of $(q + p)^n$. The sum of the probabilities is

$$\sum_{x=0}^{n} f(x) = \sum_{x=0}^{n} \binom{n}{x} p^x q^{n-x} = (q + p)^n = 1. \qquad (4.3.3)$$

Example 4.3.2 A coin is tossed four times. What is the probability of getting

a) exactly two heads?
b) at least one head?
c) more than one head?

Solution Let the random variable X be the number of heads on the four tosses. Thus the random variable X is binomial with probability function

$$f(x) = P(X = x) = \binom{4}{x} \left(\tfrac{1}{2}\right)^x \left(\tfrac{1}{2}\right)^{4-x}, \qquad x = 0, 1, 2, 3, 4.$$

a) The probability of getting exactly two heads is

$$f(2) = P(X = 2) = \binom{4}{2} \left(\tfrac{1}{2}\right)^2 \left(\tfrac{1}{2}\right)^2 = 6\left(\tfrac{1}{4}\right)\left(\tfrac{1}{4}\right) = \tfrac{3}{8}.$$

b) The probability of getting at least one head is equal to the probability of getting one or more heads. Hence

$$\begin{aligned} P(X \geq 1) &= P(X = 1) + P(X = 2) + P(X = 3) + P(X = 4) \\ &= 1 - P(X = 0) \\ &= 1 - \binom{4}{0}\left(\tfrac{1}{2}\right)^0\left(\tfrac{1}{2}\right)^4 \\ &= 1 - \tfrac{1}{16} = \tfrac{15}{16}. \end{aligned}$$

c) The probability of getting more than one head is

$$\begin{aligned} P(X > 1) &= P(X = 2) + P(X = 3) + P(X = 4) \\ &= 1 - P(X = 0) - P(X = 1) \\ &= 1 - \left(\tfrac{1}{2}\right)^4 - 4\left(\tfrac{1}{2}\right)^4 \\ &= 1 - \tfrac{1}{16} - \tfrac{4}{16} = \tfrac{11}{16}. \end{aligned}$$

Example 4.3.3 In a family of five children, what is the probability that there will be exactly two boys, assuming that the sexes are equally likely?

Solution Let X be the number of boys in a family of five children. Thus

$$n = 5, \qquad p = \tfrac{1}{2}, \qquad x = 2.$$

Hence by Equation (4.3.1)

$$P(X = 2) = f(2) = \binom{5}{2} \left(\tfrac{1}{2}\right)^2 \left(\tfrac{1}{2}\right)^3$$
$$= \frac{5!}{2!\,3!}\left(\tfrac{1}{2}\right)^2 \left(\tfrac{1}{2}\right)^3$$
$$= 10\left(\tfrac{1}{2}\right)^5 = \tfrac{10}{32}.$$

Example 4.3.4 From a lot containing 20 items, five of which are defective, four items are drawn with replacement. What is the probability of getting

a) exactly one defective item?
b) at least one defective item?

Solution Let X be the number of defective items drawn. Then the possible values of X are 0, 1, 2, 3, 4.

Let $p = P$ (defective item), then $p = \tfrac{5}{20} = \tfrac{1}{4}$. The random variable X is binomial, since the items are drawn with replacement and the probability of drawing a defective item does not change from trial to trial.

a) Using Equation (4.3.1) we have

$$P(X = 1) = f(1) = \binom{4}{1}\left(\tfrac{1}{4}\right)^1 \left(\tfrac{3}{4}\right)^3$$
$$= \frac{4!}{1!\,3!}\left(\tfrac{1}{4}\right)^1 \left(\tfrac{3}{4}\right)^3$$
$$= 4\left(\tfrac{1}{4}\right)\left(\tfrac{27}{64}\right) = \tfrac{27}{64}.$$

b) The probability of at least one defective item is

$$P(X \geq 1) = P(X = 1) + P(X = 2) + P(X = 3) + P(X = 4)$$
$$= 1 - P(X = 0)$$
$$= 1 - \binom{4}{0}\left(\tfrac{1}{4}\right)^0 \left(\tfrac{3}{4}\right)^4$$
$$= 1 - \left(\tfrac{3}{4}\right)^4$$
$$= 1 - \tfrac{81}{256} = \tfrac{175}{256}.$$

Theorem 4.3.2. Let the random variable X have the binomial distribution

$$f(x) = \binom{n}{x}p^x q^{n-x}, \qquad x = 0, 1, 2, \ldots, n.$$

Then the mean μ and the variance σ^2 of the binomial distribution are

$$\mu = E(X) = np \qquad (4.3.4)$$

and

$$\sigma^2 = E(X^2) - [E(X)]^2 = npq. \qquad (4.3.5)$$

Proof. Method I. The binomial random variable X is the sum of n independent Bernoulli random variables X_i, each of which takes the value 1 with probability p and the value 0 with probability $1 - p$. Thus

and
$$E(X_i) = p, \quad i = 1, 2, \ldots, n$$
and
$$E(X_i^2) = p, \quad i = 1, 2, \ldots, n.$$
Since $X = X_1 + X_2 + \cdots + X_n$, it follows that
$$\begin{aligned}E(X) &= E(X_1 + X_2 + \cdots + X_n)\\ &= E(X_1) + E(X_2) + \cdots + E(X_n)\\ &= p + p + \cdots + p = np\end{aligned}$$
and
$$\sigma^2 = \text{Var}(X) = \text{Var}(X_1 + X_2 + \cdots + X_n).$$
Since the X_i, $i = 1, 2, \ldots, n$ are independent, we have
$$\begin{aligned}\sigma^2 = \text{Var}(X) &= \text{Var}(X_1) + \text{Var}(X_2) + \cdots + \text{Var}(X_n)\\ &= pq + pq + \cdots + pq\\ &= npq.\end{aligned}$$

Method II. By Definition 3.6.1

$$\begin{aligned}\mu = E(X) &= \sum_{x=0}^{n} x \binom{n}{x} p^x q^{n-x}\\ &= \sum_{x=0}^{n} \frac{xn!}{x!(n-x)!} p^x q^{n-x}\\ &= \sum_{x=1}^{n} \frac{n!}{(x-1)![n-1-(x-1)]!} p^x q^{n-1-(x-1)}\\ &= np \sum_{x=1}^{n} \frac{(n-1)! p^{x-1} q^{n-1-(x-1)}}{(x-1)![n-1-(x-1)]!}\\ &= np(q+p)^{n-1}\\ &= np.\end{aligned}$$

Thus $\mu = E(X) = np$.

By Definition 3.7.1

$$\begin{aligned}E(X^2) &= \sum_{x=0}^{n} x^2 \binom{n}{x} p^x q^{n-x}\\ &= \sum_{x=0}^{n} \frac{x^2 n! \, p^x q^{n-x}}{x!(n-x)!}.\end{aligned} \qquad (4.3.6)$$

Using the identity $x^2 = x(x-1) + x$ in Equation (4.3.6), we have

$$E(X^2) = \sum_{x=0}^{n} \frac{[x(x-1) + x]n!}{x!(n-x)!} p^x q^{n-x}$$

$$= \sum_{x=0}^{n} \frac{x(x-1)n!}{x!\,(n-x)!} p^x q^{n-x} + \sum_{x=0}^{n} \frac{xn!}{x!\,(n-x)!} p^x q^{n-x}$$

$$= \sum_{x=2}^{n} \frac{n(n-1)(n-2)!}{(x-2)!\,[n-2-(x-2)]!} p^x q^{n-x} + E(X)$$

$$= n(n-1)p^2 \sum_{x=2}^{n} \frac{(n-2)!}{(x-2)!\,[n-2-(x-2)]!} p^{x-2} q^{n-x} + E(X)$$

$$= n(n-1)p^2 (q+p)^{n-2} + np$$

$$= n(n-1)p^2 + np$$

$$= n^2 p^2 - np^2 + np.$$

Thus
$$\sigma^2 = E(X^2) - [E(X)]^2$$
$$= n^2 p^2 - np^2 + np - n^2 p^2$$
$$= np - np^2$$
$$= np(1-p)$$
$$= npq.$$

Therefore $\sigma^2 = npq$.

Example 4.3.5 A coin is tossed 64 times. Find the mean and the standard deviation of the number of heads obtained.

Solution Let the random variable X be the number of heads in 64 tosses of a coin. Then X has a binomial distribution with $p = \frac{1}{2}$ and $n = 64$. Therefore
$$\mu = E(X) = np = 64(\tfrac{1}{2}) = 32$$
and
$$\sigma = \sqrt{npq} = \sqrt{64(\tfrac{1}{2})(\tfrac{1}{2})} = 4.$$

4.4 THE MULTINOMIAL DISTRIBUTION

Suppose an experiment results in K mutually exclusive outcomes $E_1, E_2, \ldots, E_K$. If an experiment is repeated n times, then the multinomial distribution is the joint distribution of the number of times each E_i, $i = 1, 2, \ldots, K$, occurs. The multinomial distribution is a generalization of the binomial distribution.

Definition 4.4.1 Let $E_1, E_2, \ldots, E_K$ be K mutually exclusive outcomes of an experiment. Let the random variable $(X_1, X_2, \ldots, X_K)$ be the number of times each E_i, $i = 1, 2, \ldots, K$, occurs in n independent trials. Let p_i, $i = 1, 2, \ldots, K$ be the probability of E_i on a single trial. Then the random variable $(X_1, X_2, \ldots, X_K)$ is called the multinomial random variable.

Example 4.4.1 The following experiments involve multinomial random variables.

1. A die is rolled n times. Let X_1 be the number of times 1 occurs, X_2 be the number of times 2 occurs, ..., X_6 be the number of times 6 occurs. Then $(X_1, X_2, X_3, X_4, X_5, X_6)$ is a multinomial random variable.
2. An urn contains N_1 black balls, N_2 red balls, and N_3 green balls. Let n balls be drawn successively, with replacement. Let X_1 be the number of black balls drawn, X_2 be the number of red balls drawn, and X_3 be the number of green balls drawn. Then (X_1, X_2, X_3) is a multinomial random variable.
3. Thirteen cards are drawn from a standard deck of 52 cards successively and with replacement. Let X_1 be the number of hearts drawn, X_2 be the number of diamonds drawn, X_3 be the number of spades drawn, and X_4 be the number of clubs drawn. Then (X_1, X_2, X_3, X_4) is a multinomial random variable.

Theorem 4.4.1 If $(X_1, X_2, \ldots, X_K)$ is a multinomial random variable in n independent trials with probabilities p_i, $i = 1, 2, \ldots, K$, on a single trial, then the probability distribution of the random variable $(X_1, X_2, \ldots, X_K)$, called the multinomial distribution, is

$$f(x_1, x_2, \ldots, x_K) = \frac{n!}{x_1! \, x_2! \ldots x_K!} p_1^{x_1} p_2^{x_2} \cdots p_K^{x_K}, \qquad (4.4.1)$$

$$x_i = 0, 1, 2, \ldots, n$$
$$i = 1, 2, \ldots, K$$

where

$$\sum_{i=1}^{K} x_i = n, \qquad \sum_{i=1}^{K} p_i = 1.$$

Proof. The probability that in n independent trials E_1 occurs x_1 times, E_2 occurs x_2 times, ..., E_K occurs x_K times in one definite order is

$$p_1^{x_1} p_2^{x_2} \cdots p_K^{x_K}, \qquad \sum_{i=1}^{K} x_i = n, \qquad \sum_{i=1}^{K} p_i = 1. \qquad (4.4.2)$$

Since we are interested in events occurring in any order, the number of mutually exclusive ways in which this can happen is

$$\frac{x_1! \, x_2! \cdots x_K!}{n!} \qquad (4.4.3)$$

Therefore, by Corollary 2 of Theorem 2.6.1 the probability function of the random variable $(X_1, X_2, \ldots, X_K)$, is

Some discrete probability distributions

$$f(x_1, x_2, \ldots, x_K) = \frac{n!}{x_1!\, x_2! \cdots x_K!}\, p_1^{x_1} p_2^{x_2} \cdots p_K^{x_K},$$
$$x_i = 0, 1, 2, \ldots, n$$
$$i = 1, 2, \ldots, K$$

where

$$\sum_{i=1}^{K} x_i = n, \qquad \sum_{i=1}^{K} p_i = 1.$$

The probability function (4.4.1) is called the multinomial distribution since it is a general term in the multinomial expansion

$$(p_1 + p_2 + \cdots + p_K)^n.$$

For $K = 2$, the multinomial distribution (4.4.1) reduces to the binomial distribution (4.3.1).

Example 4.4.2 A die is rolled 12 times. What is the probability of obtaining two 1's, three 2's, one 3, two 4's, three 5's, and one 6?

Solution Let the random variable $(X_1, X_2, \ldots, X_6)$ represent the number of times 1, 2, ..., 6 appear, respectively, when a die is rolled 12 times. Then the random variable $(X_1, X_2, \ldots, X_6)$ has a multinomial distribution with $p_i = \frac{1}{6}$, $i = 1, 2, \ldots, 6$; $x_1 = 2$, $x_2 = 3$, $x_3 = 1$, $x_4 = 2$, $x_5 = 3$, and $x_6 = 1$; $n = 12$.

By Equation (4.4.1) we get

$$P(X_1 = 2, X_2 = 3, X_3 = 1, X_4 = 2, X_5 = 3, X_6 = 1)$$
$$= f(2, 3, 1, 2, 3, 1)$$
$$= \frac{12!}{2!\, 3!\, 1!\, 2!\, 3!\, 1!} \left(\tfrac{1}{6}\right)^2 \left(\tfrac{1}{6}\right)^3 \left(\tfrac{1}{6}\right)^1 \left(\tfrac{1}{6}\right)^2 \left(\tfrac{1}{6}\right)^3 \left(\tfrac{1}{6}\right)^1 = \tfrac{11!}{12} \left(\tfrac{1}{6}\right)^{12}.$$

Example 4.4.3 An urn contains five black, four red, and two green balls. Six balls are drawn successively, with replacement. What is the probability of obtaining two black, three red, and one green ball?

Solution Let the random variable (X_1, X_2, X_3) represent the number of black, red, and green balls drawn. Then the random variable (X_1, X_2, X_3) has a multinomial distribution with $n = 6$; $p_1 = \tfrac{5}{11}$, $p_2 = \tfrac{4}{11}$, $p_3 = \tfrac{2}{11}$; $x_1 = 2$, $x_2 = 3$, $x_3 = 1$.

By Equation (4.4.1) we get

$$P(X_1 = 2, X_2 = 3, X_3 = 1) = f(2, 3, 1) = \frac{6!}{2!\, 3!\, 1!} \left(\tfrac{5}{11}\right)^2 \left(\tfrac{4}{11}\right)^3 \left(\tfrac{2}{11}\right)^1$$
$$= 60 \left(\tfrac{5}{11}\right)^2 \left(\tfrac{4}{11}\right)^3 \left(\tfrac{2}{11}\right)^1.$$

4.5 THE GEOMETRIC DISTRIBUTION

Suppose that an experiment consists of a sequence of independent Bernoulli trials. If we perform independent trials until we get a first success, then the waiting time for a first success or the number of trials needed for the occurrence of first success is a geometric random variable.

Definition 4.5.1 Let p be the probability of success in a sequence of independent Bernoulli trials. Let the random variable X be the number of trials needed for the occurrence of first success. Then X is called a geometric random variable.

Example 4.5.1 The following examples involve geometric random variables.
1. A coin is tossed until a head appears. Then X, the number of trials required to obtain a first head, is a geometric random variable.
2. A die is rolled until a "6" appears. Then X, the number of trials required to get a first 6, is a geometric random variable.
3. A box contains three defective and seven non-defective items. Items are drawn successively with replacement. Then X, the number of draws needed until we get a defective item, is a geometric random variable.

Theorem 4.5.1 If X is a geometric random variable with probability p of success and $q = 1 - p$ of failure on a single trial, then the probability function of the random variable X, called the geometric distribution, is

$$f(x) = P(X = x) = q^{x-1}p, \quad x = 1, 2, \ldots . \quad (4.5.1)$$

Proof. The number of trials X needed for the occurrence of first success may be an infinite number of positive integral values $1, 2, \ldots$. Let $x - 1$ be the number of failures preceding the first success

$$\underbrace{FF \ldots F}_{x - 1.}S$$

By Theorem 2.7.1, the probability of an unbroken sequence of $x - 1$ failures followed by a success is $q^{x-1}p$. Therefore the probability function of the random variable X is

$$f(x) = P(X = x) = q^{x-1}p, \quad x = 1, 2, \ldots .$$

The probabilities of getting a first success at trials $1, 2, 3, \ldots$ correspond to the successive terms in an infinite geometric series

$$f(x) = p + qp + q^2p + \cdots .$$

Some discrete probability distributions

The sum of the probabilities is

$$\sum_{x=1}^{\infty} f(x) = \frac{p}{1-q} = \frac{p}{p} = 1.$$

Example 4.5.2 A die is rolled until a 1 appears.
a) In a sequence of independent trials, what is the probability function of the number of trials required for the first 1 to appear?
b) What is the probability of obtaining a 1 in three trials?

Solution Let the random variable X be the number of trials required to get a first 1. Then X has a geometric distribution with $p = \frac{1}{6}$.
a) Therefore the probability function of X is

$$f(x) = \left(\tfrac{5}{6}\right)^{x-1} \left(\tfrac{1}{6}\right), \qquad x = 1, 2, \ldots.$$

b) The probability of obtaining a 1 in three trials is

$$\begin{aligned}P(X = 3) = f(3) &= \left(\tfrac{5}{6}\right)^{3-1} \left(\tfrac{1}{6}\right) \\ &= \left(\tfrac{5}{6}\right)^{2} \left(\tfrac{1}{6}\right) \\ &= \tfrac{25}{216}.\end{aligned}$$

Theorem 4.5.2 Let the random variable X have the geometric distribution

$$f(x) = q^{x-1} p \qquad x = 1, 2, \ldots.$$

Then the mean μ and the variance σ^2 of the geometric distribution are

$$\mu = E(X^{\mathbf{2}}) = \frac{1}{p} \tag{4.5.2}$$

and

$$\sigma^2 = E(X^2) - [E(X)]^2 = \frac{q}{p^2}. \tag{4.5.3}$$

Proof. By Definition 3.6.1

$$\begin{aligned}\mu = E(X) &= \sum_{x=1}^{\infty} x f(x) \\ &= \sum_{x=1}^{\infty} x q^{x-1} p \\ &= 1 \cdot p + 2 \cdot qp + 3 \cdot q^2 p + \cdots.\end{aligned}$$

Let
$$S = 1 \cdot p + 2 \cdot qp + 3 \cdot q^2 p + \cdots. \tag{4.5.4}$$

The infinite series on the right-hand side of Equation (4.5.4) is an arithmetic-geometric series. Multiplying both sides of the Equation (4.5.4) by q, the common ratio of the geometric series, we get

$$qS = qp + 2 \cdot q^2 p + \cdots. \tag{4.5.5}$$

Subtracting Equation (4.5.5) from Equation (4.5.4), we have

$$(1 - q)S = p + qp + q^2 p + \cdots. \tag{4.5.6}$$

This is an infinite geometric series with common ratio q. Thus

$$(1 - q)S = \frac{p}{1 - q} \quad \text{or} \quad S = \frac{p}{(1 - q)^2} = \frac{p}{p^2} = \frac{1}{p}.$$

Therefore $\mu = E(X) = \frac{1}{p}$.

By Definition 3.7.1

$$E(X^2) = \sum_{x=1}^{\infty} x^2 f(x)$$

$$= \sum_{x=1}^{\infty} x^2 q^{x-1} p$$

$$= 1 \cdot p + 4 \cdot qp + 9 \cdot q^2 p + \cdots. \tag{4.5.7}$$

Let

$$S = 1 \cdot p + 4 \cdot qp + 9 \cdot q^2 p + \cdots. \tag{4.5.8}$$

Multiplying both sides of the Equation (4.5.8) by q, the common ratio of the geometric series, we get

$$qS = qp + 4 \cdot q^2 p + \cdots. \tag{4.5.9}$$

Subtracting Equation (4.5.9) from Equation (4.5.8), we have

$$(1 - q)S = 1 \cdot p + 3 \cdot qp + 5 \cdot q^2 p + \cdots. \tag{4.5.10}$$

The infinite series on the right-hand side of Equation (4.5.10) is an arithmetic-geometric series. Multiplying both sides of Equation (4.5.10) by q, the common ratio of the geometric series, we get

$$q(1 - q)S = q \cdot p + 3 \cdot q^2 p + \cdots. \tag{4.5.11}$$

Subtracting Equation (4.5.11) from Equation (4.5.10), we have

$$(1 - q)^2 S = p + 2qp + 2q^2 p + \cdots. \tag{4.5.12}$$

The infinite series on the right-hand side of Equation (4.5.12) is an infinite geometric series excluding the first term. Thus

$$(1 - q)^2 S = p + \frac{2qp}{1 - q}$$

or
$$S = \frac{p}{(1-q)^2} + \frac{2qp}{(1-q)^3}$$
$$= \frac{p}{p^2} + \frac{2qp}{p^3}$$
$$= \frac{1}{p} + \frac{2q}{p^2}.$$

Therefore
$$\sigma^2 = E(X^2) - [E(X)]^2$$
$$= \frac{1}{p} + \frac{2q}{p^2} - \frac{1}{p^2}$$
$$= \frac{p + 2q - 1}{p^2}$$
$$= \frac{2q - q}{p^2} = \frac{q}{p^2}.$$

Example 4.5.3 A die is rolled until a 6 appears. What is the mean and standard deviation of the number of trials required?

Solution Let the random variable X be the number of trials required to obtain a first 6. Then X has a geometric distribution with $p = \frac{1}{6}$. Therefore
$$\mu = E(X) = 1/p = 6$$
and
$$\sigma = \sqrt{q/p^2} = \sqrt{\frac{5}{6}/(\frac{1}{6})^2} = \sqrt{30} = 5.47.$$

4.6 THE NEGATIVE BINOMIAL DISTRIBUTION

The negative binomial distribution is a generalization of the geometric distribution. Suppose that an experiment consists of a sequence of independent Bernoulli trials. If we perform this experiment until we get K successes, then the number of trials needed for the occurrence of K successes is a negative binomial random variable. In the negative binomial distribution, the number of trials is a random variable and the number of successes is fixed, whereas in the binomial distribution the number of successes is a random variable and the number of trials is fixed.

Definition 4.6.1 Let p be the probability of success in a sequence of independent Bernoulli trials. Let the random variable X be the number of trials needed for the occurrence of $K \geq 1$ successes. Then X is called the negative binomial random variable.

Example 4.6.1 The following experiments involve negative binomial random variables.

1. A coin is tossed repeatedly until we get three heads. Then X, the number of trials required to obtain three heads, is a negative binomial random variable.
2. A box contains three defective and seven nondefective items. Items are drawn successively with replacement. Then X, the number of draws needed until we get three nondefective items, is a negative binomial random variable.
3. An urn contains M white and N black balls. Balls are drawn successively with replacement. Then X, the number of draws needed until we get four black balls, is a negative binomial random variable.

Theorem 4.6.1 If X is a negative binomial random variable with probability p of success and $q = 1 - p$ of failure on a single trial, then the probability function of the random variable X, for the occurrence of K successes, called the negative binomial distribution, is

$$f(x) = \binom{x-1}{K-1} p^K (1 - p)^{x - K}, \quad x = K, \; K + 1, \ldots \quad (4.6.1)$$

Proof. The number of trials X needed for the occurrence of $K \geq 1$ successes may be an infinite number of positive integral values $K, K + 1, \ldots$ Let the $x - 1$ trials produce $K - 1$ successes and the trial x produce the Kth success. The probability of exactly $K - 1$ successes in $x - 1$ trials is $\binom{x-1}{K-1} p^{K-1} q^{x - 1 - (K - 1)}$, and the probability of the Kth success on trial x is p. Therefore by Theorem 2.7.1 the probability function of the random variable X is

$$f(x) = \binom{x-1}{K-1} p^{K-1} q^{x - 1 - (K - 1)} p$$
$$= \binom{x-1}{K-1} p^K q^{x - K}, \quad x = K, K + 1, \ldots$$

The probability function (4.6.1) is also called Pascal's distribution. If $K = 1$, the negative binomial distribution (4.6.1) reduces to the geometric distribution (4.5.1).

Example 4.6.2 A die is rolled. What is the probability of getting a 6 for the third time on the seventh trial?

Solution Using the negative binomial distribution (4.6.1) with $x = 7$, $K = 3$, and $p = \frac{1}{6}$, we have

$$P(\text{getting a 6 for the third time in seven trials})$$
$$= \binom{6}{2} \left(\frac{1}{6}\right)^3 \left(\frac{5}{6}\right)^4$$
$$= \frac{6!}{2! \, 4!} \left(\frac{1}{6}\right)^3 \left(\frac{5}{6}\right)^4$$
$$= 15 \left(\frac{1}{6}\right)^3 \left(\frac{5}{6}\right)^4.$$

Theorem 4.6.2 Let the random variable X have the negative binomial distribution

$$f(x) = \binom{x-1}{K-1}p^K q^{x-K}, \quad x = K, K+1, \ldots \quad (4.6.2)$$

Then the mean μ and the variance σ^2 are

$$\mu = E(X) = \frac{K}{p} \quad (4.6.3)$$

and

$$\sigma^2 = \frac{Kq}{p^2}. \quad (4.6.4)$$

Proof. Let X_1 = number of trials required up to the first success.

X_2 = number of trials required between the first success and up to and including the second success.

$\vdots$

X_K = number of trials required between the $(K-1)$th success and up to and including the Kth success. Thus $X = X_1 + X_2 + \cdots + X_K$ is the total number of trials required for the K successes. The X_i's ($i = 1, 2, \ldots, K$) are independent random variables, each having a geometric distribution. Therefore by Theorem 3.7.3

$$E(X) = E(X_1) + E(X_2) + \cdots + E(X_K).$$

By Theorem 4.5.2, $E(X_i) = 1/p$ ($i = 1, 2, \ldots, K$). Hence

$$E(X) = \frac{1}{p} + \frac{1}{p} + \cdots + \frac{1}{p}$$
$$= \frac{K}{p}.$$

Since X_i's ($i = 1, 2, \ldots, K$) are independent, by Theorem 3.9.4

$$\sigma^2 = \text{Var}(X) = \text{Var}(X_1) + \text{Var}(X_2) + \cdots + \text{Var}(X_K).$$

By Theorem 4.5.2, $\text{Var}(X_i) = q/p^2$ ($i = 1, 2, \ldots, K$). Therefore

$$\sigma^2 = \text{Var}(X) = \frac{q}{p^2} + \frac{q}{p^2} + \cdots + \frac{q}{p^2}$$
$$= \frac{Kq}{p^2}.$$

Example 4.6.3 A die is rolled until four 6's are obtained. What is the mean and the standard deviation of the number of trials required?

Solution Let the random variable X be the number of trials required to obtain four 6's. Then X has a negative binomial distribution with $p = \frac{1}{6}$ and $K = 4$. Therefore

$$\mu = E(X) = \frac{K}{p} = \frac{4}{\frac{1}{6}} = 24$$

and

$$\sigma = \sqrt{\frac{Kq}{p^2}} = \sqrt{\frac{(4)(\frac{5}{6})}{(\frac{1}{6})^2}} = \sqrt{120} = 10.95.$$

4.7 THE HYPERGEOMETRIC DISTRIBUTION

Consider a finite population which consists of two kinds of objects. A sample of fixed size is drawn successively, without replacement. We may like to know the number of objects of one kind or the other in the sample.

Definition 4.7.1 Let N be the number of objects in a finite population and let a be the number of these objects which are of particular type A. Let the random variable X be the number of objects of type A that occur in a sample of size n drawn at random, without replacement. Then X is called a hypergeometric random variable.

Example 4.7.1 The following experiments involve hypergeometric random variables.

1. An urn contains four white and six black balls. Three balls are drawn, without replacement. Then the random variable X, the number of black balls drawn, is a hypergeometric random variable.
2. A box contains four defective and eight nondefective items. Three items are drawn, without replacement. Then the random variable X, the number of defective items drawn, is a hypergeometric random variable.
3. A group consists of 25 men and 15 women. A sample of 15 is selected, without replacement. Then the random variable X, the number of women in the sample, is a hypergeometric random variable.

Theorem 4.7.1 Let N be the number of objects in a finite population and let a be the number of these objects which are of particular type A. Then the probability function of the random variable X, the number of objects of type A in a sample of size n, is

$$f(x) = P(X = x) = \frac{\binom{a}{x}\binom{N-a}{n-x}}{\binom{N}{n}}, \quad x = 0, 1, 2, \ldots, n. \quad (4.7.1)$$

Proof. The total number of sample points in the sample space of this experiment is $\binom{N}{n}$ and each sample point is equally likely. There are $\binom{a}{x}\binom{N-a}{n-x}$ sample points with exactly x of type A and $n - x$ of type "not A." Thus the total number of favorable sample points is

$$\binom{a}{x}\binom{N-a}{n-x}.$$

Therefore each sample point has probability

$$f(x) = P(X = x) = \frac{\binom{a}{x}\binom{N-a}{n-x}}{\binom{N}{n}}, \quad x = 0, 1, 2, \ldots, n.$$

The probability function (4.7.1) is called the hypergeometric distribution.

Example 4.7.2 A box contains three defective and seven nondefective items. Three items are drawn, without replacement. Find the probability function of the number of defective items drawn.

Solution Let the random variable X be the number of defective items drawn. Then

$$N = 10, a = 3, n = 3 \quad \text{and} \quad X = 0, 1, 2, 3.$$

Therefore by Equation (4.7.1) we have

$$f(x) = P(X = x) = \frac{\binom{3}{x}\binom{7}{3-x}}{\binom{10}{3}}, \quad x = 0, 1, 2, 3.$$

The probabilities corresponding to 0, 1, 2, and 3 defectives are

$$f(0) = P(X = 0) = \frac{\binom{3}{0}\binom{7}{3}}{\binom{10}{3}} = \frac{35}{120}$$

$$f(1) = P(X = 1) = \frac{\binom{3}{1}\binom{7}{2}}{\binom{10}{3}} = \frac{63}{120}$$

$$f(2) = P(X = 2) = \frac{\binom{3}{2}\binom{7}{1}}{\binom{10}{3}} = \frac{21}{120}$$

$$f(3) = P(X = 3) = \frac{\binom{3}{3}\binom{7}{0}}{\binom{10}{3}} = \frac{1}{120}$$

The hypergeometric distribution of X can be written in tabular form as follows:

$X = x$	0	1	2	3
$f(x) = P(X = x)$	$\frac{35}{120}$	$\frac{63}{120}$	$\frac{21}{120}$	$\frac{1}{120}$

Example 4.7.3 Five cards are drawn from a standard deck of 52 cards. What is the distribution of the number of aces drawn?

Solution Let the random variable X be the number of aces drawn. Then $N = 52$, $a = 4$, $n = 5$, and $X = 0, 1, 2, 3, 4$. Therefore by Equation (4.7.1) we have

$$f(x) = P(X = x) = \frac{\binom{4}{x}\binom{48}{5-x}}{\binom{52}{5}}, \quad x = 0, 1, 2, 3, 4.$$

Theorem 4.7.2 Let the random variable X have the hypergeometric distribution

$$f(x) = \frac{\binom{a}{x}\binom{N-a}{n-x}}{\binom{N}{n}}, \quad x = 0, 1, \ldots, n. \tag{4.7.2}$$

Then the mean μ and the variance σ^2 of the hypergeometric distribution are

$$\mu = E(X) = \frac{na}{N} \tag{4.7.3}$$

and

$$\sigma^2 = \frac{N-n}{N-1} \cdot n \cdot \frac{a}{N}\left(1 - \frac{a}{N}\right). \tag{4.7.4}$$

Proof. By Definition 3.6.1

$$\mu = E(X) = \sum_{x=0}^{n} \frac{x\binom{a}{x}\binom{N-a}{n-x}}{\binom{N}{n}}$$

$$= \frac{1}{\binom{N}{n}} \sum_{x=0}^{n} \frac{xa!}{x!(a-x)!} \cdot \binom{N-a}{n-x}$$

$$= \frac{a}{\binom{N}{n}} \sum_{x=1}^{n} \frac{(a-1)!}{(x-1)![a-1-(x-1)]!} \binom{N-a}{n-x}$$

$$= \frac{a}{\binom{N}{n}} \sum_{x=1}^{n} \binom{a-1}{x-1}\binom{N-a}{n-x}$$

$$= \frac{a}{\binom{N}{n}} \sum_{y=0}^{n-1} \binom{a-1}{y}\binom{N-a}{n-y-1} \tag{4.7.5}$$

where $x - 1 = y$.
Since

$$f(x) = \frac{\binom{a}{x}\binom{N-a}{n-x}}{\binom{N}{n}}, \quad x = 0, 1, \ldots, n$$

is a probability distribution, we have

$$\sum_{x=0}^{n} \binom{a}{x}\binom{N-a}{n-x} = \binom{N}{n}. \tag{4.7.6}$$

Now by Equation (4.7.6), Equation (4.7.5) can be written as

$$\mu = E(X) = \frac{a}{\binom{N}{n}} \cdot \binom{N-1}{n-1}.$$

Thus

$$\mu = \frac{a}{N!/n!\,(N-n)!} \frac{(N-1)!}{(n-1)!\,(N-n)!}.$$

Therefore $\mu = na/N$.

By Definition 3.7.1

$$E(X^2) = \sum_{x=0}^{n} x^2 f(x)$$

$$= \sum_{x=0}^{n} \frac{x^2 \binom{a}{x}\binom{N-a}{n-x}}{\binom{N}{n}}. \qquad (4.7.7)$$

Using the identity $x^2 = x(x-1) + x$ in Equation (4.7.7), we get

$$E(X^2) = \sum_{x=0}^{n} \frac{[x(x-1)+x]\binom{a}{x}\binom{N-a}{n-x}}{\binom{N}{n}}$$

$$= \sum_{x=0}^{n} \frac{x(x-1)\binom{a}{x}\binom{N-a}{n-x}}{\binom{N}{n}} + \sum_{x=0}^{n} \frac{x\binom{a}{x}\binom{N-a}{n-x}}{\binom{N}{n}}. \qquad (4.7.8)$$

We have already obtained the second sum in evaluating the mean. Using the same procedure on the sum of the first term of Equation (4.7.8), we get

$$E(X^2) = \frac{a(a-1)n(n-1)}{N(N-1)} + \frac{na}{N}$$

$$= \frac{a(a-1)n(n-1) + na(N-1)}{N(N-1)}$$

$$= \frac{na}{N(N-1)}[(a-1)(n-1) + N - 1].$$

Therefore

$$\sigma^2 = E(X^2) - [E(X)]^2$$

$$= \frac{na(na - a - n + N)}{N(N-1)} - \frac{n^2 a^2}{N^2}$$

$$= \frac{N-n}{N-1} \cdot n \cdot \frac{a}{N}\left(1 - \frac{a}{N}\right).$$

4.8 THE POISSON DISTRIBUTION

Many experiments consist of yielding an infinite number of possible integral values 0, 1, 2, ... in a continuous time interval or in a continuous region of space. A unit of time may be a minute, an hour, a day, a week; and a unit of space may be a length, an area, a volume. The Poisson distribution applies to experiments yielding discrete data in a continuous space.

Definition 4.8.1 Let the random variable X be the number of successes in a given time interval or in a given region of space. Then X is called the Poisson random variable if the following conditions are satisfied:
  i) The number of successes in two disjoint time intervals or regions of space are independent.
  ii) The probability of a success for a small time interval or region of space is proportional to the length of the time interval or region of space.
  iii) The probability of two or more successes in a small time interval or region of space is negligible.

Example 4.8.1 The following experiments involve Poisson random variables.
1. The number of deaths per year by a rare disease in a large city.
2. The number of misprints per page in a large volume of a printed material.
3. The number of automobile accidents per month in a large city.
4. The number of defects in a manufactured product.
5. The number of telephone calls per minute at a telephone switchboard.
6. The number of rats per acre.
7. The number of airplane arrivals per hour at an airport.

Definition 4.8.2 Let X be a Poisson random variable with the possible values $0, 1, 2, \ldots$. Then the probability function of the random variable X, called the Poisson distribution, is

$$f(x) = P(X = x) = \frac{e^{-\lambda}\lambda^x}{x!}, \quad x = 0, 1, 2, \ldots, \quad \lambda > 0. \quad (4.8.1)$$

where $e = 2.71828\ldots$ and λ is the average number of successes occurring in the given time interval or region of space.

The sum of the probabilities is

$$\sum_{x=0}^{\infty} f(x) = \sum_{x=0}^{\infty} \frac{e^{-\lambda}\lambda^x}{x!}$$

$$= e^{-\lambda} \sum_{x=0}^{\infty} \frac{\lambda^x}{x!}$$

$$= e^{-\lambda}e^{\lambda} = 1$$

since it is known in calculus that

$$e^\lambda = 1 + \frac{\lambda}{1!} + \frac{\lambda^2}{2!} + \cdots = \sum_{x=0}^{\infty} \frac{\lambda^x}{x!}. \qquad (4.8.2)$$

Example 4.8.2 A book has 200 pages and 200 misprints distributed at random. What is the probability that a page contains

a) exactly two misprints?
b) fewer than two misprints?

Solution Let the random variable X be the number of misprints per page. Then the random variable X has the Poisson distribution with $\lambda = 1$.

a) The probability that a page contains exactly two misprints is

$$P(X = 2) = f(2) = \frac{e^{-1}(1)^2}{2!}$$
$$= \frac{e^{-1}}{2}$$
$$= .1839.$$

b) The probability that a page contains fewer than two misprints is

$$P(X < 2) = \sum_{x=0}^{1} \frac{e^{-1}(1)^x}{x!}$$
$$= e^{-1}\left(1 + \frac{1}{1!}\right)$$
$$= .7356.$$

Example 4.8.3 If the number of telephone calls an operator receives from 9:00 to 9:05 follows a Poisson distribution with $\lambda = 2$, what is the probability that the operator will not receive a phone call in that same time interval tomorrow?

Solution Let X be the number of calls the operator receives from 9:00 to 9:05. Then the random variable X has the Poisson distribution with $\lambda = 2$. By Equation (4.8.1) we have

$$P(X = 0) = f(0) = e^{-2}$$
$$= .135335.$$

Theorem 4.8.1 Let the random variable X have the Poisson distribution

The Poisson distribution

$$f(x) = \frac{e^{-\lambda}\lambda^x}{x!}, \quad x = 0, 1, 2, \ldots, \quad \lambda > 0. \tag{4.8.3}$$

Then the mean μ and the variance σ^2 of the Poisson distribution are

$$\mu = E(X) = \lambda \tag{4.8.4}$$

and

$$\sigma^2 = E(X^2) - [E(X)]^2 = \lambda. \tag{4.8.5}$$

Proof. By Definition 3.6.1

$$\mu = E(X) = \sum_{x=0}^{\infty} xf(x)$$

$$= \sum_{x=0}^{\infty} \frac{xe^{-\lambda}\lambda^x}{x!}$$

$$= \sum_{x=1}^{\infty} \frac{e^{-\lambda}\lambda^x}{(x-1)!}$$

$$= \lambda e^{-\lambda} \sum_{x=1}^{\infty} \frac{\lambda^{x-1}}{(x-1)!}. \tag{4.8.6}$$

By Equation (4.8.2) the sum on the right-hand side of Equation (4.8.6) is equal to e^λ. Thus

$$\mu = E(X) = \lambda e^{-\lambda} e^\lambda = \lambda.$$

By Definition 3.7.1

$$E(X^2) = \sum_{x=0}^{\infty} x^2 f(x)$$

$$= \sum_{x=0}^{\infty} \frac{x^2 e^{-\lambda}\lambda^x}{x!}. \tag{4.8.7}$$

Using the identity $x^2 = x(x-1) + x$ in Equation (4.8.7), we get

$$E(X^2) = \sum_{x=0}^{\infty} \frac{[x(x-1) + x]e^{-\lambda}\lambda^x}{x!}$$

$$= \sum_{x=0}^{\infty} \frac{x(x-1)e^{-\lambda}\lambda^x}{x!} + \sum_{x=0}^{\infty} \frac{xe^{-\lambda}\lambda^x}{x!}$$

$$= \sum_{x=2}^{\infty} \frac{e^{-\lambda}\lambda^x}{(x-2)!} + E(X)$$

$$= \lambda^2 e^{-\lambda} \sum_{x=2}^{\infty} \frac{\lambda^{x-2}}{(x-2)!} + \lambda. \tag{4.8.8}$$

By Equation (4.8.2) the sum in the first term on the right-hand side of Equation (4.8.8) is equal to e^λ. Thus

$$E(X^2) = \lambda^2 e^{-\lambda} e^\lambda + \lambda = \lambda^2 + \lambda. \qquad (4.8.9)$$

Therefore

$$\begin{aligned}\sigma^2 &= E(X^2) - [E(X)]^2 \\ &= \lambda^2 + \lambda - \lambda^2 \\ &= \lambda.\end{aligned}$$

4.9 THE UNIFORM DISTRIBUTION

Suppose an experiment results in N mutually exclusive outcomes, all equally likely. We have considered a number of experiments of this type in Chapter 3 such as tossing a coin, rolling a die, and drawing a card.

Definition 4.9.1 A random variable X is called a discrete uniform random variable if it has N possible outcomes, all equally likely.

Example 4.9.1 The following experiments involve uniform random variables.
1. Toss a coin once. Let $x = 0$ represent the outcome tail and $x = 1$ represent the outcome head. Then X is a discrete uniform random variable.
2. Roll a die once. Then X, the number that appears on the top, is a discrete uniform random variable.
3. Draw a card from a standard deck of 52 cards. Then X, any card drawn, is a discrete uniform random variable.

Definition 4.9.2 Let $x_1, x_2, \ldots, x_N$ be the possible outcomes of the random variable X, then the probability function of the random variable X, called the discrete uniform distribution, is

$$f(x) = P(X = x) = \frac{1}{N}, \qquad x = x_1, x_2, \ldots, x_N. \qquad (4.9.1)$$

Example 4.9.2 A coin is tossed once, what is the distribution of the number of heads?

Solution Let the random variable X be the number of heads when a coin is tossed once. Then X has a discrete uniform distribution.

$$f(x) = \tfrac{1}{2}, \qquad x = 0, 1.$$

Example 4.9.3 A die is rolled. What is the probability distribution of the number that appears on the top?

Solution Let the random variable X be the number that appears on the top when a die is rolled. Then X has a discrete uniform distribution.
$$f(x) = \tfrac{1}{6}, \quad x = 1, 2, 3, 4, 5, 6.$$

Example 4.9.4 A card is drawn from a standard deck of 52 cards. What is the probability distribution of any card drawn?

Solution Let the random variable X be any card drawn from a standard deck of 52 cards. Then X has a discrete uniform distribution.
$$f(x) = \tfrac{1}{52}, \quad x = 1, 2, \ldots, 52.$$

Theorem 4.9.1 Let the random variable X have a discrete uniform distribution
$$f(x) = \frac{1}{N}, \quad x = x_1, x_2, \ldots, x_N. \tag{4.9.2}$$

Then the mean μ and the variance σ^2 of the discrete uniform distribution are
$$\mu = \frac{N+1}{2} \tag{4.9.3}$$

and
$$\sigma^2 = E(X^2) - [E(X)]^2 = \frac{N^2 - 1}{12}. \tag{4.9.4}$$

Proof. By Definition 3.6.1
$$\begin{aligned}
\mu = E(X) &= \sum_{x=1}^{N} x f(x) \\
&= \sum_{x=1}^{N} x \, \frac{1}{N} \\
&= \frac{1}{N} \sum_{x=1}^{N} x \\
&= \frac{1}{N} \frac{N(N+1)}{2}
\end{aligned} \tag{4.9.5}$$

since it is known in elementary algebra that
$$1 + 2 + \cdots + N = \frac{N(N+1)}{2}. \tag{4.9.6}$$

Thus $\mu = E(X) = \dfrac{(N + 1)}{2}$.

By Definition 3.7.1

$$E(X^2) = \sum_{x=1}^{N} x^2 f(x)$$

$$= \sum_{x=1}^{N} x^2 \cdot \frac{1}{N}$$

$$= \frac{1}{N} \sum_{x=1}^{N} x^2$$

$$= \frac{1}{N} \frac{N(N + 1)(2N + 1)}{6}.$$

Since it is known in elementary algebra that

$$1^2 + 2^2 + \cdots + N^2 = \frac{N(N + 1)(2N + 1)}{6}. \tag{4.9.7}$$

Thus $E(X^2) = \dfrac{(N + 1)(2N + 1)}{6}$.

Therefore

$$\sigma^2 = E(X^2) - [E(X)]^2$$

$$= \frac{(N + 1)(2N + 1)}{6} - \left[\frac{N + 1}{2}\right]^2$$

$$= \frac{(N + 1)(2N + 1)}{6} - \frac{(N + 1)^2}{4}$$

$$= (N + 1)\left[\frac{2N + 1}{6} - \frac{N + 1}{4}\right]$$

$$= (N + 1)\left[\frac{4N + 2 - 3N - 3}{12}\right]$$

$$= \frac{(N + 1)(N - 1)}{12}$$

$$= \frac{N^2 - 1}{12}.$$

4.10 SUMMARY

Some of the discrete probability distributions studied in this chapter have actually been introduced without explicit names in the examples and problems of Chapters 2 and 3. Several discrete probability distributions that describe most populations encountered in practice are derived. We discuss a set of assumptions that characterize a particular experiment and derive the corresponding probability distribution. Mean, variance, and useful properties of these distributions are derived. Extensive tables for these distributions now exist that make the evaluation of probabilities very simple.

EXERCISES

1. A die is tossed once and a success is recorded if a 1 is obtained; otherwise a failure is recorded. What is the probability of obtaining x successes?

2. If a random variable X has a Bernoulli distribution, given by Definition 4.2.2, show that $E(X^k) = p$ for $k = 1, 2, 3, \ldots$.

3. If $X_1, X_2, \ldots, X_n$ is a sequence of Bernoulli variables such that $E(X_i) = p$, show that

   a) $E \sum_{i=1}^{n} X_i = np$,

   b) $\text{Var}(\sum_{i=1}^{n} X_i) = npq$.

   and compare the results to the mean and variance of a Binomial distribution.

4. A die is rolled four times. What is the probability that
   a) exactly three rolls will show 2?
   b) exactly two rolls will show 2?

5. Out of 16 families with four children each, how many of them are expected to have
   a) no boys?
   b) one boy?
   c) two boys?

6. The probability that a basketball player gets a basket in a single throw is $\frac{1}{5}$. What is the probability of getting at least four baskets in 25 throws?

7. Find the binomial distribution for which the mean is 7 and the variance is $\frac{28}{5}$.

8. Three electric bulbs are drawn at random from a large consignment of electric bulbs, of which 10% are defective. What is the probability of getting
   a) no defectives?
   b) one defective?
   c) two defectives?
   d) three defectives?

9. The following table gives the number of heads obtained when six coins are tossed 64 times.

Number of heads	0	1	2	3	4	5	6
Frequency	1	9	10	22	16	5	1

Compare the results with the expected binomial frequencies.

10. Five dice are thrown 243 times. How many times do you expect three dice to show a five or a six?

11. Suppose that 40% of the people in a particular city smoke cigarettes.
    a) A group of 10 people is chosen. What is the probability that more than four are smokers?
    b) How large must the group be, in order that the probability is at least .90 that one or more of them smoke?

12. The reliability (probability of working properly) of a transistor is .9. In a radio with 10 transistors, what is the probability that a breakdown will occur?

13. The probability that a production line produces defective parts is .1. Eight pieces are selected from the line.
    a) What is the probability of at least one defective?
    b) What is the probability that all pieces are good?
    c) If 50 pieces were selected, what is the expected number of defectives?

14. A die is rolled four times. What is the probability that the results are 3, 4, 5, and 6 in any order?

15. A die is rolled six times. What is the probability that the results are 1, 2, 3, 3, 5, 6?

16. On the staff of the faculty of Mathematics of a certain university 40% are mathematicians, 40% are statisticians and 20% are computer scientists. A committee of three is selected, each person being chosen independently of others. What is the probability that the committee has one mathematician, one statistician and one computer scientist?

17. In a university, 60% are undergraduates, 30% are M.S. students, and 10% Ph.D. students. A committee of five is to be sent to the Senate. Each person is chosen independently of others. What is the probability that there will be two undergraduates, two M.S. students, and one Ph.D. student?

18. A box contains 10 balls, of which two are red, four are white, three are black, and one is yellow. A random sample of five balls is drawn successively with replacement from the box. What is the probability that the sample contains one red, two white, one black, and one yellow ball?

19. A production line produces good articles with probability .7, average ones with probability .2 and defective ones with probability .1. Ten articles are selected.

a) What is the probability of eight good ones and one defective?

b) What is the probability that there is an equal number of good and defective articles?

20. The probability that Dick misses any one shot in a game of billiards (a player continues to shoot until he misses a shot) is .30. What is the probability that

a) his turn lasts exactly four plays?

b) his turn lasts at least four plays?

21. The probability that a person exposed to a certain contagious disease will catch it is .4. Find the probability that the fifteenth person exposed to this disease is the third one to catch it.

22. The probability of John hitting a target at each shot is .2. He decides to shoot until he gets five hits. What is the probability that he can make it in 10 shots?

23. In an oral examination the probability that Harry answers any question correctly is $\frac{1}{3}$. What is the probability that the fifth question is the first one he gets correct?

24. A man decides to throw a pair of dice until he gets two sixes. What is the expected number of throws until he stops?

25. The probability is .4 that any person in a particular population smokes. A sequence of people are asked if they smoke. What is the probability that the fourth person asked is the first smoker?

26. In an oral examination a contestant answers true-false questions. He continues to answer questions until he gets eight correct answers. What is the probability that

a) he gets them right on the twentieth question?

b) he needs more than 20 questions?

(Assume that he guesses the answer in each question.)

27. It is estimated that 80% of the electric bulbs manufactured by a firm are not defective and the remaining are defective. They can be detected only by testing on a machine. A man needs two good bulbs. He selects bulbs at random, tests them on the machine, and throws away the defectives. What is the probability that he must test more than four bulbs to get two good bulbs?

28. A committee consisting of three persons A, B, and C is to be represented by one of them in a meeting. Each tosses a coin and the odd man is to represent the committee. If the coins all show tails or all show heads they are tossed again. What is the probability that a decision is reached in four or less tosses?

29. It is known that 2% of a group of students are color-blind. A committee of five students who are not color-blind is to be obtained from this group of students. Each student is selected at random and examined by some device. If he is found color-blind, then he is not included in the committee. What is the probability that 10 students from this group are to be examined?

30. An operation on laboratory animals, to artificially increase their blood pressure, is known to be 70% successful. Six animals with high blood pressure are needed for a particular experiment. What is the probability that eight animals must be operated on?

31. A man has $10, which he bets at a dollar a time, on a game which has a probability of .2 of winning. He will stop after his first win. What is the probability that he loses his $10 before a win occurs?

32. Six balls are drawn at random successively without replacement from a bag containing seven white and five red balls. What is the probability of drawing three or fewer white balls?

33. The names of six boys and six girls are written on paper slips and placed in a box. Five names are drawn at random. What is the probability that three are boys and two are girls?

34. Of 20 cups of coffee, 12 are made with instant coffee and eight are not. Five cups are selected. What is the probability that at least three are made with instant coffee?

35. Suppose that 10% of 100 articles are defective. Six articles are chosen from the 100. What is the probability that there are exactly 2 defectives? Compare this probability with that where, after each draw, the article is replaced and mixed thoroughly.

36. Eight books, all of the same size, are bound in red and nothing printed on the cover but it is known that four are dictionaries and four are encyclopedias. Four books are drawn at random, without replacement.

   a) What is the probability that all are dictionaries?
   b) What is the probability that two or more are dictionaries?

37. If four cards are drawn, without replacement, from a standard deck of 52 cards, what is the probability that

   a) exactly two will be aces?
   b) at least one of them will be an ace?

38. In a population of 150 individuals, 16% wear glasses. What is the probability of getting at most three individuals wearing glasses in a sample of 15? (Sampling without replacement.)

39. Calculate the probability of exactly two accidents at a busy intersection on a given day if on the average there are .5 accidents per day.

40. What is the probability that there will be three incoming telephone calls at a switch board during a particular two-minute time interval if on the average there are 1.5 incoming calls in the two-minute span?

41. Assume that the number of items of a certain kind purchased in a store during a week's time follows a Poisson distribution with $\mu = 2$. How large a stock should the merchant have on hand to yield a probability of .95 that he will be able to supply the demand?

42. A typist on the average makes three errors per page. What is the probability of her typing a page
   a) with no errors?
   b) with at least two errors?

43. In a city on the average there are four automobile accidents per month. What is the probability of
   a) no accidents in a given month?
   b) at most three accidents in a given month?

44. A Geiger counter records 60 counts per minute on the average for a particular radio-active substance. What is the probability of
   a) two counts in a five-second period?
   b) no counts in a two-second period?
   c) k counts in an S-second period?

45. Flaws in large plates of glass occur, on the average, one per 10 square feet. What is the probability that a 6′ × 10′ sheet will contain (a) no flaws? (b) at least one flaw?

Chapter 5

THE NORMAL DISTRIBUTION

5.1 THE NORMAL DISTRIBUTION

The normal distribution occupies the central position in probability and statistics. The normal distribution is the most frequently used of all probability distributions. The distribution was first discovered by DeMoivre in 1733 and later by Gauss in 1809. The normal distribution is also known as the "Gaussian Distribution." Its graph is called the "normal curve." The normal distribution has convenient mathematical properties and it also serves as an approximation to other distributions.

Definition 5.1.1 A continuous random variable X is said to be normally distributed if its probability density function is

$$f(x) = \frac{1}{\sigma\sqrt{2\pi}} \exp\left[-\tfrac{1}{2}\left(\frac{x-\mu}{\sigma}\right)^2\right],$$

$$-\infty < x < \infty, \ -\infty < \mu < \infty, \ \sigma > 0 \quad (5.1.1)$$

where

i) μ is the mean of the normal distribution,
ii) σ is the standard deviation of the normal distribution,
iii) $e = 2.71828$,
iv) $\pi = 3.14159$.

We abbreviate the normal distribution (5.1.1) with a mean μ and a variance σ^2 with the symbol $N(\mu, \sigma^2)$.

Since $f(x)$ is a probability density function, by Definition 3.3.1, the total area under the curve $f(x)$ above the x-axis is equal to 1. In terms of calculus

$$\int_{-\infty}^{\infty} \frac{1}{\sigma\sqrt{2\pi}} \exp\left[-\tfrac{1}{2}\left(\frac{x-\mu}{\sigma}\right)^2\right] dx = 1. \quad (5.1.2)$$

Definition 5.1.2 If a random variable X is normally distributed with a mean μ and a variance σ^2, then the probability that X lies between a and b is

$$P(a < X < b) = \int_a^b \frac{1}{\sigma\sqrt{2\pi}} \exp\left[-\tfrac{1}{2}\left(\frac{x-\mu}{\sigma}\right)^2\right] dx \tag{5.1.3}$$

= Area under the curve $f(x)$ above the x-axis and between the ordinates $x = a$ and $x = b$.

The graph of three normal distributions with the same standard deviation but different means is shown in Fig. 5.1.1, and the graph of three normal distributions with the same mean but different standard deviations is shown in Fig. 5.1.2.

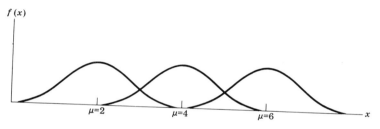

Fig 5.1.1 Normal distributions with $\sigma = 1$ and $\mu = 2, 4, 6$.

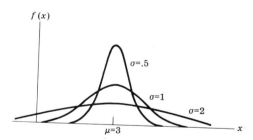

Fig. 5.1.2 Normal distributions with $\mu = 3$ and $\sigma = .5, 1, 2$.

It can be seen from the graphs of the normal distributions shown in Figs. 5.1.1 and 5.1.2 that the normal distribution has the following properties:

1. The normal distribution is symmetrical about the point $x = \mu$; that is, the graph of the normal distribution to the left of $x = \mu$ is the same as to the right.

2. The mean is in the middle and divides the area in half.
3. The total area under the curve $f(x)$ above the x-axis is equal to 1.

The normal distribution possesses many more interesting properties. We state below some of these properties without proofs as proofs are dependent on advanced mathematical procedures beyond the scope of this book.

Theorem 5.1.1 If X has the distribution $N(\mu, \sigma^2)$, then $Y = cX + d$ has the distribution $N(c\mu + d, c^2\sigma^2)$.

Example 5.1.1 If X has the normal distribution with mean 3 and variance 2, then $Y = 5X + 4$ has the normal distribution with mean $5(3) + 4 = 19$ and variance $25(2) = 50$.

Theorem 5.1.2 If $X_1, X_2, \ldots, X_N$ are independent random variables with the distributions $N(\mu_i, \sigma_i^2)$, $i = 1, 2, \ldots, N$, then $S = X_1 + X_2 + \cdots + X_N$ has the distribution

$$N\left(\sum_{i=1}^{N} \mu_i, \sum_{i=1}^{N} \sigma_i^2\right).$$

Example 5.1.2 If X_1, X_2 are independent normal random variables with means 2, 3 and variances 1, 2, respectively, then $S = X_1 + X_2$ has the normal distribution with

$$\text{mean} = 2 + 3 = 5$$

and

$$\text{variance} = 1 + 2 = 3.$$

Theorem 5.1.3 If X_1, X_2 are independent random variables with the distributions $N(\mu_i, \sigma_i^2)$, $i = 1, 2$, then $D = X_1 - X_2$ has the distribution $N(\mu_1 - \mu_2, \sigma_1^2 + \sigma_2^2)$.

Example 5.1.3 If X_1, X_2 are independent normal random variables with means 15, 12 and variances 10, 8, respectively, then $D = X_1 - X_2$ has the normal distribution with

$$\text{mean} = 15 - 12 = 3$$

and

$$\text{variance} = 10 + 8 = 18.$$

5.2 THE STANDARD NORMAL DISTRIBUTION

Definition 5.2.1 A normal distribution with $\mu = 0$ and $\sigma^2 = 1$ is called a standard normal distribution or a unit normal distribution if its probability density function is

$$f(x) = \frac{1}{\sqrt{2\pi}} e^{-(x^2/2)}, \qquad -\infty < x < \infty. \tag{5.2.1}$$

We abbreviate the normal distribution (5.2.1) with the symbol $N(0, 1)$.

Since $f(x)$ is a probability density function, by Definition 3.3.1, the total area under the curve $f(x)$ above the x-axis is equal to 1. In terms of calculus

$$\int_{-\infty}^{\infty} \frac{1}{\sqrt{2\pi}} e^{-(x^2/2)} \, dx = 1. \tag{5.2.2}$$

Definition 5.2.2 If a random variable X is normally distributed with a mean 0 and variance 1, then the probability that X lies between a and b is

$$P(a < X < b) = \int_a^b \frac{1}{\sqrt{2\pi}} e^{-(x^2/2)} \, dx \tag{5.2.3}$$

$= $ Area under the curve $f(x)$ above the x-axis and between the ordinates $x = a$ and $x = b$.

This area cannot be evaluated by ordinary means. The area under the density function of the standard normal distribution (5.2.1) can be evaluated by numerical integration. The table of areas or probabilities under the standard normal distribution are given in Appendix Table 2.

Theorem 5.2.1 If X has the distribution $N(\mu, \sigma^2)$, then $Z = (X - \mu)/\sigma$ has the distribution $N(0, 1)$.

Proof. The mean of Z is

$$E(Z) = E\left(\frac{X - \mu}{\sigma}\right)$$
$$= \frac{1}{\sigma} E(X - \mu)$$
$$= \frac{1}{\sigma} [E(X) - \mu]$$
$$= \frac{1}{\sigma} (\mu - \mu)$$
$$= 0.$$

The variance of Z is

$$\text{Var}(Z) = \text{Var}\left(\frac{X - \mu}{\sigma}\right)$$

$$= \frac{1}{\sigma^2} \text{Var}(X - \mu)$$
$$= \frac{1}{\sigma^2} \text{Var}(X)$$
$$= \frac{1}{\sigma^2} \sigma^2$$
$$= 1.$$

Hence by Theorem 5.1.1 Z is normally distributed with mean 0 and variance 1.

Example 5.2.1 If X is normally distributed with mean 3 and variance 4, what is the probability that X lies between 3 and 5?

Solution

$$z_1 = \frac{3-3}{2} = 0, \quad z_2 = \frac{5-3}{2} = 1.$$

Thus
$$P(3 < X < 5) = P(0 < Z < 1)$$
$$= \text{Area between } z = 0 \text{ and } z = 1.$$

From Appendix Table 2, the shaded area (Fig. 5.2.1) is .3413. Therefore $P(3 < X < 5) = .3413$.

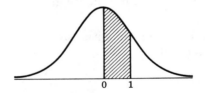

Figure 5.2.1

Example 5.2.2 Let X have a normal distribution with mean μ and variance σ^2. Find

a) $P(\mu - \sigma < X < \mu + \sigma)$,
b) $P(\mu - 2\sigma < X < \mu + 2\sigma)$,
c) $P(\mu - 3\sigma < X < \mu + 3\sigma)$.

Solution

a) $z_1 = \dfrac{(\mu - \sigma) - \mu}{\sigma} = -1, \quad z_2 = \dfrac{(\mu + \sigma) - \mu}{\sigma} = 1.$

The standard normal distribution

Thus

$$P(\mu - \sigma < X < \mu + \sigma) = P(-1 < Z < 1)$$
$$= \text{Area between } z = -1 \text{ and } z = 1$$
$$= \text{Area between } z = -1 \text{ and } z = 0$$
$$+ (\text{Area between } z = 0 \text{ and } z = 1).$$

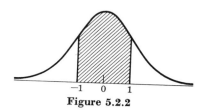

Figure 5.2.2

By symmetry the area between $z = -1$ and $z = 0$ (Fig. 5.2.2) is the same as the area between $z = 0$ and $z = 1$. Therefore

$$P(\mu - \sigma < X < \mu + \sigma) = .3413 + .3413$$
$$= .6826.$$

This means that 68.26% of the area falls within one standard deviation of the mean.

b) $z_1 = \dfrac{(\mu - 2\sigma) - \mu}{\sigma} = -2, \quad z_2 = \dfrac{(\mu + 2\sigma) - \mu}{\sigma} = 2.$

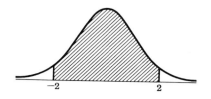

Figure 5.2.3

Thus

$$P(\mu - 2\sigma < X < \mu + 2\sigma) = P(-2 < Z < 2)$$
$$= (\text{Area between } z = -2 \text{ and } z = 0)$$
$$+ (\text{Area between } z = 0 \text{ and } z = 2)$$
$$= .4773 + .4773$$
$$= .9546.$$

112 The normal distribution

(See Fig. 5.2.3.) Therefore $P(\mu - 2\sigma < X < \mu + 2\sigma) = .9546$. This means that 95.46% of the area falls within two standard deviations of the mean.

c) $z_1 = \dfrac{(\mu - 3\sigma) - \mu}{\sigma} = -3, \quad z_2 = \dfrac{(\mu + 3\sigma) - \mu}{\sigma} = 3.$

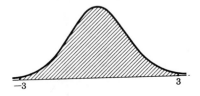

Figure 5.2.4

Thus

$$P(\mu - 3\sigma < X < \mu + 3\sigma) = P(-3 < Z < 3)$$
$$= \text{(Area between } z = -3 \text{ and } z = 0)$$
$$+ \text{(Area between } z = 0 \text{ and } z = 3)$$
$$= .4987 + .4987 = .9974.$$

(See Fig. 5.2.4.) Therefore $P(\mu - 3\sigma < X < \mu + 3\sigma) = .9974$. This means that 99.74% of the area falls within three standard deviations of the mean.

Example 5.2.3 The heights of male students at a certain university are normally distributed with a mean of 68.50 inches and a standard deviation of 2.3 inches.

a) What is the probability that any one male student at this university is over six feet tall?

b) What is the percentage of male students at this university who are between 70 and 72 inches tall?

Solution Let the random variable X be the height of male students at a certain university. Then X is normally distributed with $\mu = 68.50$ and $\sigma = 2.3$.

a) $z = \dfrac{72.0 - 68.5}{2.3} = 1.52.$

(See Fig. 5.2.5.) Thus

$$P(X > 72) = P(Z > 1.52)$$
$$= \text{Area to the right of } z = 1.52$$

$$= .5000 - .4357$$
$$= .0643.$$

Therefore $P(X > 72) = .0643$.

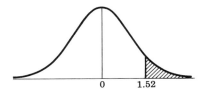

Figure 5.2.5

b) $z_1 = \dfrac{70.0 - 68.5}{2.3} = \dfrac{1.5}{2.3} = .65,$

$z_2 = \dfrac{72.0 - 68.5}{2.3} = \dfrac{3.5}{2.3} = 1.52.$

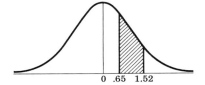

Figure 5.2.6

(See Fig. 5.2.6.) Thus

$$\begin{aligned} P(70 < X < 72) &= P(.65 < Z < 1.52) \\ &= \text{Area between } z = .65 \text{ and } z = 1.52 \\ &= (\text{Area between } z = 0 \text{ and } z = 1.52) \\ &\quad - (\text{Area between } z = 0 \text{ and } z = .65) \\ &= .4357 - .2422 \\ &= .1935. \end{aligned}$$

Therefore $P(70 < X < 72) = .1935$.

Example 5.2.4 Let the weights of one pound coffee jars filled by an automatic machine be normally distributed with $\mu = 1.03$ pounds and $\sigma = .02$ pounds. Find the probability that
a) the weight of any coffee jar is less than 1 pound.
b) the weight is more than 1.06 pounds.

Solution Let the random variable X be the true weight of coffee jars filled by an automatic machine. Then X is normally distributed with $\mu = 1.03$ and $\sigma = .02$.

a) $z = \dfrac{1.00 - 1.03}{.02} = -1.5.$

(See Fig. 5.2.7.) Thus
$$P(X < 1) = P(Z < -1.5)$$
$$= \text{Area to the left of } z = -1.5$$
$$= .5000 - .4332$$
$$= .0668.$$

Therefore $P(X < 1) = .0668$.

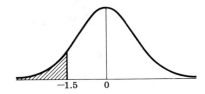

Figure 5.2.7

b) $z = \dfrac{1.06 - 1.03}{.02} = 1.5.$

(See Fig. 5.2.7.) Thus
$$P(X > 1.06) = P(Z > 1.5)$$
$$= \text{Area to the right of } z = 1.5$$
$$= .5000 - .4332$$
$$= .0668.$$

Therefore $P(X > 1.06) = .0668$.

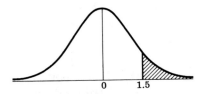

Figure 5.2.8

Example 5.2.5 The lifetimes of a certain make of battery are normally distributed. Suppose that 33% of the batteries have a lifetime under 45 hours and 10% of the batteries over 65 hours. Find the mean and standard deviation of the distribution.

Solution Let the random variable X be the lifetime of a certain make of battery. Let μ and σ be the mean and the standard deviation of the distribution of X.

Since the area lying to the left of $x = 45$ is .33, the area between $z = 0$ and $z_1 = (\mu - 45)/\sigma$ is $.5000 - .3300 = .1700$, from the normal tables. The value of z_1 corresponding to this area is .44. Hence

$$\frac{45 - \mu}{\sigma} = -.44. \qquad (5.2.4)$$

Area to the right of $x = 65$ is .10, the area between $z = 0$ and $z_2 = (65 - \mu)/\sigma$ is $.50 - .10 = .40$. The value of z_2 corresponding to this area is 1.28. Hence

$$\frac{65 - \mu}{\sigma} = 1.28. \qquad (5.2.5)$$

Solving Equations (5.2.4) and (5.2.5), we get

$$\mu = 50.12, \qquad \sigma = 11.63.$$

Example 5.2.6 In a certain examination the grades are normally distributed with an average grade of 75 and a standard deviation of 8. The instructor gave a grade A to students with marks 90 and over. If 12 students received a grade A, how many students took the examination?

Solution Let the random variable X be the grades of students in a certain examination. Then X is normally distributed with $\mu = 75$ and $\sigma = 8$. Then

$$z = \frac{90 - 75}{8} = \frac{15}{8} = 1.87.$$

(See Fig. 5.2.9.) Thus

$$P(X > 90) = P(Z > 1.87)$$
$$= \text{Area to the right of } z = 1.87$$
$$= .5000 - .4693$$
$$= .0307.$$

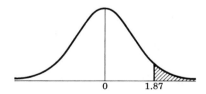

Figure 5.2.9

Therefore the total number of students, N, who took the examination

116 The normal distribution

is given by $\dfrac{3N}{100} = 12$ or $N = 400$.

5.3 THE NORMAL APPROXIMATION TO THE BINOMIAL DISTRIBUTION

In Section 4.3 we defined a binomial random variable as the sum of n independent Bernoulli random variables. By Theorem 4.3.1, the number of successes X in n independent trials is given by

$$P(X = x) = f(x) = \binom{n}{x} p^x q^{n-x}, \qquad x = 0, 1, 2, \ldots, n \qquad (5.3.1)$$

where p is the probability of success and $q = 1 - p$ is the probability of failure on a single trial. The computation of binomial probabilities (5.3.1) becomes laborious and difficult as the number of trials n becomes large. However, binomial probabilities can be approximated by a normal distribution if the number of trials n is large. The accuracy of the normal approximation to the binomial distribution also depends upon p. The approximation is more rapid if n is large and p and q close to $\frac{1}{2}$. If p is close to 0 or 1, the normal approximation to the binomial distribution becomes poorer for any given n. The normal approximation can be applied to every distribution we have studied so far under a wide range of conditions.

Example 5.3.1 A fair coin is tossed 12 times. Find the probability of getting from four to six heads by using (a) the binomial distribution, (b) the normal approximation to the binomial distribution.

Solution a) Let the random variable X be the number of heads in 12 tosses of a fair coin. Then X has a binomial distribution with $n = 12$, $p = \frac{1}{2}$, $q = \frac{1}{2}$. Hence by Equation (5.3.1)

$$P(X = x) = f(x) = \binom{12}{x} \left(\tfrac{1}{2}\right)^x \left(\tfrac{1}{2}\right)^{12-x}, \qquad x = 0, 1, \ldots, 12. \qquad (5.3.2)$$

Therefore

$$\begin{aligned}
f(0) &= .000244 & f(7) &= .193248 \\
f(1) &= .002728 & f(8) &= .120780 \\
f(2) &= .016104 & f(9) &= .053680 \\
f(3) &= .053680 & f(10) &= .016104 \\
f(4) &= .120780 & f(11) &= .002728 \\
f(5) &= .193248 & f(12) &= .000244 \\
f(6) &= .225436
\end{aligned}$$

Thus

$$\begin{aligned}
P(4 \le X \le 6) &= f(4) + f(5) + f(6) \\
&= .120780 + .193248 + .225436 \\
&= .539464.
\end{aligned}$$

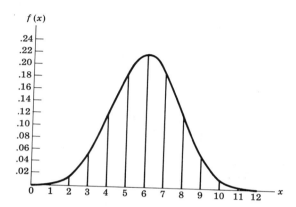

Fig. 5.3.1 Probability distribution for the number of heads and a normal curve.

b) The probability distribution for the number of heads in 12 tosses of the fair coin is shown graphically in Fig. 5.3.1. In the figure we superimpose a normal distribution with mean

$$\mu = np = 12(\tfrac{1}{2}) = 6$$

and

$$\sigma = \sqrt{npq} = \sqrt{12(\tfrac{1}{2})(\tfrac{1}{2})} = \sqrt{3} = 1.732.$$

Since the binomial distribution is discrete and we want to use the normal approximation as if the data were continuous, we make a correction for the continuity by subtracting $\tfrac{1}{2}$ from the lower value and adding $\tfrac{1}{2}$ to the upper value. Thus

$$z_1 = \frac{3.5 - 6.0}{1.732} = -1.44, \qquad z_2 = \frac{6.5 - 6.0}{1.732} = .28.$$

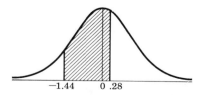

Figure 5.3.2

(See Fig. 5.3.2.) Therefore
$$\begin{aligned} P(3.5 < X < 6.5) &= P(-1.44 < Z < .28) \\ &= (\text{Area between } z = 0 \text{ and } z = -1.44) \\ &\quad + (\text{Area between } z = 0 \text{ and } z = .28) \\ &= .4251 + .1103 \\ &= .5354. \end{aligned}$$

The exact probability obtained in part (a) is .539464.

The above result can be stated in Theorem 5.3.1, the proof of which is beyond the scope of this book.

Theorem 5.3.1 Let the random variable X be the number of successes in n independent trials. Let p be the probability of success and $q = 1 - p$ be the probability of failure on a single trial, then the standardized binomial random variable

$$Z = \frac{X - np}{\sqrt{npq}}$$

is approximately normally distributed with mean 0 and variance 1 if the number of trials n is sufficiently large.

Example 5.3.2 Find the probability that a student can guess from 50 to 60 correct answers out of 100 questions on a true-false examination.

Solution Let the random variable X be the number of correct answers guessed by a student out of 100 questions, on a true-false examination. Then X is a binomial random variable with $n = 100$ and $p = \frac{1}{2}$.

We require the probability that the number of correct answers be 50 to 60. Using the normal approximation to the binomial with

$$\mu = (100)(\tfrac{1}{2}) = 50 \quad \text{and} \quad \sigma = \sqrt{npq} = \sqrt{100(\tfrac{1}{2})(\tfrac{1}{2})} = 5,$$

we have

$$z_1 = \frac{49.5 - 50}{5} = -.01$$

$$z_2 = \frac{60.5 - 50}{5} = 2.1.$$

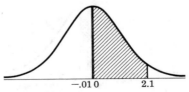

Figure 5.3.3

(See Fig. 5.3.3.) Thus

$$\begin{aligned}
P(50 \leq X \leq 60) &= P(-.01 < Z < 2.1) \\
&= (\text{Area between } z = 0 \text{ and } z = -.01) \\
&\quad + (\text{Area between } z = 0 \text{ and } z = 2.1) \\
&= .0040 + .4821 \\
&= .4861.
\end{aligned}$$

Therefore $P(50 \leq X \leq 60) = .4861$.

Example 5.3.3 If 3% of the light bulbs made by a certain manufacturer are defective, what is the probability that out of 600 light bulbs selected at random 20 or more are defective?

Solution Let the random variable X be the number of defective bulbs in a sample of 600. Then X is a binomial random variable with

$$\mu = np = 600 \left(\tfrac{3}{100}\right) = 18$$

and

$$\sigma = \sqrt{npq} = \sqrt{600 \left(\tfrac{3}{100}\right) \left(\tfrac{97}{100}\right)} = 4.18.$$

We require the probability that the number of defective bulbs is more than 20. Using the normal approximation to the binomial, we have

$$z = \frac{19.5 - 18}{4.18} = \frac{1.5}{4.18} = .36.$$

(See Fig. 5.3.4.) Thus

$$\begin{aligned}
P(X > 20) &= P(Z > .36) \\
&= \text{Area to the right of } z = .36 \\
&= .5000 - .1406 \\
&= .3594.
\end{aligned}$$

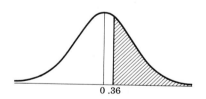

Figure 5.3.4

Therefore $P(X > 20) = .3594$.

5.4 SUMMARY

The normal distribution occupies the central position in probability and statistics. The normal distribution is the most frequently used of all probability distributions. The normal distribution possesses many convenient mathematical properties and also serves as an approximation to other distributions.

EXERCISES

1. Find the area under the standard normal distribution
    a) between $z = 0$ and $z = .7$,
    b) between $z = -.5$ and $z = 0$,
    c) between $z = -.6$ and $z = 2.2$,
    d) between $z = .4$ and $z = 1.4$,
    e) to the right of $z = -1.1$,
    f) to the left of $z = -1.2$.

2. Find the values of z in the following cases where the area refers to that under standard normal distribution:
    a) area between 0 and z is 0.4032,
    b) area to the left of z is 0.7580,
    c) area between -1.2 and z is 0.1269,
    d) area between -1.2 and z is 0.6429.

3. Let a random variable Z be normally distributed with mean 0 and variance 1. Find
    a) $P(Z \geq 1.96)$,
    b) $P(-3 \leq Z \leq 3)$,
    c) $P(Z \leq -1)$,
    d) $P(|Z| \leq 1)$,
    e) $P(|Z| \geq 3)$.

4. A random variable X is normally distributed with $\mu = 2$ and $\sigma = 1$. Find
    a) $P(X > 4)$,
    b) $P(0 < X < 2)$.

5. A random variable X is normally distributed with $\mu = 3$ and $\sigma = 6$. Find a number X_0 such that
    a) $P(X > X_0) = .05$,
    b) $P(X > -X_0) = .95$.

6. The heights of 500 students are normally distributed with mean 70.0 inches and standard deviation of 4.0 inches. How many students have their heights
    a) greater than 75 inches?
    b) less than or equal to 60 inches?
    c) between 68 inches and 72 inches, both inclusive?

7. The marks in a certain statistics examination are normally distributed with a mean of 70 and a standard deviation of 10. The top 5% of the students are to receive grade A. What is the minimum mark a student must get in order to obtain a grade A?

8. In a certain college it was found, after an I.Q. test, that the students' I.Q.'s are normally distributed with a mean 105 and standard deviation 5. Find the probability that an individual selected at random has an I.Q. between 95 and 115.

9. In a recruitment center 60% pass the minimum height requirement for selection to the army. The heights of the people present for the examination are normally distributed with mean 62 inches and standard deviation 3 inches. Find the minimum height requirement for a pass in the test.

10. The mean grade of a group of 500 school grades is 75% and the standard deviation 8%. Assuming the distribution to be normal find
    a) the number of grades above 85%,
    b) the highest grade of the lowest 20,
    c) the limits within which the middle 400 lie.

11. In a normal distribution 30% of the items are under 40 and 10% are over 65. Find the mean μ and standard deviation σ of the distribution.

12. In an examination 60% of the candidates obtained marks below 75% and 5% got above 85% and the rest between 75% and 85%. If the marks follow a normal distribution, find the mean and the standard deviation of the marks.

13. The diameters of ball bearings produced by a certain firm are normally distributed with mean of .5 inches and standard deviation of .05 inches. Ball bearings with a diameter between .45 and .55 are accepted and those with a diameter outside these limits will be termed defective and rejected. Find the percentage of defective ball bearings manufactured by the company.

14. It is known that the life of a certain make of transistor is normally distributed with mean 30 months. If a purchaser requires that 95% of them have lives exceeding 20 months, what is the largest value the standard deviation can have and still satisfy the purchaser?

15. The average seasonal rainfall in a certain city is 25 inches with a standard deviation of 5 inches. Out of a period of 100 years how many would you expect to have rainfall between 20 and 30 inches of rain? Assume that the seasonal rainfall for this city follows a normal distribution.

16. In a certain construction work the average wage is $4.00 per hour and the standard deviation is $.50. If the wages are assumed to follow a normal distribution, what percentage of the workers receive wages between $2.50 and $3.00 per hour?

17. A normal distribution with mean 75 and standard deviation 10 has 300 variates between 65 and 95. How many variates are there in the whole distribution?

18. In a certain normal distribution with a mean of 100 and a standard deviation 20 there are 500 variates between 110 and 150. How many variates are there between 90 and 120?

19. The weight of oranges from a large shipment averages 10.0 ounces with a standard deviation of 1.5 ounces. If these weights are normally distributed, what percent of all these oranges would be expected to weigh between 7.9 ounces and 12.4 ounces?

20. A random variable X is normally distributed. If $P(X > 15) = .0082$ and $P(X < 5) = .6554$, find the mean μ and the variance σ^2 of this distribution.

21. Find the number k, so that for a normally distributed random variable X, $P[\mu - k\sigma < X < \mu + k\sigma] = p$ where
    a) $p = .95$,
    b) $p = .99$,
    c) $p = .90$.

22. Let X and Y be two normal random variables with means 10 and 15 and standard deviations 3 and 4, respectively. Find
    a) $P[X + Y > 25]$,
    b) $P[10 < X + Y < 30]$,
    c) $P[X + Y < 10]$.

23. In a certain town the weekly wages of men are normally distributed with mean 100 and variance 12, while women's wages are normally distributed (independently of the men's) with mean 80 and variance 4. For a man and woman chosen randomly, what is the probability that the sum of their wages is
    a) greater than 190?
    b) between 170 and 180?
    c) less than 175?

24. If the temperature of a chemical solution is normally distributed with mean 60 and variance 25 when expressed in degrees centigrade, what is its distribution when expressed in degrees Fahrenheit?

25. If X has a normal distribution with mean 0 and variance 1, find k so that $P[X \leq k] = 3P[X > k]$.

26. If 60% of the population favors person A for president and a sample of 100 people is taken, what is the probability that at least 50 people in the sample favor A?

27. A fair coin is tossed 200 times. What are the probabilities of obtaining
    a) less than 110 heads?
    b) an equal number of heads and tails occur?
    c) more than 90 tails?

28. A single die is thrown 108 times and the number of 6's is denoted by X. Use the normal approximation to the binomial to find
    a) $P[20 < X < 30]$,
    b) $P[X \leq 7]$,
    c) $P[X > 20]$.

29. If a true-false test has 100 questions, what is the probability that a student who tosses a coin (heads = true, tails = false) will get 65 or more correct?

30. If in Exercise 29, there are two wrong and one correct answers to each problem and the student throws a die (1 or 2 = correct, 3, 4, 5 or 6 = incorrect), what is the probability of getting 50 or more correct answers?

31. If it is known that 5% of a batch of light bulbs are defective, then what is the probability, in a sample of 200, of finding
    a) exactly 5 defective?
    b) more than 10?
    c) between 0 and 5 (inclusive)?

32. If a drug is effective on 60% of patients, what is the probability that of 30 patients 3 or fewer will have no effect?

Chapter 6

THE IDEA AND CHOICE OF A SAMPLE

6.1 THE CONCEPT OF SAMPLING

We stated in Chapter 1 that statistics is a discipline concerned with the application of mathematics to the study of data obtained from observations, and that the roots of statistics lie in the theory of probability. In Chapters 2 through 5 we saw that the theory of probability was based on deductive reasoning; that is, starting from some axioms and definitions we were able to develop the theory based purely on reason. Thus we calculated the probability that the throw of a die would result in a one without actually performing the experiment. For the remainder of the book we return to our original concern of applying mathematics to the study of observed data. As we shall see in due course, the theory of probability provides many of the techniques for this study.

In order to study observed data, it is necessary to collect them, and do so in such a way that valid scientific conclusions can be based on the results of the study. Before discussing how this should be done, let us reconsider some of the examples of Chapter 1. In Example 1.1.1 a comparison is to be made between a new drug and a neutral substance, by means of two sets of ten patients. The twenty patients themselves are not of primary interest; that is, the experiment is not to be performed solely for their benefit. Instead the experimenter wants to generalize the results to a larger group, for example all hypertensive patients, or all hypertensive patients in a geographical region. In Example 1.1.2 only 30 children are examined, but the experimenter wants to generalize the results to a larger group (perhaps all the children in the school or town).

These examples illustrate the central problem (some would say the only problem) of statistics, namely, that of generalizing results on the basis of limited experiments or observations.

Definition 6.1.1 The total group under discussion, or the group to which the results will be generalized, is called the population.

Definition 6.1.2 A characteristic of a population is a variable which can be measured for each member of the population.

Definition 6.1.3 A parameter is a number which describes a population.

Example 6.1.1 In a study concerning heights of people in Canada, the population would be all the people in Canada, the characteristic would be the height of each person, and a parameter might be the average height of the population.

Example 6.1.2 If we were concerned with the speeds of cars made in 1970, then the population would be all cars made in 1970, and the characteristic would be the speed of each car. A parameter might be the speed of the fastest car in the population.

Example 6.1.3 In Example 1.1.1, the population is all hypertensive patients in some geographical region, for example, New York State, while the population characteristic is the blood pressure of each member of the population.

Example 6.1.4 For the purpose of the U.S. Census Bureau, the population would be every person in the United States. The characteristic of the population might be whether they have a telephone or not.

Example 6.1.5 In Example 1.1.4, there are two populations, namely, the iron ore in region A and in region B. The population characteristics are the melting points of the ore.

It is important to define clearly both the population and the characteristic being measured before any experimentation begins. In practical situations one is faced with populations of which he cannot examine every member. Thus we could not examine the blood pressure of every hypertensive patient in New York State, nor could we give an experimental drug to each one. In Example 1.1.4 it is impossible to measure the melting point of all iron ore in regions A and B. In cases such as these an experimenter must be satisfied with measurements of the characteristics of a limited number of members of the population, that is, he must do with only a limited number of hypertensive patients or a limited number of iron-melting-point measurements.

Definition 6.1.4 A sample is a subset of a population, usually chosen because measurements on the entire population cannot or will not be made.

Definition 6.1.5 A statistic is a number which describes the sample.

Example 6.1.6. If in Example 1.1.1 the population is all hypertensive patients in New York State, then the set of 20 on whom measurements were made constitutes the sample. A statistic might be the average blood pressure of the sample.

Example 6.1.7 If in Example 6.1.4 the population consists of all people in the United States, and the characteristic is whether or not they have telephones, then a sample might consist of 1000 people. A statistic might be the number of people with telephones in the sample.

Example 6.1.8 In Example 1.1.4 there are two populations, the iron ore in regions A and B. The two samples consist of 10 and 6 determinations of the melting points, respectively. The statistics might be the average melting points of the two samples.

Definition 6.1.6 We denote the values obtained from a sample by x. x_1 corresponds to the first value, x_2 to the second, etc. If the sample has N observations, their values are $x_1, x_2 \ldots, x_N$.

Example 6.1.9 In Example 1.1.3 $x_1 = 13.696$, $x_2 = 13.699$, ..., $x_{10} = 13.707$.

Before discussing how to choose a sample that enables us to reach scientific conclusions we discuss why we sample in the first place instead of measuring the whole population. The three main reasons are given as follows.

I. One reason has already been alluded to, and that is that for reasons of time, resources or patience one may be unable (for practical purposes) to measure the population.

Example 6.1.10 In judging the popularity of a particular television program, it would take too long and be too expensive to ask the opinion of every viewer in the country. Sampling their opinions is therefore the only reasonable way of performing the job.

Example 6.1.11 As a counterexample however, note that when the census is taken (every ten years in the United States), the whole population is measured, presumably because the time and resources are adequate to do so.

II. A second reason for sampling is because some testing procedures destroy the sample. For instance in testing a batch of ammunition it would be foolish to insist on sampling the entire batch. In testing light bulbs for life length, it is obviously only practical to test a subset of the population.

III. A third reason is that sampling may be the only option open to us. For instance an anthropologist interested in knowing the skull dimensions of prehistoric man, has at his disposal only the few results unearthed by archeologists. The question of measuring the population is of course impossible.

In the case of medical experiments involving human patients there is a fourth reason based on ethical considerations; namely, that new information, on for instance the effects of medication, should be obtained without causing suffering or damage to the patient. For new medications this is best obtained by using a carefully chosen sample.

6.2 SELECTION OF THE SAMPLE

The importance of the question of how to sample should not be underrated. If the sample is chosen badly, then no amount of sophisticated mathematical and statistical techniques will render it acceptable. The technique of sampling is the very cornerstone of all modern scientific inquiry.

The main requirement of a sample is that it must be representative of the population. Intuitively, samples are supposed to be microcosmic representations of their population. If in order to measure the average height of Canadians, we choose a sample consisting of basketball players, it will not be representative and therefore conclusions made for the sample will not be applicable to the population.

One of the more infamous cases of a non-representative sample concerns the poll taken in 1936 by the Literary Digest to predict the winner of the presidential election between Landon and Roosevelt. Although the sample was carefully chosen to be representative, the results depended on voluntary responses. Since apparently there was a stronger tendency to respond to the question by conservatives than by liberals, Landon was predicted as the winner, whereas in fact Roosevelt won by getting about two-thirds of the votes. Note however that since then, the method has been improved upon and modern election forecasts can be made quite accurately, on the basis of small well chosen samples.

Basically the concept of sampling is to select a sample from a population in such a way that properties of the sample will correspond, as closely as possible, with those of the population. Before proceeding we may note that a population may consist of discrete units such as ball bearings or radio tubes or of continuous substances such as water or oil. In the former it is called discrete and in the latter, continuous. From a discrete population the number of samples may be finite or infinite. For instance if the population consists of all hypertensive patients in New York, then the total number of different possible samples is finite. If the population is the number of throws of a balanced coin, it is discrete yet infinite. Further it exists only in the imagination. One might call such a population hypothetical rather than real. The number of samples from such a population is also infinite, e.g. one sample might be one throw of a coin, a second sample two throws, etc. From populations of continuous substances such as liquids, the total number of possible samples is infinite; for example, if the population consists of all iron ore in the region, then there are an infinite number of measurements which could be made to obtain the melting points of iron.

Definition 6.2.1 If a population has a finite or a countably infinite number of units or elements it is called a discrete population.

Example 6.2.1 In Example 6.1.3 the population is finite (and therefore discrete) since it would be possible (although not necessarily feasible) to measure the blood pressure of every hypertensive patient.

Definition 6.2.2. If a population is not discrete then it is called continuous.

Example 6.2.2 In measuring the specific gravity of oil, there are no natural units; that is, there are no natural elements of the population. It is therefore continuous.

Besides the nature of the population is the nature of the measurements being taken, which may be discrete or continuous.

Example 6.2.3 In sampling from a batch of oil (the population), if the measurement X is

a) the percent of impurity, then X is continuous; and
b) either acceptable or non-acceptable, then X is discrete.

Example 6.2.4 The number of people who smoke brand x cigarettes is discrete, while the length of time it takes to smoke a brand x cigarette is continuous.

Definition 6.2.3 If the set of values which the measured variable can assume is finite or countably infinite, the variable will be called discrete, otherwise it will be called continuous.

Example 6.2.5 In Example 1.1.7 experimental mice are infected with cancer cells and then examined for tumors. The number of tumors a mouse can have is a discrete measurement while the time taken for the tumors to develop is a continuous variable.

The process of actually selecting the sample depends both on the nature of the population and on the type of measurements being made, but it is generally easier to discuss for discrete populations. For this reason we will speak of the elements or members of a population as though it were discrete. Comments on continuous populations will be made later.

Definition 6.2.4 A sample will be called randomly selected if every member of the population has an equal chance of being included in the sample. By the same token if a sample is randomly selected and has N elements, then all possible samples of size N have an equal chance of being chosen.

If a sample is randomly selected then the laws of probability theory can be applied; for this reason random selection, according to Definition 6.2.4, is desirable. The word random should not be confused with haphazard. If we want to choose five numbers from the integers 1, 2, ..., 20, we might choose the first two small and then, to compensate, choose the next two large, with perhaps one in the middle. Such a selection would not be random in that biases and prejudices came into play. Similarly in choosing six rats from a cage of 36 rats for measuring hormone levels in pharmacology, one might be inclined to close one's eyes and reach into the cage and pull out six rats. However, they would most likely be the less agile ones and if this docility were due to physical or hormonal qualities in the rats, this could adversely affect the results. Sampling methods which either fail to eliminate the sampler's biases or choose samples on the basis of a characteristic thought to be related to the characteristic of interest, for example, choosing rats on the basis of agility which might be related to hormone deficiencies, are called biased samples, and any conclusion or inferences made about a population on the basis of a biased sample should be viewed with suspicion. The following example illustrates how a sample may be randomly selected.

Example 6.2.6 How can an experimenter randomly select a sample of six rats from a group of 36?

Solution Each rat could first be numbered from 1 to 36 (by attaching a small disc to its ear or tail). Then each number could be put on a slip of paper, the slips mixed together in a cardboard box and then six slips drawn out without looking. Those rats corresponding to the six drawn slips would form the sample. A simpler way, using random number tables, is given in the next section.

Before discussing that technique, however, we should just mention how to sample from continuous populations. In general there are no hard and fast rules, and experience and common sense are relied on heavily; however, the concepts of random selection should be borne in mind. For example, when sampling a bulk material like sand for, let us say, ash content, the amount sampled must be large enough to detect the ash yet not too large for laboratory analysis. To ensure that the samples are representative, one might mix the sand thoroughly after each amount is removed or take amounts from different parts of the batch. When sampling agricultural crops for yield figures (usually in terms of yield per acre) the population is in fact discrete, there being a finite number of plants. It is generally more convenient, however, to divide the areas being sampled into lots and measure the yield of all plants in each lot. The question then concerns the size of the lots. Should we have a large number of small lots or fewer large lots? In an unpublished study on the effect of fertilizers and pesticides on rice yield, one of the authors found that if the lots are too small, that is, less than about 100 square feet, then the overall estimated yield tends to be too high. Further, if lots are too close to each other, their results seem to be related. The point here is that experience of similar sampling studies must be used when sampling from continuous populations.

6.3 THE USE OF RANDOM NUMBER TABLES

The procedure of giving each member of the population an equal chance of being included in the sample by numbering slips of paper and drawing them from a box would become tedious if repeated often and cumbersome for large populations. The use of random number tables is a more sophisticated way of drawing a random sample. Appendix Table 3 is a short table of random numbers. Their use is illustrated by means of the following examples.

Example 6.3.1 (Pharmacology) Consider the situation in Example 1.1.7. Suppose that we have 100 mice from which we wish to choose 10 for control A and 10 for drug B. First the animals are labeled by a marker (usually attached to the ear or the tail) from 1 to 100. One turns to Appendix Table 3 and selects any place as a starting point. The numbers are arranged in columns of five digits each. Since we want numbers from 1 to 100, choose any column of two-digit numbers. The first number chosen corresponds to the first mouse in the control group, and the second number, that is the one directly below the first, corresponds to the first mouse in the drug group. The third number allocates that mouse to the control group, etc., until twenty numbers have been selected. For illustrative purpose suppose that we started in the upper left-hand corner of Appendix Table 3, then the following allocations would be made.

Control group	93, 21, 9, 1, 59, 2, 19, 83, 14, 71
Drug group	28, 22, 90, 6, 56, 74, 69, 78, 63, 78

Note that there is no number 100 in the table, but if the number 00 is drawn, we will say that this corresponds to 100.

Example 6.3.2 (Education) From Example 1.1.2 suppose we wish to select 30 children from the 150 available in a fifth grade class to take comparative tests in English and mathematics. We first assign a number from 1–150 to each child. Select a starting point in Appendix Table 3, and choose one column of two-digit numbers plus the left-hand digits of the column to its left. This will correspond to three digits; that is, any number between 001 and 999 will be eligible. We ignore numbers greater than 150 and those which recur. Read down the column until 30 numbers have been selected and choose the corresponding children.

The random number table of Appendix Table 3 is necessarily brief in this book. For repeated use or large experiments it would be advisable to obtain a book of extended tables. To be highly recommended is the Rand Corporation's *A Million Random Digits with 100,000 Normal Deviates*, The Free Press of Glencoe, New York 1955.

Note that whether one uses a random number table or some other method of randomly selecting a sample, we speak of the method as being random and not the sample. It is possible, although not likely, that in answer to the question "Do you smoke brand x cigarettes?", we might get all "yes" or all "no" answers. The results might not appear to have come from a representative or randomly selected sample; yet if the ideas above had been incorporated, we would have to say that the selection was random.

Since one can only approximate the conditions of real life, no procedure or table is going to guarantee complete random sampling; however, the approximation is generally accurate enough that valid conclusions can be drawn from the results; that is, one can validly infer about the population values from the sample values. The following definition is needed to introduce the ideas of the next chapter.

Definition 6.3.1 A sample consisting of N observations chosen by the techniques of random sampling is called a random sample of size N. The word "random" is often dropped when it is clearly understood.

Example 6.3.3 If twenty patients of Example 1.1.1 are chosen from the population by random sampling, they will be called a random sample of size 20.

6.4 STRATIFIED RANDOM SAMPLING

The type of sampling discussed so far gives each member of the population an equal chance of being included in the sample. Sometimes it is intuitively more reasonable to divide the population into levels or strata, and to sample randomly from each stratum. For instance, in taking a sample of salaries, it might be reasonable to group them first into $0–5,000 per year, $5,000–10,000 per year, etc. In sampling manufacturing plants, it might be reasonable to group into steel industry, oil industry, plastics industry, etc., or into large, medium, or small depending on the variable being measured. In socioeconomic studies it is sometimes convenient to choose a number of areas and sample randomly within them. Such a method of sampling is called stratified random sampling.

The procedure of stratifying is not discussed in any detail here, and it is not possible at this stage to compare the two methods to say which is better. However, in general, stratified random sampling is more efficient; that is, it gets the same information for smaller sample sizes and smaller costs, than unstratified random sampling. This is not obvious, but it becomes clearer when one thinks about sampling the population of the United States for (say) television program preferences. Dividing the country into areas and sampling in those areas seems much more reasonable than simply randomly sampling the whole country.

Although various problems are better handled with stratified random sampling, the strata themselves are sampled randomly. Thus a thorough knowledge of random sampling is prerequisite for studying stratified sampling. All sampling techniques in the remainder of this text will be assumed to be random sampling techniques.

6.5 SUMMARY

In order to draw from a sample conclusions which can be generalized, one must collect the data from a random sample. Basically this means that every member of a discrete population has an equal chance of being included in the sample. This is most conveniently done using a set of random number tables. Continuous populations must be handled on a more *ad hoc* basis. There is great danger in trying to generalize conclusions obtained from samples which have not been randomly selected. A brief reference is made to stratified random sampling.

EXERCISES

1. Give an example of a population and its associated characteristic. Name a parameter of the population.

2. For the experiment in Example 1.1.5 suggest (a) the population, (b) a characteristic, (c) a parameter.

3. Consider an experiment to determine the probability of a head of a biased coin. Describe (a) the population, (b) the characteristic, (c) the parameter, (d) the sample, (e) the statistic.

4. Classify the following populations as discrete or continuous:
   a) a truck-load of cement,
   b) the entire population of the United States considered for census purposes,
   c) the city water supply,
   d) all the transistors in a television set,
   e) the number of milk cows in Ontario,
   f) the amount of milk produced in Ontario,
   g) the number of people who have degrees in statistics.

5. Classify the following measurements as discrete or continuous:
   a) the percentage of impurities in a truckload of cement,
   b) the number of telephones in New York City,
   c) the weights of new born babies,
   d) the sex of new born babies,
   e) the yield of rice per acre,
   f) the number of rice plants per acre,
   g) the specific gravity of iron,
   h) the age of an experimental rat,
   i) the I.Q. of retarded children,
   j) the number of defectives in a sample of 20 objects.

6. Consider a quality control problem where a batch should be rejected if the proportion defective is greater than $\frac{1}{36}$. Each day 10 items are selected from the batch, and if one or more of the 10 is defective, the whole batch is rejected. Simulate this procedure by throwing two dice 10 times, and calling a two "defective." Repeat the simulation several times and comment on the results.

7. Suppose you know that in a particular university of 2,000 students, 10% of the student body is lefthanded. In classes of size 20 you arrange to have three chairs with writing areas on the left. How often are you likely to be short of chairs for left-handed people? Simulate the problem by drawing 20 random numbers from four-digit numbers, that is 0001–2000 with 0001–0200 corresponding to lefthanded. Repeat often (say 100 times) and comment on the results.

8. In Example 1.1.2 what would you think of choosing 30 volunteers to form the sample rather than randomly selecting them.

9. In Section 6.2 we said that in measuring the yield of rice per lot, we overestimate the total yield if the lots are too small. Why do you think this is so?

10. Suppose that a population consists of the numbers 1 through 25. Use the random number tables to select 20 samples of size 5, and find the average of each sample. See how the central value of each of the 20 samples compares with the central value of the population.

11. Give two examples of destructive sampling.

12. Suppose that in a school of 200 children, two samples of size 15 are to be chosen and assigned to two separate tasks in order to compare dexterity. Describe how you would choose and assign the students.

Chapter 7
ORGANIZATION AND ANALYSIS OF DATA

7.1 INTRODUCTION

We have discussed why a sample is taken from a population and how it is chosen so as to adequately represent the population. We now consider the problem of how to organize, summarize, and analyze the observations in such a way that they can be described by a few numbers or statistics. Consider the following example.

Example 7.1.1 Suppose 10,000 television viewers are asked if they watch a particular program, to which they may answer Y (yes) or N (no). Rather than present the results as a sequence of letters Y, Y, N, Y, etc., one may simply say how many values of Y and N there are. Suppose that 6000 answer "yes" and 4000 answer "no," then we may summarize the results of the sample by saying that p, the proportion who watch the program is 0.6, or the percentage who watch is 60%.

If the number of observations N is large, it might facilitate easier handling and easier calculations if they are grouped into a relatively small number of classes or intervals. We may then present them pictorially or graphically. This grouping of sample results is not mandatory but can be done at the option of the experimenter.

Definition 7.1.1 Observations will be said to be grouped if they are arranged into a relatively small number of classes or intervals. Otherwise they will be called ungrouped.

Example 7.1.2 If the results of Example 7.1.1 are presented as a sequence of 10,000 Y's and N's they are ungrouped, but if they are presented as 6000 Y's and 4000 N's they are grouped.

The calculations of the statistics are slightly different for grouped and ungrouped observations, and so, from Section 7.5 on we will consider calculations for both grouped and ungrouped observations.

7.2 FREQUENCY TABLES

We will be concerned in this section with methods for grouping observations. The most straightforward way of grouping is by means of a frequency table, where the observations are divided into classes or inter vals. The resulting table shows the number of observations in each class or interval.

Definition 7.2.1 A frequency table is a table in which the observations in a sample of size N are grouped into classes or intervals, so that the frequency of observations in each class can be ascertained.

The first requirement necessary for the construction of a frequency table is the range of the observations; that is, what are the largest and smallest values? This prompts the following definition.

Definition 7.2.2 The range R of a set of N observations is the largest observation minus the smallest, that is

$$R = \text{largest observation} - \text{smallest observation}.$$

Example 7.2.1 In Example 1.1.3 the largest observation is 13.707 and the smallest is 13.683. The range is

$$R = 13.707 - 13.683 = 0.024.$$

To make a frequency table from a random sample of size N, one calculates the range R of the sample first. R is then divided into a number of intervals of equal length. Each observation x_i is assigned to one and only one of the intervals. The number of observations in each interval is recorded. The intervals into which R is divided are called class intervals, and the number of observations in each interval is called the class frequency.

Definition 7.2.3 The intervals into which R, the range of a random sample of size N, is divided to form a frequency table are called class intervals.

Definition 7.2.4 The two endpoints of a class interval are called the class limits.

Definition 7.2.5 The center of a class interval, or the point half way between its class limits, is called the class mark.

Definition 7.2.6 If a frequency table has h class intervals, we shall denote their class marks by $y_1, y_2, \ldots, y_h$.

Definition 7.2.7 The length of a class interval, that is the difference between the class limits, is called the class width.

Definition 7.2.8 The number of observations in a class interval is called the class frequency. If there are h classes we denote the class frequencies by $f_1, f_2, \ldots, f_h$.

The problems of how to choose the number of class intervals and their width are discussed subsequently, but the reader will note that these problems arise only for summarizing continuous observations, since discrete observations are naturally classified. The following example will be used to illustrate the calculations of this section.

Example 7.2.2 Rats are common subjects for making preliminary investigations of newly developed pharmacological agents. For instance diuretic agents are tried first on rats and later on dogs before being used on humans. In Table 7.2.1 are the weights (in grams) of 97 forty-day-old rats to be used for a particular experiment.

TABLE 7.2.1
The weights (in grams) of 97 forty-day-old male rats

143	133	156	123	143	136	125	173	163	127
127	139	164	149	163	174	159	181	167	158
132	156	167	162	171	128	155	164	155	142
135	128	155	144	129	154	139	145	145	131
164	155	163	132	159	163	177	180	153	147
137	149	133	148	133	150	149	170	147	145
141	187	130	160	165	155	136	149	167	152
130	180	147	117	154	145	145	145	158	
139	159	128	149	144	169	164	133	149	
143	157	139	130	131	173	181	152	132	

The first problem in making a frequency table is determining the class width. Once an observation has been placed in a class interval, it loses its individuality so to speak and assumes as its value, the class mark. Thus all observations in the same class interval are assumed to have the same value, namely, the class mark of the interval.

Since for continuous variables there are no natural class intervals, their choice is a matter of judgement. One is guided by the following conditions:

a) that no serious error is incurred by substituting the class mark for the observed value

b) that for convenience the class width is made as large as possible subject to condition (a).

These two conditions imply that the larger the variation between observations the wider is the class interval and that, for a fixed range of variation, the greater the number of observations the narrower is the class interval. This last point follows intuitively, since for too few observations and too narrow intervals, each observation would fall into a separate class and no classification would have been affected. As the number of observations increases, several may fall into each class and thus the class interval lengths may be narrowed. From this it follows that the two figures needed to form a frequency table are

N, the number of observations, and

R, the range of the observations.

There is no general method of calculating the number or the width of class intervals, but a few rules of thumb are available. If N is the number of observations then generally (a) and (b) above will be satisfied if the number of class intervals is an integer approximately equal to the square root of N. In practice one will rarely try to form a frequency table with fewer than 50 or 60 observations. One rarely has more than 500–600 observations, so that the number of class intervals will usually lie between 8 and 25. If m is the number of class intervals, then the range divided by m will give the approximate width of the interval. The starting point of the intervals, that is the left-hand point of the first interval, may be chosen so as to position the intervals conveniently; for example, so that the class marks correspond to integers. Once this starting point has been chosen then the other intervals fall naturally into place. Care should be taken however to avoid ambiguities. Suppose that we wish to make a frequency table of the data in Example 7.2.1. We note that the smallest observation is 117, the largest is 187, and the range R is 70. Suppose that we choose a starting point at 117 and a class width of 10, then we could define succeeding class intervals as (a) 117–127, 127–137, 137–147, etc.; however, it is not clear into which class intervals the numbers 127, 137, 147, etc. should go. We could avoid such ambiguities by defining the class intervals as (b) 117–126, 127–136, 137–146, etc. so that there is no overlap. In Example 7.2.1 such a definition of intervals would be sufficient to classify all the observations, but if one measured the weights to the first decimal place, for example 132.7, there would be no class interval for numbers such as 126.5, 136.2, etc. Both of the above difficulties are overcome by the simple expedient of choosing class limits one half unit beyond the accuracy of the measurements so that the intervals neither create ambiguities nor leave "gaps" in the data. In the above discussion such intervals might be (c) 116.5–126.5, 126.5–136.5, 136.5–146.5, etc.

Example 7.2.3 If we wish to make a frequency table of the observations in Example 7.2.2, then

i) $N = 97$,
ii) the number of class intervals is approximately $\sqrt{97}$ or 10,
iii) the range is $R = 187 - 117 = 70$,
iv) the class width is $R/10 = 7$,
v) if the starting point is chosen to be 116.5, then the class limits of the first class are 116.5 and 123.5,
vi) the class mark of the first class is the point half-way between the class limits, or 120,
vii) the remaining class limits and class marks are shown in Table 7.2.2.

TABLE 7.2.2

Class no.	Class limits	Class mark
1	116.5–123.5	120
2	123.5–130.5	127
3	130.5–137.5	134
4	137.5–144.5	141
5	144.5–151.5	148
6	151.5–158.5	155
7	158.5–165.5	162
8	165.5–172.5	169
9	172.5–179.5	176
10	179.5–186.5	183
11	186.5–193.5	190

Note that in order to conveniently position the class marks at integer values, it is necessary to have 11 instead of 10 intervals; but this in no way contradicts our previous discussion.

Once the number and size of the class intervals have been decided upon, one is ready to make the frequency table. This is most simply done by writing vertically on the left-hand side of a page, the class limits, and next to them the class marks. The observations may be recorded by marking a 1 on the line corresponding to any class interval for each entry assigned thereto. It saves time in totalling if every fifth entry in a class is denoted by a diagonal line across the preceding four. It is important to go through the observations in a systematic way, usually by reading down columns or across rows. The markings themselves are referred to as tally marks.

The major disadvantage of this method is that it provides no means for rechecking the results, so that if a second tabulation is made with the same observations, and different results are obtained, then there is no way of knowing which is correct. For a large number of observations requiring considerable accuracy, it is better to transfer each observation to a card or slip of paper, which can then be subdivided into piles corresponding to class intervals; discrepancies can thus be checked fairly quickly. When all observations have been assigned to their class intervals, the tally marks may be totalled to give the class frequencies.

Example 7.2.4 The observations of Example 7.2.2 were examined in Example 7.2.3, and the class limits and class marks shown in Table 7.2.2 were established. We now assign each observation of Example 7.2.2 to its class interval. For convenience, suppose that we start with the number 143 and read down the columns. The first number, 143, lies between 137.5 and 144.5, and is therefore assigned to class 4. The next number, 127, is assigned to class 2. This is continued until all 97 values have been assigned. The results are shown in Table 7.2.3.

TABLE 7.2.3

Construction of a frequency table from the observations of Example 7.2.2.

Class no.	Class limits	Class marks (y)	Tally marks	Class frequency (f)	Cumulative frequency up to upper class limit
1	116.5–123.5	120	‖	2	2
2	123.5–130.5	127	⦀⦀	10	12
3	130.5–137.5	134	⦀⦀ ‖‖	13	25
4	137.5–144.5	141	⦀⦀ ‖	11	36
5	144.5–151.5	148	⦀⦀⦀ ‖	17	53
6	151.5–158.5	155	⦀⦀⦀	15	68
7	158.5–165.5	162	⦀⦀ ‖‖‖	14	82
8	165.5–172.5	169	⦀ ‖	6	88
9	172.5–179.5	176	‖‖‖	4	92
10	179.5–186.5	183	‖‖‖	4	96
11	186.5–193.5	190	‖	1	97
Total				97	

The final column headed "cumulative frequency up to upper class limit" is for later discussion. In the notation of Definitions 7.2.6 and 7.2.8, $y_1 = 120$, $y_2 = 127$, ..., $y_{11} = 190$; $f_1 = 2$, $f_2 = 10$, ..., $f_{11} = 1$.

7.3 GRAPHICAL REPRESENTATION OF THE FREQUENCY TABLE

It is often intuitively appealing if the results from the frequency table can be presented so as to convey to the eye the general run of the observations. The two most convenient methods for doing this are by means of (a) the frequency polygon and (b) the histogram or bar graph.

For both, one employs an appropriate piece of graph paper and writes the class limits and corresponding class marks on a horizontal scale. The class frequencies are measured on the vertical scale. One then plots points defined by the intersection of the class mark and its corresponding class frequency.

The frequency polygon is obtained by using straight line segments to join each adjacent point as illustrated in Fig. 7.3.1. The left-hand and right-hand lines are drawn from the last points at each end to the base at the center of the next class interval. The histogram is formed by placing vertical bars with width centered at the class mark, height equal to the frequency and whose width equals the class interval width as in Fig. 7.3.2.

Definition 7.3.1 A frequency polygon is a graphical representation of a frequency table in which class frequencies are plotted against class marks. These are then joined by straight line segments.

Definition 7.3.2 A histogram or bar graph is a graphical representation of a frequency table in which class frequencies for each class interval are represented by vertical bars, whose height represents the class frequency f.

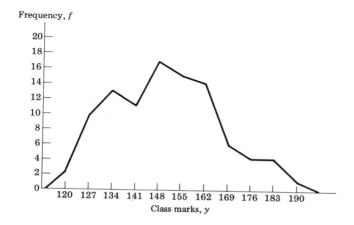

Fig. 7.3.1 Frequency polygon of weights (grams) of 97 forty-day-old rats.

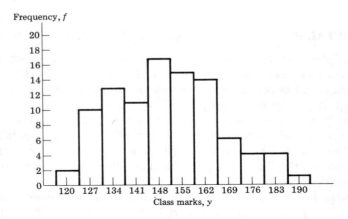

Fig. 7.3.2 Histogram of weights (grams) of 97 forty-day-old rats.

Example 7.3.1 The observations of Example 7.2.2, having been put into a frequency table in Table 7.2.3, can now be represented by either a frequency polygon as in Fig. 7.3.1 or a histogram as in Fig. 7.3.2. In both cases the horizontal and vertical axes are plotted in the same way. From Table 7.2.3 the highest class frequency is seen to be 17 ($f_5 = 17$). Thus the vertical axis need go only from 0 to 17. On the horizontal axis, the class marks (y) are plotted. For class one, $y_1 = 120$, $f_1 = 2$, and a dot is recorded on the graph representing that point. Other points (y_i, f_i) are similarly plotted. The frequency polygon is then completed by joining, with a straight edge, each adjacent pair of dots. The polygon is completed by joining the endpoints to the base at the center of the next class interval. The histogram is completed by drawing a horizontal line, through each dot, the length of the class interval. Vertical lines are then dropped to the base line.

In both the polygon and the histogram the total areas under the graphs are proportional to the total number of observations. In the polygon the ratio of areas of two class intervals is not proportional to the ratio of their number of observations; whereas for the histogram, the area over any fractional part of an interval appears to be the same, no matter what part, but this is in general not so. For comparative purposes the polygon and histogram are superimposed in Fig. 7.3.3. The areas under both graphs are easily seen to be the same.

Finally a few general notes on the presentation of frequency polygons and histograms. They should explain themselves as far as possible, so that their title should include information concerning what the measurements are, what material or subjects were used, and what restrictions apply. The horizontal axis should be clearly labeled showing what is measured and the units.

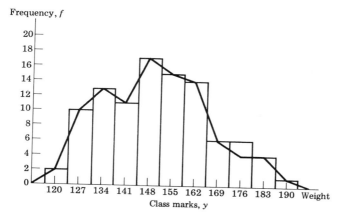

Fig. 7.3.3 Superimposed histogram and polygon of weights (grams) of 97 forty-day-old rats.

7.4 CUMULATIVE FREQUENCIES

The two most important variables which result from a frequency table are the class marks $y_1, y_2, \ldots, y_h$ and the class frequencies $f_1, f_2, \ldots, f_h$. For instance y_i is the value which every observation in class interval i is assumed to have, and f_i is the number of observations in the ith class interval. It is sometimes desirable or necessary to know the frequency up to and including the ith class interval, and this can be done by obtaining the cumulative frequency from the frequency table.

Definition 7.4.1 The cumulative frequency for the ith class interval is the frequency of the observations up to and including the ith class; that is, it is the sum of frequencies from the first through the ith class interval.

The cumulative frequency of the first class interval is f_1. For the second class interval it is $f_1 + f_2$. In general for the ith class interval, the cumulative frequency is $\sum_{j=1}^{i} f_j$. We note when $i = h$, that $\sum_{j=1}^{h} f_j = N$.

Graphing of the cumulative frequencies proceeds in the same way as for the polygon or histogram except that on the horizontal axis, one plots the upper class limit for each class interval rather than the class mark, and the vertical scale must go from zero to N. The resulting graph, sometimes called a step function, is similar to Fig. 7.4.1.

Example 7.4.1 In Table 7.2.3, the cumulative frequency for the first class interval is 2, for the second $2 + 10 = 12$, for the third $2 + 10 + 13 = 25$, etc. The cumulative frequencies up to the upper class limits of the class intervals, that is up to and including that class interval, are listed in the right-hand column of Table 7.2.3, and graphed in Fig. 7.4.1.

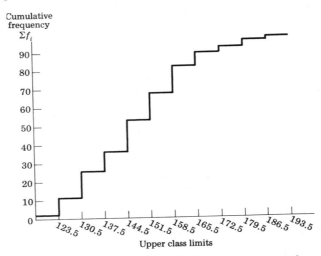

Fig. 7.4.1 A typical cumulative-frequency graph.

7.5 MEASURES OF LOCATION

We stated in Section 7.1 that we wished to describe sample results by means of a few statistics. The methods of Sections 7.2, 7.3, and 7.4 are mainly concerned with summarizing the sample so that it may be presented visually. We now return to the problem of how to calculate the statistics. There are two very useful measures of a set of sample observations. The first is the "center" of the observations, called the measure of location. The second, to be discussed later, is a measure of spread or dispersion.

Definition 7.5.1 By measure of location we mean a number which describes in some manner the "center" of a set of observations.

We put the word center in quotation marks to show that there are various ways of defining it.

Example 7.5. The center of a billiard cue may be (a) the point half-way between the ends or (b) the point at which the cue balances. These are usually not the same.

The measure of location of a set of observations with which the student is probably most familiar is the arithmetic mean, or the sum of the N observations divided by N. We now consider this measure in more detail.

Definition 7.5.2 The arithmetic mean of a set of observations (often called the sample mean), denoted by $\bar{x}$ for ungrouped observations and $\bar{y}$ for grouped observations, is defined as follows.

i) For ungrouped observations

$$\bar{x} = \frac{x_1 + x_2 + \cdots + x_N}{N} = \sum_{i=1}^{N} x_i/N. \qquad (7.5.1)$$

ii) For grouped observations with class marks $y_1, \ldots, y_h$ and class frequencies $f_1, \ldots, f_h$

$$\bar{y} = \frac{1}{N} \sum_{i=1}^{h} f_i y_i. \qquad (7.5.2)$$

Note that Equation (7.5.2) reduces to Equation (7.5.1) if the observations are ungrouped, since then $h = N$ and $f_i = 1$ and $y_i = x_i$.

Example 7.5.2 The arithmetic mean or the sample mean of the ungrouped observations of Example 7.2.2 is

$$\bar{x} = (143 + 127 + 132 + \cdots + 145 + 152)/97 = 14529/97$$
$$= 149.8 \text{ (to one decimal)}.$$

For the grouped observations of the same numbers shown in Table 7.2.3, the arithmetic mean is

$$\bar{y} = \tfrac{1}{97} [(2)(120) + (10)(127) + \cdots + (1)(190)]$$
$$= \tfrac{1}{97} \cdot (14552) = 150.0 \text{ (to one decimal)}.$$

An important property of the arithmetic mean is one's ability to simplify the arithmetic of calculating $\bar{x}$ and $\bar{y}$ by adding, subtracting, multiplying by, or dividing by a convenient constant.

Theorem 7.5.1 (a) If a and b are constants ($a \neq 0$), $x_1, x_2, \ldots, x_N$ are observations, and $z_i = ax_i + b$, then $\bar{z} = a\bar{x} + b$. (b) If $y_1, y_2, \ldots, y_h$ are class marks and $z_i = ay_i + b$, then $\bar{z} = a\bar{y} + b$.

Proof.

a) By Definition 7.5.2

$$\bar{z} = \sum_{i=1}^{N} z_i/N.$$

But $z_i = ax_i + b$, so

$$\bar{z} = \frac{1}{N} \sum_{i=1}^{N} (ax_i + b) = \frac{a}{N} \sum_{i=1}^{N} x_i + b$$
$$= a\bar{x} + b.$$

b) For grouped data, $z_i = ay_i + b$, so

$$\bar{z} = \frac{1}{N} \sum_{i=1}^{h} f_i(ay_i + b)$$

$$= \frac{1}{N} \sum_{i=1}^{h} a f_i y_i + \frac{1}{N} \sum_{i=1}^{h} f_i b$$

$$= a\bar{y} + b \text{ (remembering that } \sum_{i=1}^{h} f_i = N\text{).}$$

Example 7.5.3 We illustrate the use of the previous theorem just for the grouped observations of Table 7.2.3. The class marks are $y_1 = 120$, $y_2 = 127, \ldots, y_{11} = 190$. If we form the z-values by subtracting 155 from each y-value and divide by 7, we get

$$z_i = \frac{y_i - 155}{7} = \frac{1}{7} y_i - \frac{155}{7}$$

so that $a = \frac{1}{7}$, $b = -\frac{155}{7}$. The resulting z-values are $z_1 = -5$, $z_2 = -4$, $z_3 = -3, \ldots, z_{11} = 5$,

$$\bar{z} = \frac{1}{N} \sum_{i=1}^{h} f_i z_i = \frac{1}{97} [(2)(-5) + \cdots + (1)(5)] = -0.71.$$

Now $\bar{z} = a\bar{y} + b$ implies that $\bar{y} = (\bar{z} - b)/a$ or

$$\bar{y} = \left[-0.71 - \left(\frac{-155}{7} \right) \right] \Big/ \frac{1}{7}$$
$$= 150.0.$$

The usefulness of Theorem 7.5.1 is apparent. It permits the arithmetic to be performed on the relatively small z-values instead of the larger y-values.

A property of the arithmetic mean which we will need later is stated in the following theorem.

Theorem 7.5.2 *The sum of deviations of the observed values of a sample from their mean is zero.*

Proof. For ungrouped data, the sum of observations about their mean is

$$\sum_{i=1}^{N} (x_i - \bar{x}) = \sum_{i=1}^{N} x_i - N\bar{x}. \tag{7.5.3}$$

From Definition 7.5.2

$$N\bar{x} = \sum_{i=1}^{N} x_i$$

and Equation (7.5.3) is zero as stated. For grouped observations, the sum of deviations about their mean is

Measures of location   147

$$\sum_{i=1}^{h} f_i(y_i - \bar{y}) = \sum_{i=1}^{h} f_i y_i - N\bar{y}$$

which by Definition 7.5.2 is zero.

For combining the results of two sets of observations we have the following theorem.

Theorem 7.5.3   Let the results of a first set of N observations be x_{11}, x_{12}, x_{13}, ..., x_{1N} and of a second set of M be x_{21}, x_{22}, x_{23}, ..., x_{2M}. The respective means (in the ungrouped case) are

$$\bar{x}_1 = \frac{1}{N} \sum_{i=1}^{N} x_{1i}, \qquad \bar{x}_2 = \frac{1}{M} \sum_{i=1}^{M} x_{2i}. \tag{7.5.4}$$

The overall mean $\bar{x}$ of the entire set of $M + N$ observations is

$$\bar{x} = \frac{N\bar{x}_1 + M\bar{x}_2}{N + M}. \tag{7.5.5}$$

Proof.   By Definition 7.5.2 the overall mean is

$$\bar{x} = \left[\sum_{i=1}^{N} x_{1i} + \sum_{i=1}^{M} x_{2i} \right] \bigg/ (N + M). \tag{7.5.6}$$

But from Equation (7.5.1) above

$$\sum_{i=1}^{N} x_{1i} = N\bar{x}_1, \qquad \sum_{i=1}^{M} x_{2i} = M\bar{x}_2.$$

Substituting these in Equation (7.5.6) we obtain the desired result.

It might seem reasonable in certain circumstances to regard the center of a set of observations as that value for which half the observations are larger than it and half smaller. Such a value, if it exists, is called the median.

Definition 7.5.3   The central value of a set of observations which have been ordered according to magnitude is called the median. If there are an even number of observations, the median is the arithmetic mean of the two central values. In symbols, if $x_1, x_2, \ldots, x_N$ is a set of N observations arranged in increasing order of magnitude, then the median denoted by Md is defined as

$$Md = \begin{cases} x_{(N+1)/2} & \text{if } N \text{ is odd,} \\ \dfrac{x_{N/2} + x_{(N+1)/2}}{2} & \text{if } N \text{ is even.} \end{cases}$$

For observations arranged in a frequency table, the median is the value which divides the area of their histogram in half.

Example 7.5.4

a) The median of the numbers 1, 1, 2, 2, 2, 6, 6, 7, 9, 10, 10 is $Md = 6$.
b) The median of the numbers 1, 1, 2, 2, 2, 6, 6, 7, 9, 10 is $Md = (2 + 6)/2 = 4$.

For ungrouped data the median can usually be found easily enough by inspection as in Example 7.5.4. We consider next the calculation of the median in the case of grouped data. Basically, we find which class interval contains the median, and then, assuming that the values in that class are evenly distributed, we linearly interpolate within that class interval.

Example 7.5.5 For the observations of Example 7.2.2, arranged in the frequency table of Table 7.2.3, there are 97 observations, so the area of the histogram is divided in half if we find the value which has $97/2 = 48.5$ observations above it and 48.5 below it. From the discussion of the cumulative frequency, we see from Table 7.2.3, that the median lies in class number 5; that is, the median lies between 144.5 and 151.5. The length of the class interval is 7, and the class frequency is 17. Linearly interpolating between 144.5 and 151.5, the median is

$$Md = 144.5 + \tfrac{7}{17} \cdot (48.5 - 36) = 149.7.$$

More generally we give the following definition.

Definition 7.5.4 If for observations arranged in a frequency table
i) there are N observations,
ii) a and b are the lower and upper class limits of the class interval containing the median,
iii) the class width is Δ,
iv) N_1 and N_2 are the cumulative frequencies up to a and b, respectively; then

$$Md = a + \frac{(N/2 - N_1)(\Delta)}{N_2 - N_1}.$$

Note that $N_2 - N_1$ is the class frequency of the class interval containing Md.

The median in general does not have the nice algebraic properties of the arithmetic mean. In particular the median of the combination of two sets of observations with respective medians Md_1 and Md_2 is not easily definable. It does have one very desirable property, however; namely, that it is not affected by the size of the observations, only their number. Thus for an economic population with many low-earners and a few high-earners, the arithmetic mean might be inflated to give a wrong impression;

Measures of location

whereas the median is the point for which half the people earn more and half earn less.

Example 7.5.6 The following are 10 artificial figures of yearly wage earnings in thousands of dollars 3, 3, 2, 4, 3, 5, 8, 10, 6, 100. The median is 4.5, while the mean is 14.4. The median clearly gives a better picture of the true situation than the arithmetic mean does.

Perhaps the easiest measure of location of a set of ungrouped observations is that value which occurs most often. Such a value, if it exists, is called the mode.

Definition 7.5.5 The most frequently occurring value of an observation is called the mode, and denoted by M_0.

Example 7.5.7 The mode of the following set of numbers 1, 1, 2, 2, 2, 6, 6, 7, 8, 9, 10 is $M_0 = 2$, since 2 occurs most often.

Definition 7.5.6 A set of observations with two modes is called bimodal.

Example 7.5.8 The following set of observations 1, 1, 2, 2, 2, 6, 7, 7, 7, 10 is bimodal since 2 and 7 occur an equal number of times.

When data are arranged in a frequency table, the mode is hardly ever used, since although it is easy to find the class interval which contains the mode, it is not certain how one should locate the mode within that interval. An approximate guide is the following:

$$\text{Mode} = 3 \cdot (\text{Median}) - 2 \cdot (\text{arithmetic mean}).$$

For practical purposes, when dealing with continuous observations, the usual procedure is to avoid using the mode.

A descriptive statistic rarely used but sometimes of interest is the geometric mean which is the Nth root of the product of the N sample values $x_1, x_2, \ldots, x_N$.

Definition 7.5.7 The geometric mean of a set of N values $x_1, x_2, \ldots, x_N$, denoted by G, is

$$G = \sqrt[N]{x_1 \cdot x_2 \cdot \cdots \cdot x_N}$$
$$= (x_1 \cdot x_2 \cdot x_3 \cdot \cdots \cdot x_N)^{1/N}. \qquad (7.5.7)$$

Example 7.5.9

a) The geometric mean of the numbers 2, 4, 4, 8 is

$$G = (2 \cdot 4 \cdot 4 \cdot 8)^{\frac{1}{4}} = (256)^{\frac{1}{4}} = 4.$$

b) The geometric mean of the numbers 1, 2, 4, 7, 11 is

$$G = (1 \cdot 2 \cdot 4 \cdot 7 \cdot 11)^{\frac{1}{5}} = (616)^{\frac{1}{5}} = 3.6.$$

Using the property of logarithms that

$$\log(x_1 \cdot x_2 \cdot \cdots \cdot x_N) = \log x_1 + \log x_2 + \cdots + \log x_N$$

we note that

$$\log G = \frac{1}{N}(\log x_1 + \log x_2 + \cdots + \log x_N)$$

$$= \frac{1}{N}\sum_{i=1}^{N} \log x_i \qquad (7.5.8)$$

and that G is found using antilogarithms. Equation (7.5.8) is usually more convenient than Equation (7.5.7). The two drawbacks of the geometric mean are that G is not defined if any observation is zero or negative, and that its use is not as intuitively appealing as the arithmetic mean or median.

A final measure of location mentioned in passing is the harmonic mean.

Definition 7.5.8 The harmonic mean, denoted by H, of a set of observations $x_1, x_2, \ldots, x_N$ is

$$H = \left[\frac{1}{N}\sum_{i=1}^{N}\frac{1}{x_i}\right]^{-1}.$$

Example 7.5.10 Calculate the harmonic mean for the following observations: 2, 5, 10.

Solution

$$H = [\tfrac{1}{3}(\tfrac{1}{2} + \tfrac{1}{5} + \tfrac{1}{10})]^{-1} = (\tfrac{8}{30})^{-1} = \tfrac{30}{8} = 3.75.$$

The study of the harmonic mean seems hardly worthwhile at present, and it is almost never used in elementary analyses.

7.6 COMPARISON OF MEASURES OF LOCATION

In order to make comparisons between

a) the arithmetic mean,
b) the median,
c) the mode,
d) the geometric mean, and
e) the harmonic mean,

we proceed by a process of elimination. The harmonic mean appears to be the least useful. The mode is only really a useful statistic when used to describe data from a popular or layman's point of view, as, for instance, in the newspaper; for analysis of scientific data, however, it is usually ignored. There are a few times when the geometric mean can be useful; most notable is in the analysis of random variables whose mean and variance are related, but such applications are beyond the scope of this book. The only two serious contenders for the role of a "good" measure of location are the median and the arithmetic mean. The median is easy to calculate, and, as in Example 7.5.6, sometimes gives a truer picture of reality than the arithmetic mean. However, it is generally more difficult to manage mathematically and is therefore deemed less desirable. The arithmetic mean is also easy to calculate and easy to study mathematically. For these reasons the arithmetic mean is the measure of location used in most circumstances.

7.7 MEASURES OF DISPERSION

It is generally not sufficient to characterize a set of observations by a measure of location only. One generally wants to state how spread out or dispersed they are too.

Definition 7.7.1 By measure of dispersion we mean a number which describes in some manner how spread out or dispersed the observations are.

There are three measures which we will consider:
i) the sample variance,
ii) the sample mean absolute deviation,
iii) the sample quartiles.

We consider first the sample variance.

Definition 7.7.2 The sample variance of a set of N observations is denoted by s^2 and defined as

i) $\quad s_x^2 = \dfrac{1}{N-1} \sum\limits_{i=1}^{N} (x_i - \bar{x})^2$ for ungrouped data $\hspace{2em}$ (7.7.1)

ii) $\quad s_y^2 = \dfrac{1}{N-1} \sum\limits_{i=1}^{h} f_i(y_i - \bar{y})^2$ for grouped data. $\hspace{2em}$ (7.7.2)

The reasons for dividing by $N-1$ instead of N are discussed in Section 8.2. Note that Equation (7.7.2) reduces to Equation (7.7.1) when the data are ungrouped, that is, when $f_i = 1$, $y_i = x_i$ and $h = N$. The word "sample" is dropped when its meaning is understood.

The units in which s^2 is measured are the square of the original units.

The desire to obtain a measure of dispersion based on the sample variance, but in terms of the original units of measurement leads to the following definition.

Definition 7.7.3 The square root of the sample variance is called the sample standard deviation and denoted by s.

Example 7.7.1 Calculate the sample variance and sample standard deviation from the numbers 1, 2, 3, 4, 5.

Solution The arithmetic mean of the numbers is

$$\bar{x} = \frac{1 + 2 + 3 + 4 + 5}{5} = 3.$$

The sample variance is

$$s_x^2 = \frac{(1-3)^2 + (2-3)^2 + (3-3)^2 + (4-3)^2 + (5-3)^2}{4} = 2.5.$$

The sample standard deviation is $s_x = \sqrt{2.5} = 1.58$.

The use of formulas (7.7.1) and (7.7.2) can be clumsy if N is large, and simpler arthimetic is facilitated by the following theorem.

Theorem 7.7.1 The sample variance of a set of N observations can be written

i) $\quad s_x^2 = \dfrac{1}{N-1}\left\{\displaystyle\sum_{i=1}^{N} x_i^2 - N\bar{x}^2\right\}\quad$ or

$$\frac{1}{N-1}\left\{\sum_{i=1}^{N} x_i^2 - \frac{\left(\sum_{i=1}^{N} x_i\right)^2}{N}\right\} \tag{7.7.3}$$

for ungrouped observations, and

ii) $\quad s_y^2 = \dfrac{1}{N-1}\left\{\displaystyle\sum_{i=1}^{h} f_i y_i^2 - N\bar{y}^2\right\}\quad$ or

$$\frac{1}{N-1}\left\{\sum_{i=1}^{h} f_i y_i^2 - \frac{\left(\sum_{i=1}^{h} f_i y_i\right)^2}{N}\right\} \tag{7.7.4}$$

for grouped observations.

Proof. We prove only the first formula in cases (i) and (ii).

a) Consider the expression $\sum_{i=1}^{N}(x_i - \bar{x})^2$. By expansion we may write

$$\sum_{i=1}^{N}(x_i - \bar{x})^2 = \sum_{i=1}^{N}(x_i^2 - 2x_i\bar{x} + \bar{x}^2)$$
$$= \sum_{i=1}^{N} x_i^2 - 2\bar{x}\sum_{i=1}^{N} x_i + N\bar{x}^2.$$

Using Equation (7.5.1) $\sum_{i=1}^{N} x_i = N\bar{x}$, so

$$\sum_{i=1}^{N}(x_i - \bar{x})^2 = \sum_{i=1}^{N} x_i^2 - 2N\bar{x}^2 + N\bar{x}^2$$
$$= \sum_{i=1}^{N} x_i^2 - N\bar{x}^2.$$

Putting this in Equation (7.7.1) gives the result.

b) Consider

$$\sum_{i=1}^{h} f_i(y_i - \bar{y})^2 = \sum_{i=1}^{h} f_i(y_i^2 - 2y_i\bar{y} + \bar{y}^2)$$
$$= \sum_{i=1}^{h} f_i y_i^2 - 2\bar{y}\sum_{i=1}^{h} f_i y_i + N\bar{y}^2.$$

Using Equation (7.5.2) $\sum_{i=1}^{h} f_i y_i = N\bar{y}$, so

$$\sum_{i=1}^{h} f_i(y_i - \bar{y})^2 = \sum_{i=1}^{h} f_i y_i^2 - N\bar{y}^2.$$

Putting this in Equation (7.7.2) gives the desired result.

Equations (7.7.3) and (7.7.4) are often called the computational forms of the sample variance.

Example 7.7.2 Using the observations of Example 7.7.1, 1, 2, 3, 4, 5, with arithmetic mean $\bar{x} = 3$, calculate the sample variance using Equation (7.7.3).

Solution $\sum_{i=1}^{n} x_i^2 = 1^2 + 2^2 + 3^2 + 4^2 + 5^2 = 55$, thus

$$s_x^2 = \tfrac{1}{4}\{55 - (5)(3^2)\} = \tfrac{1}{4}(55 - 45) = 2.5.$$

Analogous to Theorem 7.5.1, we may simplify the calculations of s^2 by performing a linear transformation on the data.

Theorem 7.7.2 a) If a and b are constants ($a \neq 0$), $x_1, x_2, \ldots, x_N$ are observations (ungrouped) $z_i = ax_i + b$, then $s_z^2 = a^2 s_x^2$.
b) If $y_1, y_2, \ldots, y_h$ are the class marks of grouped data and $z_i = ay_i + b$, then $s_z^2 = a^2 s_y^2$.

154   Organization and analysis of data

Proof.
a) By Definition 7.7.2

i) $s_z^2 = \dfrac{1}{N-1} \sum_{i=1}^{N} (z_i - \bar{z})^2.$

But $z_i = ax_i + b$ and $\bar{z} = a\bar{x} + b$, so

$$s_z^2 = \frac{1}{N-1} \sum_{i=1}^{N} (ax_i + b - a\bar{x} - b)^2$$

$$= \frac{a^2}{N-1} \sum_{i=1}^{N} (x_i - \bar{x})$$

$$= a^2 s_x^2.$$

b) By Definition 7.7.2

ii) $s_z^2 = \dfrac{1}{N-1} \sum_{i=1}^{h} f_i(z_i - \bar{z})^2$, so

$$s_z^2 = \frac{1}{N-1} \sum_{i=1}^{h} f_i(ay_i + b - ay_i - b)^2$$

$$= \frac{a^2}{N-1} \sum_{i=1}^{h} f_i(y_i - \bar{y})^2$$

$$= a^2 s_y^2.$$

Example 7.7.3   Use Theorem 7.7.2 to calculate the sample variance of the numbers 5, 7, 9, 11, 13.

Solution   If $x_1 = 5$, $x_2 = 7$, ..., $x_5 = 13$, and $z_i = (x_i - 3)/2$, then $z_1 = 1$, $z_2 = 2$, ..., $z_5 = 5$. From Example 7.7.1 we know that $s_z^2 = 2.5$. Now $z_i = (x_i - 3)/2$ implies that $x_i = 2z_i + 3$, so that in terms of Theorem 7.7.1, $a = 2$, $b = 3$, and $s_x^2 = 4s_z^2 = 10$. This can of course be verified by direct calculation.

If we wish to combine two sample variances to obtain their arithmetic average, called the pooled sample variance, then we may use the idea of Theorem 7.5.3.

Definition 7.7.4   If s_1^2 and s_2^2 are two independent sample variances computed from samples of size N and M, respectively, then s_p^2, the pooled sample variance, is

$$s_p^2 = \frac{(N-1)s_1^2 + (M-1)s_2^2}{N+M-2}.$$

Example 7.7.4   Using the observations in Example 1.1.4

$$\sum_{i=1}^{10} x_{iA}^2 = 22{,}702{,}503, \qquad \bar{x}_A = 1506.7,$$

$$\sum_{i=1}^{6} x_{iB}^2 = 13{,}488{,}680, \qquad \bar{x}_B = 1499.3.$$

From Theorem 7.7.1

$$s_A^2 = \frac{22{,}702{,}503 - (10)(150.7)^2}{9} = 117.12,$$

$$s_B^2 = \frac{13{,}488{,}680 - (6)(1499.3)^2}{5} = 135.46.$$

Now from Definition 7.7.4 the pooled sample variance is

$$s_p^2 = \frac{(9)(117.12) + (5)(135.46)}{14} = \frac{1731.38}{14} = 123.7.$$

We mention, as a second measure of dispersion, the sum of the absolute values of deviations from the mean, and call it the mean absolute deviation.

Definition 7.7.5 The mean absolute deviation is denoted by m.a.d. and is

a) $\dfrac{1}{N} \sum\limits_{i=1}^{N} |x_i - \bar{x}|$ for ungrouped observations

and

b) $\dfrac{1}{N} \sum\limits_{i=1}^{h} f_i |y_i - \bar{y}|$ for grouped observations.

Note that without the absolute value signs, the sums (a) and (b) would be zero, from Theorem 7.5.2.

Example 7.7.5 The mean absolute deviation of the numbers 1, 2, 5, 4, 3 (whose arithmetic mean is 3) is

$$\text{m.a.d.} = \tfrac{1}{5}[|1-3| + |2-3| + |5-3| + |4-3| + |3-3|]$$
$$= \tfrac{6}{5} = 1.2.$$

The mean absolute deviation does not in general have convenient mathematical properties and is therefore rarely used in practice.

As an extension of the idea of median, which divides the observations into two groups with equal frequencies, one might define two more values, Q_1 and Q_3, such that one quarter of the observations lie below Q_1 and one quarter above Q_3. The numbers Q_1 and Q_3 are called quartiles.

Definition 7.7.6 The two quartiles, denoted by Q_1 and Q_3, of a set of observations are numbers such that one quarter of the observations lie below Q_1 and one quarter above Q_3.

Note that the two quartiles along with the median divide the observations into four groups each with equal frequencies.

Example 7.7.6 For the numbers 1, 2, 3, 4, 4, 6, 7, 8, 9, 10, 10 the values of Q_1 and Q_3 are 3 and 9, while the median is 6.

We recall that for ungrouped observations the median was one of the observations only if the number of observations was odd. Similarly Q_1 and Q_3 are observation values only if the number of observations is one less than a multiple of 4. For other numbers of observations Q_1 and Q_3 could be defined as some sort of mean of two observations, but the infrequent use of quartiles hardly justifies the time spent.

To find the quartiles from grouped observations, we proceed as for the median. First find the class intervals within which Q_1 and Q_3 lie, and linearly interpolate in them. We shall not give a general formula, but instead give an example.

Example 7.7.7 From the 97 grouped observations in Table 7.2.3, Q_1 and Q_3 lie in the class intervals containing 24.25 and 72.75, respectively, that is in class number 3 and 7. To find Q_1 proceed as follows:

i) The upper and lower boundaries of the interval containing Q_1 are 130.5 and 137.5.

ii) The class width is 7.0.

iii) The frequencies up to 130.5 and 137.5 are 12 and 25, respectively.

By linear interpolation

$$Q_1 = 130.5 + (7)\left(\frac{24.25 - 12}{25 - 12}\right) = 130.5 + 6.59 = 137.09.$$

Similarly to find Q_3, note that

i) the upper and lower class boundaries of the interval containing Q_3 are 158.5 and 165.5,

ii) the class width is 7.0,

iii) the frequencies up to 158.5 and 165.5, respectively, are 68 and 82.

By linear interpolation

$$Q_3 = 158.5 + (7)\left(\frac{72.75 - 68}{82 - 68}\right) = 160.87.$$

The idea of quartiles can be extended to deciles and percentiles as follows.

Definition 7.7.7 The rth decile, denoted by d_r, is the number below which $10r\%$ of the observations lie.

Definition 7.7.8 The rth percentile, denoted by p_r, is the number below which $r\%$ of the observations lie.

The calculation of deciles and percentiles follows naturally from the ideas of median and quartiles.

The comparison of the three measures of dispersion is soon taken care of. Quartiles, deciles, and percentiles are, like the median, simple to calculate, but in general are difficult to manage mathematically. The mean absolute deviation, also simple to compute, has the same failing, namely, that its characteristics cannot be studied with ease.

The variance, although slightly more difficult to calculate, has mathematical properties which are deemed desirable. Hence it has become generally accepted and used as the "usual" measure of dispersion.

7.8 SUMMARY

In order to make a set of data visually presentable so that an idea of the population shape can be seen at a glance, the data can be arranged into a frequency table, that is, into a number of adjacent class intervals (usually of equal width). The middle of each class, called the class mark, can then be plotted against the class frequency and presented in the form of a histogram or frequency polygon. Associated with the frequency table is the cumulative frequency, whose graph is a non-decreasing step function.

The data may then be quantitatively summarized by calculating measures of location and dispersion. Five measures of location are discussed; the arithmetic mean, the median, the mode, the geometric mean and the harmonic mean. In most cases the arithmetic mean is the most useful of these. Of the measures of dispersion discussed, only the sample variance and the sample standard deviation are of great practical interest, although the quartiles are of marginal interest. Also mentioned are deciles and percentiles.

EXERCISES

Problems 1, 2, 3, 4, 5, and 6 refer to the following set of 119 observations which are the I.Q.'s of schoolchildren.

158   Organization and analysis of data

111	117	112	127	113	94
104	93	102	96	102	98
128	99	128	109	107	104
120	110	136	141	144	114
156	132	80	123	97	95
114	140	107	119	109	127
102	98	113	127	117	129
110	141	111	98	122	117
113	131	116	103	99	88
132	120	110	119	121	100
136	100	140	113	119	122
111	96	93	131	114	91
99	123	106	103	126	147
106	93	120	124	90	124
127	111	99	77	76	133
119	91	112	136	116	101
122	123	119	87	123	92
98	130	121	133	103	117
115	134	108	116	156	124
96	127	150	129	139	

1. a) Calculate the range of the observations.
   b) Make a frequency table.
   c) Draw the histogram and frequency polygon.
   d) Plot the cumulative frequencies.

2. a) Calculate the median for grouped and ungrouped data.
   b) Calculate the arithmetic mean for grouped and ungrouped data.
   c) For the grouped data, in which class interval does the mode lie?

3. a) Simplify the observations by subtracting 110 and dividing by 10.
   b) Do likewise for the grouped data.

4. Calculate the mean for the grouped and ungrouped cases using the new simplified data of Exercise 3.

5. Let the first two columns correspond to school A and the last four correspond to school B; calculate their ungrouped means and verify that Theorem 7.5.3 holds true.

6. a) Calculate the variance and standard deviation.
   b) Calculate the quartiles.

7. The following data are the number of hours worked in a day by four workers: 6, 8, 4, 7. Compute (a) the median, (b) the arithmetic mean, (c) the geometric mean, (d) the harmonic mean, (e) the mode, (f) the range, (g) the variance, (h) the standard deviation, (i) the mean absolute deviation.

8. Throw a single die ten times and record the outcomes. Then calculate (a) the median, (b) the arithmetic mean, (c) the geometric mean, (d) the mode, (e) the range, (f) the variance, (g) the standard deviation.

9. Using the observations of Example 1.1.1, calculate (a) the arithmetic mean, (b) the variance, (c) the standard deviation.

10. Using the ideas of Example 7.5.5 and Definition 7.5.5, construct a generalized formula for calculating the quartiles.

11. Using Exercise 10, find the upper and lower quartiles for the first two columns of data in the table of I.Q.'s shown on p. 158.

12. What can be said about a set of data, if the variance is zero?

13. Suggest a problem where the median would be a more appropriate measure of location than the arithmetic mean.

14. The marks of a student in eight examinations are 83, 92, 69, 65, 89, 74, 70, 82. Find the median of these marks.

15. Given the frequency distribution of the times at which infants first walk in a certain township.

Age in months	Number of infants
7.5– 8.4	1
8.5– 9.4	8
9.5–10.4	15
10.5–11.4	30
11.5–12.4	52
12.5–13.4	58
13.5–14.4	25
14.5–15.4	10
15.5–16.4	1

Find (a) the mean, (b) the median, (c) the mode. Verify the empirical relation Mean − Mode = 3 (Mean − Median).

16. Find the quartiles of the following frequency distribution of the scores obtained by a group of 200 students in a certain test.

Marks	Number of students
30–40	5
40–50	15
50–60	40
60–70	65
70–80	55
80–90	17
90–100	3

17. Find the mean deviation and the standard deviation of the heights of 100 students at a certain college.

Height in inches	Number of students
58–60	2
61–63	8
64–66	16
67–69	35
70–72	25
73–75	10
76–78	4

18. After calculating the mean and the standard deviation of ten observations, a copying mistake is detected. In copying the sample values 456 was written for 546 by mistake. If the incorrect average and standard deviation calculated with the wrong entry are 250.1 and 45.2, find the correct value of the standard deviation.

19. The arithmetic mean of a set of marks of 200 students is found to be 55.2, and the arithmetic mean of marks of another set of 300 students is found to be 60.5. Find the arithmetic mean of marks of the combined set of 500 students.

20. The mean and standard deviation of wages of a group of 100 workers is found to be $120.00 and $2.5, respectively. In another group of 200 workers the mean and standard deviation of wages is found to be $125.00 and $3.00 respectively. What is the mean and the standard deviation of wages of the combined group of 300 workers?

21. The mean of two samples of sizes 40 and 110, respectively, are 50.2 and 53.7 and the standard deviations are 7 and 9. Find the mean and the standard deviation of the combined sample of size 150.

22. The mean score of 55 students of sections A and B in a certain statistics examination is 67. The mean score of 30 students in Section A is 60. Find the mean score of 25 students in Section B.

23. The mean of a set of 15 observations is 2 and the sum of their squares is 90. Find the standard deviation of the set.

24. Find two numbers whose arithmetic mean is 5 and geometric mean is 4.

25. Calculate the geometric mean of the following frequency distribution.

Class limits	Frequencies
0–4	8
5–9	10
10–14	20

(continued)

Table continued—

Class limits	Frequencies
15–19	35
20–24	15
25–29	10
30–34	2

26. Let x and y be two non-negative numbers and let H, G, and A be their harmonic mean, geometric mean, and arithmetic mean, respectively. Prove that $H \leq G \leq A$.

27. Calculate the harmonic mean of the weekly incomes in dollars of eight families: 150, 100, 200, 125, 80, 300, 75, 110.

28. On a certain farm a unit of work can be done by John, Bill, Tom, Harry and, Dick in 3, 6, 7, 8, and 10 hours, respectively. How many hours on the average do they take to do one unit of work?

29. Three cities A, B, and C are equidistant. A taxi driver travels from A to B at 45 miles per hour, and from B to C at 55 miles per hour, and from C to A at 65 miles per hour. Determine the average speed of the taxi for the entire trip.

Chapter 8
ESTIMATION

8.1 INTRODUCTION

In Chapter 6 we stated that the central problem of statistics was to generalize from a sample to a population. In this chapter we consider one facet of that generalization, namely, how to estimate the unknown population parameters by using the sample statistics.

Example 8.1.1 If a random variable X is described by a normal density with mean μ and variance σ^2, then we might want to estimate μ and σ^2 on the basis of a random sample $x_1, x_2 \ldots, x_N$.

Example 8.1.2 Let X be a random variable with a binomial probability function with parameters n and p. The parameter n is usually known while p is unknown. We may desire to estimate p using n and the value x, that X takes in n trials. Since p the probability corresponds to relative frequency, it is natural to estimate p as x/n. If x is the number of defective transistors in a sample of size n, then p is estimated as the proportion of defectives in the sample.

Example 8.1.3 If the birth records in a particular city show that of 1000 babies born, 510 were boys, then the estimate of p, the probability that the next baby is a boy, is .510.

It should be clear that, using a sample, the information about the parameters will be incomplete and that any statements about the parameters will contain an element of uncertainty. This uncertainty is measured in terms of probability.

Example 8.1.4 If the population is the students in a particular university and we wish to know their average height μ, on the basis of a sample, we will never know μ exactly. The best that we will be able to do is to estimate μ from a sample of size N. If $\bar{X}$ estimates μ then we know that

Some properties of the sample mean and variance   163

as N is increased our degree of uncertainty concerning μ will decrease until, if the sample consisted of the whole population, $\bar{X}$ and μ would be one and the same. This is another way of saying that $E(\bar{X}) = \mu$, so that we may think of the expectation of a statistic as its average over every member of the population. Not every estimator has the property that its expectation equals the parameter being estimated, as we shall see later.

8.2 SOME PROPERTIES OF THE SAMPLE MEAN AND VARIANCE

The arithmetic mean, often called the sample mean, is defined in Definition 7.5.2 and the sample variance in Definition 7.7.2. These two statistics are particularly important for further study because of their mathematical properties, which we now examine in more detail. If $X_1, X_2, \ldots, X_N$ are N random variables then the sample mean and the sample variance are also random variables.

Theorem 8.2.1   If $X_1, X_2, \ldots, X_N$ each have expectation μ, then $E(\bar{X}) = \mu$.

Proof. $E(\bar{X}) = E\{(1/N)(X_1 + X_2 + \cdots + X_N)\}$.
Using Theorem 3.7.1 with $a = 1/N$, $b = 0$, we may write

$$E(\bar{X}) = (1/N) E(X_1 + X_2 + \cdots + X_N).$$

Using Theorem 3.7.3 this is

$$E(\bar{X}) = (1/N) \{E(X_1) + E(X_2) + \cdots + E(X_N)\}.$$

But $E(X_i) = \mu$, so

$$E(\bar{X}) = (1/N) \{\mu + \mu + \cdots + \mu\}$$
$$= (1/N) (N\mu)$$
$$= \mu.$$

Example 8.2.1   If X_1, X_2 are the number of dots on the upturned faces of two dice, respectively, then show that $E(\bar{X}) = \frac{7}{2}$.

Solution   In Example 3.6.1 we showed that

$$E(X_1) = \tfrac{7}{2} = \mu, \; E(X_2) = \tfrac{7}{2} = \mu.$$

Thus $E(\bar{X}) = \mu = \frac{7}{2}$.

Theorem 8.2.2   For N independent random variables $X_1, X_2, \ldots, X_N$, if $E(X_i) = \mu$, $\text{Var}(X_i) = \sigma_X^2$, $i = 1, 2, \ldots, N$, then

$$\text{Var}(\bar{X}) = \sigma_X^2/N.$$

Proof. From Definition 3.8.1 $\text{Var}(\bar{X}) = E[\bar{X} - \mu]^2$, so

$$\text{Var}(\bar{X}) = E\left\{\frac{X_1 + X_2 + \cdots + X_N}{N} - \mu\right\}^2$$
$$= E\{(1/N)[(X_1 - \mu) + (X_2 - \mu) + \cdots + (X_N - \mu)]\}^2$$
$$= (1/N^2) E[(X_1 - \mu) + (X_2 - \mu) + \cdots + (X_N - \mu)]^2$$
by Theorem 3.7.1
$$= (1/N^2) E[(X_1 - \mu)^2 + (X_2 - \mu)^2 + \cdots + (X_N - \mu)^2 + \text{cross-products}]$$

where each cross-product term is of the form $(X_i - \mu)(X_j - \mu)$, $i \neq j$. By Theorem 3.7.3, the expectation operator E may operate on each term separately. The cross-product terms will be of the form $E[(X_i - \mu) \times (X_j - \mu)]$, $i \neq j$. By Theorem 3.7.4, since X_i, X_j are independent

$$E[(X_i - \mu)(X_j - \mu)] = E(X_i - \mu)E(X_j - \mu)$$
$$= (\mu - \mu)(\mu - \mu)$$
$$= 0$$

and
$$\text{Var}(\bar{X}) = (1/N^2)\{E(X_1 - \mu)^2 + \cdots + E(X_N - \mu)^2\}.$$

Now from Definition 3.8.1, $E(X_i - \mu)^2 = \sigma_X^2$, so
$$\text{Var}(\bar{X}) = (1/N^2)[\sigma_X^2 + \sigma_X^2 + \cdots + \sigma_X^2]$$
$$= \sigma_X^2/N.$$

σ_X^2/N is denoted by $\sigma_{\bar{X}}^2$ and called the variance of the sample mean.

Example 8.2.2 Continuing Example 8.2.1 with two dice with outcomes X_1 and X_2, show that $\text{Var}(\bar{X}) = \frac{35}{24}$.

Solution In Example 3.8.3 we found that
$$\sigma_{X_1}^2 = \sigma_{X_2}^2 = \sigma_X^2 = \tfrac{35}{12}.$$
By Theorem 8.2.2
$$\text{Var}(\bar{X}) = \sigma_X^2/N = \frac{35/12}{2} = \tfrac{35}{24}.$$

Example 8.2.3 Two fair coins are tossed three times each. If $\bar{X}$ is the average number of heads, then $E(\bar{X}) = \frac{3}{2}$ and $\text{Var}(\bar{X}) = \frac{3}{8}$.

Solution Suppose that X_1 and X_2 are the number of heads in the three tosses of each coin separately. Then in Example 3.6.2 we showed that
$$E(X_1) = E(X_2) = \mu = \tfrac{3}{2}$$
and in Example 3.8.2 that

$$\text{Var}(X_1) = \text{Var}(X_2) = \tfrac{3}{4}.$$
So using Theorems 8.2.1 and 8.2.2
$$E(\bar{X}) = \tfrac{3}{2} \quad \text{and} \quad \text{Var}(\bar{X}) = \tfrac{3}{8}.$$
Theorems 8.2.1 and 8.2.2 prove that if the random variables $X_1, X_2 \ldots, X_N$ have mean μ and variance σ^2, then $\bar{X}$ has mean μ and variance σ^2/N.

If the density from which $X_1, X_2, \ldots, X_N$ came is normal, then we may present an extremely important result concerning $\bar{X}$. The proof is beyond the scope of this book.

Theorem 8.2.3 If $X_1, X_2, \ldots, X_N$ have a normal distribution with mean μ and variance σ^2, then $\bar{X}$ is also normal, with mean μ and variance σ^2/N.

Further, from Theorem 5.2.1, we know that $\sqrt{N}(\bar{X} - \mu)/\sigma$ has a standard normal distribution. Theorem 8.2.3 and the above result will prove very useful to us in due course. Before passing on to the study of the sample variance, we state without proof an extremely important and remarkable theorem.

Theorem 8.2.4 (Central Limit Theorem) If $X_1, X_2, \ldots, X_N$ have any density with mean μ and variance σ^2, then the density of $Z = \sqrt{N}(\bar{X} - \mu)/\sigma$ will be approximately distributed as a standard normal for N large.

We now consider the problem of obtaining the overall mean of two sets of observations.

Theorem 8.2.5 If from two separate distributions with means μ_1 and μ_2 and variance σ_1^2 and σ_2^2, we draw two samples of sizes N and M with means $\bar{X}_1$ and $\bar{X}_2$, respectively, then
a) $E(\bar{X}_1 - \bar{X}_2) = \mu_1 - \mu_2$,
b) $\text{Var}(\bar{X}_1 - \bar{X}_2) = \sigma_1^2/N + \sigma_2^2/M$.

Proof. Using Theorem 3.7.2 we may write conclusion (a) immediately, thus
$$E(\bar{X}_1 - \bar{X}_2) = E(\bar{X}_1) - E(\bar{X}_2) = \mu_1 - \mu_2.$$
To prove (b), we write, using the results of Exercise 48, of Chapter 3, that
$$\text{Var}(\bar{X}_1 - \bar{X}_2) = \text{Var}(\bar{X}_1) + \text{Var}(\bar{X}_2).$$
Then using Theorem 8.2.2 we have
$$\text{Var}(\bar{X}_1 - \bar{X}_2) = \sigma_1^2/N + \sigma_2^2/M.$$
As an extension of Theorem 8.2.5, it can be shown that if $\bar{X}_1$ and $\bar{X}_2$

are normal, then $\bar{X}_1 - \bar{X}_2$ is also normal. The following theorems concern the sample variance s_X^2 defined as

$$s_X^2 = 1/(N-1) \sum_{i=1}^{N} (X_i - \bar{X})^2.$$

Theorem 8.2.6 If $X_1, X_2, \ldots, X_N$ are independent random variables with $E(X_i) = \mu$, $\text{Var}(X_i) = \sigma^2$, $i = 1, 2, \ldots, N$, then $E(s_X^2) = \sigma^2$.

Proof. Using Theorem 3.7.1

$$E(s_X^2) = E\{1/(N-1) \sum_{i=1}^{N} (X_i - \bar{X})^2\} = 1/(N-1) E \sum_{i=1}^{N} (X_i - \bar{X})^2.$$

Writing $X_i - \bar{X}$ as $(X_i - \mu) + (\mu - \bar{X})$ we have

$$E(s_X^2) = 1/(N-1) E \sum_{i=1}^{N} [(X_i - \mu) + (\mu - \bar{X})]^2.$$

Using the result of Exercise 6 of this chapter we have

$$E(s_X^2) = 1/(N-1) E\{\sum_{i=1}^{N} (X_i - \mu)^2 - N(\bar{X} - \mu^2)\}$$

$$= 1/(N-1) \{\sum_{i=1}^{N} E(X_i - \mu)^2 - NE(\bar{X} - \mu)^2\}.$$

From Definition 3.8.1 $E(X_i - \mu)^2 = \sigma^2$ and from Theorem 8.2.2 $E(\bar{X} - \mu)^2 = \sigma^2/N$, so

$$E(s_X^2) = 1/(N-1) \{N\sigma^2 - N\sigma^2/N\} = \sigma^2.$$

8.3 POINT ESTIMATION

If we use a single statistic to estimate a single parameter, for example the sample mean to estimate the population mean, we are using a point estimator of that parameter.

Definition 8.3.1 A point estimator is a single statistic which is used to estimate a population parameter.

Example 8.3.1 In Example 1.1.3, to estimate the specific gravity of mercury, we could use $\bar{x} = 13.895$ as the point estimate.

Note that if we are talking of a statistic in general we call it an estimator, but once it takes a specified value it is called an estimate. The sample mean is not the only estimator of the population mean. We could for instance have used the median. We need some criteria concerning desirability of estimators to help us choose between them. We would first like to ask that our estimator be near the parameter. This would be difficult to ensure all the time since the value of the estimator changes from sample to sample. We could compromise by asking that the estimator should equal the parameter "on the average." But from the discussion at the end of Section 8.1, this is equivalent to asking that the expectation of the estimator should equal the parameter. An estimator with this property is called unbiased. Otherwise it is biased.

Definition 8.3.2 The statistic T is called an unbiased estimator of a parameter θ, if $E(T) = \theta$.

Example 8.3.2 The statistic $\bar{X}$, the sample mean, is an unbiased estimator of the parameter μ, the population mean.

Example 8.3.3 Show that $\bar{X}^2$ is a biased estimator of μ^2.

Solution $\sigma^2/N = E(\bar{X} - \mu)^2$

Therefore $\sigma^2/N = E(\bar{X}^2) - \mu^2$, so
$$E(\bar{X}^2) = \sigma^2/N + \mu^2, \qquad E(\bar{X}^2) \neq \mu^2.$$
Therefore $\bar{X}^2$ is a biased estimator of μ^2.

Example 8.3.4 The sample variance is an unbiased estimator of σ^2, the population variance, since by Theorem 8.2.6 $E(s_X^2) = \sigma^2$.

Example 8.3.5 Show that the sample standard deviation is not an unbiased estimator of σ.

Solution Since the variance is always non-negative from its definition, we have that
$$\text{Var}(s_X) = E[s_X - E(s_X)]^2 \geq 0, \quad \text{or} \quad E(s_X^2) - [E(s_X)]^2 \geq 0.$$
But from Theorem 8.2.6 $E(s_X^2) = \sigma^2$, so $\sigma^2 \geq [E(s_X)]^2$ and thus $\sigma \geq E(s_X)$ so that s_X has expectation less than or equal to σ and is therefore biased.

It should now be clear to the student that we divide by $N - 1$ instead of N in calculating the sample variance, in order to avoid an unbiased estimator of σ^2. In general, if given the choice between a biased and an unbiased estimator we would choose the latter. But what if we have a choice between two unbiased estimators? This brings us to the second criterion for discriminating between estimators. We would like an estimator to have a high probability of being close to the parameter.

This is another way of saying that we want an estimator with a small variance. Given a choice of two unbiased estimators, we would prefer the one with the smaller variance.

Example 8.3.6 In Fig. 8.3.1 we graph the distribution of three estimators T_1, T_2 and T_3. On the vertical axis the probability density of the estimator is plotted. On the horizontal axes are the values of T_1, T_2, T_3, and θ which is the parameter being estimated. T_2 is preferred to T_1 since T_1 is biased, while T_3 is preferred to T_2, since T_3 has smaller variance.

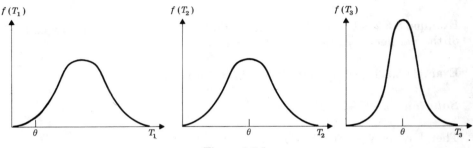

Figure 8.3.1

Example 8.3.7 In a sample $X_1, X_2, \ldots, X_N$, both X_1 and $\bar{X}$ are unbiased estimators of μ, since $E(X_1) = \mu$, $E(\bar{X}) = \mu$ However

$$\text{Var}(X_1) = \sigma^2, \qquad \text{Var}(\bar{X}) = \sigma^2/N$$

so $\bar{X}$ has a smaller variance and is therefore more desirable than X_1.

Combining the two criteria of being unbiased and having small variance leads to the following definition.

Definition 8.3.3 A statistic T which is an unbiased estimator of θ and has smaller variance than any other unbiased estimator is called a minimum-variance unbiased estimator of θ.

One can show that if $X_1, X_2, \ldots, X_N$ have a normal distribution with unknown mean μ and unknown variance σ^2, then $\bar{X}$ and $s_{\bar{X}}^2$ are minimum-variance unbiased estimators of μ and σ^2, respectively.

8.4 INTERVAL ESTIMATION OF THE POPULATION MEAN WITH KNOWN VARIANCE

Even if we have a minimum-variance unbiased estimate of a parameter, it is still only a point estimate and gives no indication of how far away the parameter lies. A more useful method of estimation is to compute an interval which has a high probability of containing μ.

Interval estimation of the population mean with known variance

Suppose that we want to estimate the mean μ of a population. We might well base our interval on $\bar{X}$, since it has desirable properties. If $\bar{X}$ is formed from N random variables from a normal density with known variances σ^2, then from Theorem 8.2.3 we know that $\bar{X}$ has a normal density and that $Z = \sqrt{N}(\bar{X} - \mu)/\sigma$ has a unit normal distribution.

Definition 8.4.1 Z_α is the point on the unit normal curve which has proportion α of the area under the curve to its left as shown in Fig. 8.4.1.

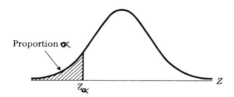

Fig. 8.4.1 Unit normal distribution showing Z_α.

Example 8.4.1 (a) $Z_{.05}$ has 5% of the area of the unit normal curve to its left. (b) $Z_{.95}$ has 95% of the area to the left and therefore 5% to the right.

By the symmetry of the normal curve $Z_\alpha = -Z_{1-\alpha}$. From Definition 8.4.1

$$P[-Z_{1-\alpha/2} \leq Z \leq Z_{1-\alpha/2}] = 1 - \alpha.$$

If $\alpha = .05$, then

$$P[-1.96 < Z < 1.96] = .95$$

or

$$P\left[-1.96 < \frac{\bar{X} - \mu}{\sigma/\sqrt{N}} < 1.96\right] = .95.$$

We may manipulate the inequalities inside the square bracket to give

$$P\left[\bar{X} - (1.96)\frac{\sigma}{\sqrt{N}} < \mu < \bar{X} + (1.96)\frac{\sigma}{\sqrt{N}}\right] = .95. \qquad (8.4.1)$$

The distance between $\bar{X} - (1.96)\sigma/\sqrt{N}$ and $\bar{X} + (1.96)\sigma/\sqrt{N}$ is an interval which, from Equation (8.4.1) has a probability of .95 of covering or including μ. Note that μ is a constant, hence we cannot talk of "the probability of μ being in the interval." It is the interval that changes because each new sample gives a new value to $\bar{X}$. When a sample has been taken and $\bar{x}$, the value of $\bar{X}$, calculated, then the interval is completely fixed, and it either contains μ or it does not. It therefore no longer makes sense to say that the probability of containing μ is .95. Since $\bar{X}$

could have had many other values, each producing a corresponding interval, and 95% of those intervals would have contained the unknown μ, we say that we are 95% confident that the particular interval calculated contains μ.

Definition 8.4.2 The interval based on Equation (8.4.1), $(\bar{X} - 1.96 \, \sigma/\sqrt{N}, \bar{X} + 1.96 \, \sigma/\sqrt{N})$ is called a 95% confidence interval. The value of .95 or 95% is called the confidence level, or confidence coefficient. In general, for confidence coefficient $1 - \alpha$, the equivalent of Equation (8.4.1) is

$$P[\bar{X} - Z_{1-\alpha/2}\sigma/\sqrt{N} < \mu < \bar{X} + Z_{1-\alpha/2}\sigma/\sqrt{N}] = 1 - \alpha$$

and the confidence interval is

$$(\bar{X} - Z_{1-\alpha/2}\sigma/\sqrt{N}, \quad \bar{X} + Z_{1-\alpha/2}\sigma/\sqrt{N}).$$

Example 8.4.2 A 90% confidence interval for μ is $(\bar{x} - 1.65 \, \sigma/\sqrt{N}, \bar{x} + 1.65 \, \sigma/\sqrt{N})$, where $\bar{x}$ is a specific value of $\bar{X}$.

Example 8.4.3 Construct a .95 and a .99 interval from a sample of size 25 for which $\bar{x}$ is calculated as 50 and for which σ is known to be 10.

Solution

a) 95% confidence interval
   i) $z_{1-\alpha/2} = z_{.95} = 1.96$.
   ii) $\bar{x} - z_{1-\alpha/2}\sigma/\sqrt{N} = 50 - (1.96)(10)/\sqrt{25} = 46.08$,
   $\bar{x} + z_{1-\alpha/2}\sigma/\sqrt{N} = 50 + (1.96)(10)/\sqrt{25} = 53.92$.

The 95% confidence interval is between 46.08 and 53.92.

b) 99% confidence interval
   i) $z_{1-\alpha/2} = z_{.99} = 2.58$.
   ii) $\bar{x} - z_{1-\alpha/2}\sigma/\sqrt{N} = 50 - (2.58)(10)/\sqrt{25} = 44.84$,
   $\bar{x} + z_{1-\alpha/2}\sigma/\sqrt{N} = 50 + (2.58)(10)/\sqrt{25} = 55.16$.

The 99% confidence interval is between 44.84 and 55.16.

To appreciate the fundamental idea of a confidence interval, it is well to think of probability as a relative frequency. If we took 100 different samples of size N, we would have 100 different confidence intervals, with, let us say, confidence coefficients .95. We could expect that 95 of these 100 intervals would include μ and that 5 of them would not include it. Of course, we have in practice only one interval which we say has a .95 probability of including μ. To illustrate this meaning more clearly, we construct the following artificial example.

Example 8.4.4 Consider a population of students whose I.Q.'s have a normal density. We wish to estimate μ, the population mean using the results of a sample. Pretend for the sake of illustration that we know μ to be 110 (of course in reality this would not be so, otherwise we would have no need to estimate it.) If σ, the population standard deviation is known to be 30 and we take 20 samples each of size 100, then a 95% confidence interval will be ($\bar{X} \pm (1.96)(30)/10$) or ($\bar{X} - 5.88, \bar{X} + 5.88$). The mean $\bar{X}$ and the upper and lower limits are shown in Table 8.4.1.

TABLE 8.4.1

The upper and lower limits of 95% confidence intervals for 20 samples of size 100

Mean ($\bar{X}$)	Upper limit ($\bar{X} + 5.88$)	Lower limit ($\bar{X} - 5.88$)
107.1	112.98	101.22
110.7	116.58	104.82
110.3	116.18	104.42
105.5	111.38	99.62
113.6	119.48	107.72
107.1	112.98	101.22
105.7	111.58	99.82
111.2	117.08	105.32
110.8	116.68	104.92
108.4	114.28	102.52
113.9	119.78	108.02
111.8	117.68	105.92
107.5	113.38	101.62
110.9	116.78	105.02
104.1	109.98	98.22
112.3	118.18	106.42
110.6	116.48	104.72
109.2	115.08	103.32
112.9	118.78	107.02
110.7	116.58	104.82

We may note that 19 of the 20 include μ while one does not. In general, we prefer a short confidence interval to a long one since it confines the probable values of μ to a smaller set of numbers. The length of the interval is the upper limit minus the lower limit or $2(Z_{1-\alpha/2})\sigma/\sqrt{N}$. This can be made shorter by decreasing σ, increasing N, or decreasing $Z_{1-\alpha/2}$, that is, increasing α.

Decreasing σ is often beyond the control of the experimenter. Increasing N, the sample size, usually has the disadvantage of increasing the costs of the experiment. Increasing α and therefore decreasing $Z_{1-\alpha/2}$ means that the level of confidence decreases. In most experiments the value of α is rarely larger than .1. An experimenter must decide for himself which of σ, N, α he will manipulate, and this will depend on the nature of the experiment and the costs of sampling.

One should note that the development of confidence intervals depends heavily on the density of $(\bar{X} - \mu)\sqrt{N}/\sigma$ being normal. We know from Theorem 8.2.3 that if $X_1, X_2, \ldots, X_N$ are normal then $(\bar{X} - \mu)\sqrt{N}/\sigma$ will be exactly normal. If the original density is not normal however, then Theorem 8.2.4 assures us that $\bar{X}$ is approximately normal (this of course means that $(\bar{X} - \mu)\sqrt{N}/\sigma$ is approximately normal too) for N large. The size that N must be, depends on how different from normality the density is. As a rule of thumb, Definition (8.4.2) can be used for samples where $N > 30$.

8.5 INTERVAL ESTIMATION OF THE POPULATION MEAN WITH UNKNOWN VARIANCE

In Section 8.4 we assumed that σ^2, the population variance, was known and this permitted us to say that $(\bar{X} - \mu)\sqrt{N}/\sigma$ had a unit normal density. More often than not, in real problems the value of σ^2 is unknown and has to be estimated by s_X^2. Then the expression $(\bar{X} - \mu)\sqrt{N}/\sigma$ becomes $(\bar{X} - \mu)\sqrt{N}/s_X$ which is given the name t_{N-1}, that is

$$t_{N-1} = \frac{\bar{X} - \mu}{s_X/\sqrt{N}}.$$

The statistic t_{N-1} no longer has a normal density since it has a random variable, $\bar{X}$, in the numerator and a random variable, s_X, in the denominator. In fact t_{N-1} is called a Student's t statistic and its density is called a Student's t distribution. The name Student was a pseudonym for the discoverer of the density, W. S. Gosset. Until his discovery in 1908, it was common practice to treat the statistic t_{N-1} as an approximate unit normal variable. The mathematical form for the Student's t distribution can be shown to depend on $N - 1$, where N is the sample size. The quantity $N - 1$ is usually referred to as the degrees of freedom.

Definition 8.5.1 If $X_1, X_2, \ldots, X_N$ are N normal random variables with unknown mean μ and variance σ^2, respectively, then

$$t_{N-1} = \frac{\bar{X} - \mu}{s_X/\sqrt{N}}$$

Interval estimation of the population mean with unknown variance

is called a Student's t statistic with $N - 1$ degrees of freedom, and its density is called the Student's t distribution.

The graph of the Student's t distribution changes for different values of $N - 1$ as shown in Fig. 8.5.1, but basically it is an upside-down bell-shaped curve similar to the normal, except slightly wider. Intuitively one might expect that as N increased, the difference between the t and the normal would diminish. This is indeed so and for $N > 30$ they are almost indistinguishable. The areas under the curve of the Student's t distribution have been numerically calculated (just as for the normal density). However, a separate table would be required for each value of $N - 1$. For this reason the complete table of areas is not given. Instead, some of the more important and commonly used areas are tabulated in Appendix Table 4.

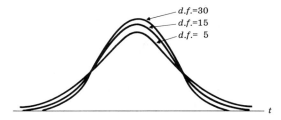

Fig. 8.5.1 Student's t distribution for degrees of freedom 5, 15, and 30.

Definition 8.5.2 t_α is the point on the Student's t curve which has proportion α of the area of the curve to its left, as shown in Fig. 8.5.2.

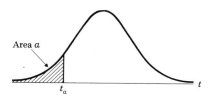

Fig. 8.5.2 t distribution showing t_α.

Example 8.5.1 (a) For $N - 1 = 10$ degrees of freedom, we find from Appendix Table 4 that $t_{.05} = -1.812$, $t_{.95} = 1.812$.

As with the normal curve, the symmetry of the Student's t curve implies that $t_\alpha = -t_{1-\alpha}$. The construction of a confidence interval for μ when σ^2 is unknown is similar to the case when σ^2 is known except that

t rather than Z tables are used. As an example if $N = 10$, $N - 1 = 9$, $\alpha = .05$ and we require a 95% confidence interval then

$$P[t_{.025} < t < t_{.975}] = .95$$

or

$$P\left[-2.26 < \frac{\bar{X} - \mu}{s_X/\sqrt{N}} < 2.26\right] = .95.$$

The inequalities may be manipulated as in Section 8.4 to give

$$P\left[\bar{X} - (2.26)\frac{s_X}{\sqrt{N}} < \mu < \bar{X} + (2.26)\frac{s_X}{\sqrt{N}}\right] = .95 \qquad (8.5.1)$$

where, of course, here $N = 10$, $\sqrt{10} = 3.16$. The same comments as before apply here; once $\bar{X}$ and s_X are calculated, it makes no sense to talk of probability. Instead one says that he is 95% confident that the interval $[\bar{X} - (2.26 s_X)/\sqrt{N}, \bar{X} + (2.26 s_X)/\sqrt{N}]$ contains μ. The generalization of Equation (8.5.1) is as follows.

Definition 8.5.3 If $X_1, X_2, \ldots, X_N$ is a sample of N random variables from a normal density with unknown mean μ and unknown variance σ^2, then based on the expression

$$P[\bar{X} - t_{1-\alpha/2} s_X/\sqrt{N} < \mu < \bar{X} + t_{1-\alpha/2} s_X/\sqrt{N}] = 1 - \alpha$$

we define a $1 - \alpha$ level confidence interval for μ, as

$$(\bar{X} - t_{1-\alpha/2} s_X/\sqrt{N}, \bar{X} + t_{1-\alpha/2} s_X/\sqrt{N}).$$

Example 8.5.2 (Physics) Calculate a 95% confidence interval for the specific gravity of mercury, using the ten observations of Example 1.1.3.

Solution

i) $\bar{x} = 13.695$.
ii) By Equation (7.7.3)
  $s_x^2 = \frac{1}{9}\{1875.585522 - (10)(13.695)^2\} = .006102$.
iii) $s_x = \sqrt{.006102} = .078$.
iv) $N = 10$, $N - 1 = 9$ = degrees of freedom.
v) $t_{.025} = -2.26$, $t_{.975} = 2.26 = t_{1-\alpha/2}$.
vi) $t_{1-\alpha/2} s_x/\sqrt{N} = (2.26)(.078)/3.17 = .056$.
vii) The lower limit $= \bar{x} - t_{1-\alpha/2} s_x/\sqrt{N} = 13.639$.
  The upper limit $= \bar{x} + t_{1-\alpha/2} s_x/\sqrt{N} = 13.751$.
viii) The required confidence interval is between 13.639 and 13.751, so that we are 95% confident that the specific gravity of mercury is included in that interval.

Notice that the length of the confidence interval is the difference of the two endpoints or $2(t_{1-\alpha/2})s_X/\sqrt{N}$ so that not only the center of the interval varies as in Section 8.4, but also the length of the interval changes from sample to sample. The ideas of Example 8.4.4 are still relevant here, and one can think of a proportion $1 - \alpha$ of the intervals containing the unknown parameter μ. The same comments regarding the dependency of Equation (8.5.1) on the normal density of the X's are valid here.

8.6 INTERVAL ESTIMATION OF THE POPULATION VARIANCE AND STANDARD DEVIATION

If we desire to estimate σ^2 and σ then we might start constructing an interval by considering the minimum variance unbiased point estimator of σ^2.

Definition 8.6.1 If $X_1, X_2, \ldots, X_N$ are N normal random variables with variance σ^2 and

$$s_X^2 = \frac{1}{N-1} \sum_{i=1}^{N} (X_i - \bar{X})^2$$

then the statistic

$$\chi_{N-1}^2 = \frac{(N-1)s_X^2}{\sigma^2}$$

is called a chi-square statistic with $N - 1$ degrees of freedom. Like the t density, the χ^2 density changes for different values of $N - 1$ as illustrated in Fig. 8.6.1. Since χ^2 is non-negative, the density is bounded by zero to the left, but is unbounded to the right.

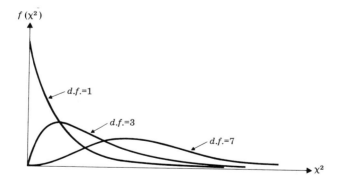

Figure 8.6.1
The chi square distribution for degrees of freedom 1, 3, and 7.

176   Estimation

The areas under the curve of the χ^2 density have been numerically calculated. As for the Student's t density, it is only worthwhile listing the more commonly used and more important areas. These are found in Appendix Table 5.

Definition 8.6.2 χ^2_α is the point on the χ^2 density which has proportion α of the area to its left as shown in Fig. 8.6.2.

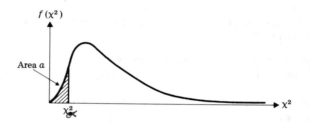

Figure 8.6.2

Example 8.6.1 For $N - 1 = 10$ degrees of freedom, we find from Appendix Table 5 that $\chi^2_{.05} = 3.94$, $\chi^2_{.95} = 18.31$.

The chi-square curve is clearly not symmetric and $\chi^2_\alpha \neq -\chi^2_{1-\alpha}$. To calculate a 95% confidence interval then

$$P[\chi^2_{.025} < \chi^2_{N-1} < \chi^2_{.975}] = .95.$$

If for example $N = 10$, $N - 1 = 9$, then

$$P[2.70 < \chi^2_{N-1} < 19.02] = .95$$

or

$$P\left[2.70 < \frac{(N-1)s^2_X}{\sigma^2} < 19.02\right] = .95.$$

The inequalities may be turned about to give

$$P\left[\frac{(N-1)s^2_X}{19.02} < \sigma^2 < \frac{(N-1)s^2_X}{2.70}\right] = .95.$$

Once the sample has been taken and s^2_x, the numerical value of s^2_X calculated, we talk of the interval as a 95% confidence interval for σ^2. In general for confidence level $1 - \alpha$, we have the following definition.

Definition 8.6.3 If $X_1, X_2, \ldots, X_N$ is a set of N random variables from a normal density with unknown mean μ and unknown variance σ^2, then based on the expression

Interval estimation of the population variance and standard deviation

$$P\left[\frac{(N-1)s_X^2}{\chi_{1-\alpha/2}^2} < \sigma^2 < \frac{(N-1)s_X^2}{\chi_{\alpha/2}^2}\right] = 1 - \alpha, \qquad (8.6.1)$$

we define a $1 - \alpha$ level confidence interval for σ^2 as

$$\left(\frac{(N-1)s_X^2}{\chi_{1-\alpha/2}^2}, \frac{(N-1)s_X^2}{\chi_{\alpha/2}^2}\right).$$

Example 8.6.2 Calculate a 95% confidence interval for the variance of the method used to determine the specific gravity of mercury, using the 10 observations of Example 1.1.3.

Solution

i) $s_x^2 = .006102$, from Example 8.5.2.
ii) Degrees of freedom $= N - 1 = 9$.
iii) $\chi_{.025}^2 = 2.70$, $\chi_{.975}^2 = 19.02$.
iv) $\dfrac{(N-1)s_x^2}{\chi_{.975}^2} = \dfrac{(9)(.006102)}{19.02} = .002887.$

$\dfrac{(N-1)s_x^2}{\chi_{.025}^2} = \dfrac{(9)(.006102)}{2.70} = .020340.$

v) The confidence interval for σ^2 is between .002887 and .020340, so that we are 95% confident that the true unknown value of σ^2 is included in that interval.

The student may note that we have rather arbitrarily constructed the confidence interval based on Equation (8.6.1) so that the area in the left-hand and right-hand tails of the χ^2 density are equal. We could just as well have used $\chi_{1-\alpha/3}^2$ and $\chi_{2\alpha/3}^2$ or any other pair which have proportion $1 - \alpha$ between them. The interval using $\chi_{1-\alpha/2}^2$ and $\chi_{\alpha/2}^2$ is called an "equal-tail" confidence interval. Other shorter confidence intervals could be constructed, but the gain obtained is hardly worth the effort at this level.

Unlike the confidence intervals of Sections 8.4 and 8.5, those for the variance are rather sensitive to departures of the original density from normality. If the experimenter believes that the population is not normal, then he would be well advised not to calculate confidence intervals based on Equation (8.6.1) for σ^2. Finally if we take the square root of each term inside the square bracket of Equation (8.6.1.) we get

$$P\left[\sqrt{\frac{(N-1)s_X^2}{\chi_{1-\alpha/2}^2}} < \sigma < \sqrt{\frac{(N-1s_X^2}{\chi_{\alpha/2}^2}}\right] = 1 - \alpha$$

giving a confidence interval

178   Estimation

$$\left(\sqrt{\frac{(N-1)s_X^2}{\chi_{1-\alpha/2}^2}}, \sqrt{\frac{(N-1)s_X^2}{\chi_{\alpha/2}^2}}\right)$$

for σ.

Example 8.6.3 As a continuation of Example 8.6.2, calculate a 95% confidence interval for σ, using the 10 evaluations for the specific gravity of mercury.

Solution

i) $s_x^2 = .006102$.

ii) $\dfrac{(N-1)s_x^2}{\chi_{.975}^2} = .002887$.

$\dfrac{(N-1)s_x^2}{\chi_{.025}^2} = .020304$.

iii) $\sqrt{\dfrac{(N-1)s_x^2}{\chi_{.975}^2}} = \sqrt{.002887} = .053$.

$\sqrt{\dfrac{(N-1)s_x^2}{\chi_{.025}^2}} = \sqrt{.020340} = .143$.

iv) The 95% confidence interval for σ is between .053 and .143.

8.7 INTERVAL ESTIMATION OF THE DIFFERENCE OF TWO NORMAL MEANS

In many experimental situations one wishes to compare the means of two normal populations. For instance in Example 1.1.3 we wish to compare the iron ore melting points in regions A and B. An engineer might want to compare two different machines or an agriculturist might want to compare the growth rate of chickens using two different feeds.

Suppose that from normal populations 1 and 2 with unknown means μ_1 and μ_2, we draw two samples of sizes N and M, respectively. If $\bar{X}_1$ and $\bar{X}_2$ are their sample means, then an obvious point estimator of $\mu_1 - \mu_2$ is $\bar{X}_1 - \bar{X}_2$. To obtain an interval estimator of $\mu_1 - \mu_2$ we first consider the case where the variances σ_1^2 and σ_2^2 of the two populations are known. From Theorem 8.2.3 both $\bar{X}_1$ and $\bar{X}_2$ are normal, and therefore from Theorem 8.2.5 and its extension $\bar{X}_1 - \bar{X}_2$ is normal with mean $\mu_1 - \mu_2$ and variance equal to $\sigma_1^2/N + \sigma_2^2/M$. Thus

$$Z = \frac{(\bar{X}_1 - \bar{X}_2) - (\mu_1 - \mu_2)}{\sqrt{\sigma_1^2/N + \sigma_2^2/M}} \quad (8.7.1)$$

has a unit normal density, so

Interval estimation of the difference of two normal means

$$P\left[-Z_{1-\alpha/2} < \frac{(\bar{X}_1 - \bar{X}_2) - (\mu_1 - \mu_2)}{\sqrt{\sigma_1^2/N + \sigma_2^2/M}} < Z_{1-\alpha/2}\right] = 1 - \alpha \quad (8.7.2)$$

or

$$P[(\bar{X}_1 - \bar{X}_2) - (Z_{1-\alpha/2})\sqrt{\sigma_1^2/N + \sigma_2^2/M} < \mu_1 - \mu_2 \\ < (\bar{X}_1 - \bar{X}_2) + (Z_{1-\alpha/2})\sqrt{\sigma_1^2/N + \sigma_2^2/M}] = 1 - \alpha$$

defines a $1 - \alpha$ level confidence interval for $\mu_1 - \mu_2$.

Example 8.7.1 (Industry) Construct 95% and 99% confidence intervals for the difference of the melting points in regions A and B, using the data of Example 1.1.4, assuming in both cases that the variance σ^2 is known to be 100.

Solution

i) $\bar{x}_A = 1506.7$, $\bar{x}_B = 1499.3$, $\bar{x}_A - \bar{x}_B = 7.4$.

ii) $\sigma_A^2 = \sigma_B^2 = 100$.

iii) $\dfrac{\sigma_A^2}{10} + \dfrac{\sigma_B^2}{6} = 100\left(\dfrac{1}{10} + \dfrac{1}{6}\right) = 26.67$.

iv) $\sqrt{\dfrac{\sigma_A^2}{10} + \dfrac{\sigma_B^2}{6}} = \sqrt{26.67} = 5.16$.

For 95% confidence intervals,

v) $z_{1-\alpha/2} = z_{.975} = 1.96$.

vi) $z_{1-\alpha/2}\sqrt{\dfrac{\sigma_A^2}{10} + \dfrac{\sigma_B^2}{6}} = (1.96)(5.16) = 10.11$.

iii) The lower limit $= (\bar{x}_A - \bar{x}_B) - z_{1-\alpha/2}\sqrt{\dfrac{\sigma_A^2}{10} + \dfrac{\sigma_B^2}{6}} = -2.71$.

The upper limit $= (\bar{x}_A - \bar{x}_B) + z_{1-\alpha/2}\sqrt{\dfrac{\sigma_A^2}{10} + \dfrac{\sigma_B^2}{6}} = 17.51$.

viii) A 95% confidence interval for the difference of the two melting points, $\mu_A - \mu_B$, is between -2.71 and 17.51.

For 99% confidence intervals, proceed as in steps (v) through (vii) except that $z_{1-\alpha/2} = 2.58$. The 99% confidence interval is between -5.91 and 20.71.

If σ_1^2 and σ_2^2 are unknown but assumed to be equal, and have estimators s_1^2 and s_2^2, respectively, then from Definition 7.7.4 a pooled estimator s_P^2 is

$$s_P^2 = \frac{(N-1)s_1^2 + (M-1)s_2^2}{N + M - 2}. \quad (8.7.3)$$

180   Estimation

In Equation (8.7.1), if $\sigma_1^2 = \sigma_2^2 = \sigma^2$, then

$$Z = \frac{(\bar{X}_1 - \bar{X}_2) - (\mu_1 - \mu_2)}{\sigma\sqrt{1/N + 1/M}}.$$

If we now replace σ by s_P, then the result as in Section 8.5, is a Student's t statistic, but with $N + M - 2$ degrees of freedom. Thus

$$t = \frac{(\bar{X}_1 - \bar{X}_2) - (\mu_1 - \mu_2)}{s_P\sqrt{1/N + 1/M}} \qquad (8.7.4)$$

and

$$P\left[-t_{1-\alpha/2} < \frac{(\bar{X}_1 - \bar{X}_2) - (\mu_1 - \mu_2)}{s_P\sqrt{1/N + 1/M}} < t_{1-\alpha/2}\right] = 1 - \alpha$$

from which we obtain

$$P[(\bar{X}_1 - \bar{X}_2) - t_{1-\alpha/2}s_P\sqrt{1/N + 1/M} < \mu_1 - \mu_2$$
$$< (\bar{X}_1 - \bar{X}_2) + t_{1-\alpha/2}s_P\sqrt{1/N + 1/M}] = 1 - \alpha \quad (8.7.5)$$

which defines a $1 - \alpha$ level confidence interval for $\mu_1 - \mu_2$.

Example 8.7.2 Construct 95% and 99% confidence intervals for the difference of the two melting points in Example 1.1.4, assuming that the variances, although both are unknown, are equal.

Solution

i)   From Example 8.7.1 $\bar{x}_A - \bar{x}_B = 7.4$.
ii) From Example 7.7.4 $s_A^2 = 117.12$, $s_B^2 = 135.46$.
iii) $s_p^2 = \dfrac{(9)(117.12) + (5)(135.46)}{14} = 123.67$.
iv) $s_p = \sqrt{123.67} = 11.12$.
v) $s_p\sqrt{1/N + 1/M} = (11.12)\sqrt{\tfrac{1}{10} + \tfrac{1}{6}} = (11.12)(.516) = 5.7$.

For 95% confidence intervals

vi) $t_{1-\alpha/2} = t_{.975} = 2.15$ (with 14 degrees of freedom).
vii) $t_{1-\alpha/2}s_p\sqrt{1/N + 1/M} = (2.15)(5.74) = 12.34$.
viii) The upper limit is $7.4 + 12.34 = 19.74$.
      The lower limit is $7.4 - 12.34 = -4.94$.
ix)   A 95% confidence interval for the difference of the two melting points, $\mu_A - \mu_B$, when the variances are unknown (but equal) is between -4.94 and 19.74.

For 99% confidence intervals proceed as in steps (vi) through (viii), except that $t_{1-\alpha/2} = 2.98$. The 99% confidence interval is between -9.71 and 24.51.

If σ_1^2 and σ_2^2 are unknown and cannot be assumed to be equal, then the problem of setting confidence intervals for $\mu_1 - \mu_2$ becomes difficult indeed. No generally accepted solution to the problem, often known as the Behrens-Fisher problem, exists although various approaches have been advanced. We discuss what is called the Welch approximation as follows.

In Equation (8.7.1) replace σ_1^2 and σ_2^2 by their estimators, obtaining

$$t = \frac{(\bar{X}_1 - \bar{X}_2) - (\mu_1 - \mu_2)}{\sqrt{s_1^2/N + s_2^2/M}} \tag{8.7.6}$$

where t has an approximate Student's t density with ν degrees of freedom where

$$\nu = \frac{[(s_1^2/N) + (s_2^2/M)]^2}{\dfrac{(1/N)^2 (s_1^2)^2}{N+1} + \dfrac{(1/M)^2 (s_2^2)^2}{M+1}} - 2. \tag{8.7.7}$$

The approximate $1 - \alpha$ level confidence interval is found from

$$P[(\bar{X}_1 - \bar{X}_2) - t_{1-\alpha/2} \sqrt{s_1^2/N + s_2^2/M} < \mu_1 - \mu_2 \\ < (\bar{X}_1 - \bar{X}_2) + t_{1-\alpha/2} \sqrt{s_1^2/N + s_2^2/M}] = 1 - \alpha \tag{8.7.8}$$

where $t_{1-\alpha/2}$ is found from the Student's t table with ν degrees of freedom. Since ν is generally not an integer, one must interpolate in the table. Linear interpolation is generally sufficiently accurate.

Welch shows in the discussion of his approximation, that if $N = M$, then the assumption of equal variances is not too critical. Thus in practice, if the equality of the two variances cannot be assumed, one should try to keep the sample sizes, N and M, equal.

Example 8.7.3 (Industry) Suppose that nine determinations of iron melting points from region C are as follows: 1500, 1481, 1498, 1504, 1473, 1502, 1525, 1508. Construct a 95% confidence interval for the difference in the melting points of region C and region A (from Example 1.1.4). Do not assume that the variances are equal.

Solution
i) $\bar{x}_A = 1506.7$, $\bar{x}_C = 1499.2$, $\bar{x}_A - \bar{x}_C = 7.5$.
ii) $s_A^2 = 117.12$, $s_C^2 = 300.16$.
iii) $\sqrt{\dfrac{s_A^2}{N} + \dfrac{s_C^2}{M}} = \sqrt{\dfrac{117.12}{10} + \dfrac{300.16}{9}} = \sqrt{45.063} = 6.71$.

iv) $\nu = \dfrac{[(s_A^2/N) + (s_C^2/M)]^2}{\dfrac{(s_A^2/N)^2}{N+1} + \dfrac{(s_C^2/M)^2}{M+1}} - 2 = \dfrac{[11.712 + 33.351]^2}{12.470 + 111.229} - 2$

$= \dfrac{2030.674}{123.699} - 2 = 14.42.$

v) $t_{1-\alpha/2} = t_{.975} = 2.14.$

vi) The lower limit $= (\bar{x}_A - \bar{x}_C) - t_{1-\alpha/2} \sqrt{s_A^2/N + s_C^2/M} = -6.99.$

The upper limit $= (\bar{x}_A - \bar{x}_C) + t_{1-\alpha/2} \sqrt{s_A^2/N + s_C^2/M} = 21.99.$

vii) A 95% confidence interval for the difference of the melting points in regions A and C is between -6.99 and 21.99.

In many situations we wish to compare two population means, but have observations which occur in pairs. For instance, "before" and "after" measurements on the same patients, or test 1 and test 2 on the same child in education. Since each pair of measurements are related, the methods of the previous sections do not apply. In fact an experimenter may purposely want to pair the measurements in order to eliminate extraneous variation.

If the pairs of observations are (X_{1i}, X_{2i}) for $i = 1, 2, \ldots, N$, where the first observation in the pair is from a normal density with mean μ_1 and the second is from a normal density with mean μ_2, then the differences $D_i = X_{1i} - X_{2i}$ are normal with mean $\mu_1 - \mu_2$ and variance σ_D^2.

If $\overline{D} = \sum_{i=1}^{N} D_i/N$, then $\overline{D}$ is normal with mean $\mu_D = \mu_1 - \mu_2$ and variance σ_D^2/N, and

$$Z_D = \dfrac{\overline{D} - \mu_D}{\sigma_D/\sqrt{N}}$$

has a unit normal density, so that

$$P[\overline{D} - Z_{1-\alpha/2}\, \sigma_D/\sqrt{N} < \mu_1 - \mu_2 < \overline{D} + Z_{1-\alpha/2}\, \sigma_D/\sqrt{N}] = 1 - \alpha \qquad (8.7.9)$$

if σ_D^2 is known. If σ_D^2 is unknown, then

$$s_D^2 = \dfrac{1}{N-1} \sum_{i=1}^{N} (D_i - \overline{D})^2$$

from which

$$t_D = \dfrac{\overline{D} - \mu}{s_D/\sqrt{N}}$$

has a Student's t distribution; in which case

$$P\left[\overline{D} - t_{1-\alpha/2}\frac{s_D}{\sqrt{N}} < \mu_1 - \mu_2 < \overline{D} + t_{1-\alpha/2}\frac{s_D}{\sqrt{N}}\right] = 1 - \alpha \quad (8.7.10)$$

defines a $1 - \alpha$ level confidence interval for $\mu_1 - \mu_2$.

Example 8.7.4 Consider Example 1.1.2. The scores are paired for each child. Assuming that the scores are normally distributed, calculate a 95% confidence interval for the true difference $\mu_E - \mu_M$ between the mathematics and English scores.

Solution

i) One first requires the differences d_i for each child. These are as follows: 8, 4, 5, 19, 3, 7, −5, −12, −19, 5, 8, −4, 17, −5, 6, 4, −13, 6, 11, −9, 24, 9, 23, −12, 10, 5, 9, 0, 7, −8.

ii) $\bar{d} = 3.43$.
$s_d^2 = 111.29$.
$s_d^2/N = 111.29/30 = 3.71$.
$s_d/\sqrt{N} = \sqrt{3.71} = 1.93$.

iii) $t_{1-\alpha/2} = t_{.975} = 2.05$ (with 29 degrees of freedom).

iv) The lower limit is $\bar{d} - t_{1-\alpha/2}\, s_d/\sqrt{N} = -.53$.
The upper limit is $\bar{d} + t_{1-\alpha/2}\, s_d/\sqrt{N} = 7.39$.

v) A 95% confidence interval for the difference $\mu_E - \mu_M$ between mathematics and English scores is between −.53 and 7.39.

8.8 INTERVAL ESTIMATION OF THE RATIO OF TWO NORMAL VARIANCES

We often require the comparison of two normal variances. If σ_1^2 and σ_2^2 are the unknown variances of the two populations, then s_1^2/s_2^2 is a point estimate of their ratio σ_1^2/σ_2^2. We need the following definition in order to set confidence limits for the ratio.

Definition 8.8.1 If $X_{11}, X_{12}, \ldots, X_{1N}$ and $X_{21}, X_{22}, \ldots, X_{2M}$ are N and M random variables from populations 1 and 2 with means μ_1 and μ_2 and variances σ_1^2 and σ_2^2, respectively, then

$$F_{N-1, M-1} = \frac{\sum_{i=1}^{N}(x_{1i} - \bar{x}_1)^2/\sigma_1^2(N-1)}{\sum_{i=1}^{M}(x_{2i} - \bar{x}_2)^2/\sigma_2^2(M-1)} = \frac{s_1^2/\sigma_1^2}{s_2^2/\sigma^2}$$

is called an F statistic with $N - 1$ degrees of freedom in the numerator and $M - 1$ degrees of freedom in the denominator.

184   Estimation

The graph of the F density changes for different values of $N - 1$ and $M - 1$, but its shape is similar to the χ^2 curve (see Fig. 8.6.1). As for the Student's t and χ^2 densities, the area under the F-density has been numerically calculated. The more important values are tabulated in Appendix Table 6.

Definition 8.8.2 $F_{N-1,\,M-1,\,\alpha}$ is the point on the F density which has proportion α of the area to its left as shown in Fig. 8.8.1.

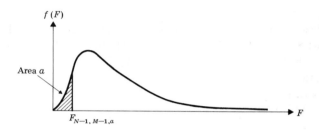

Figure 8.8.1

Since both N and M can vary, the resulting table could be extremely large. For this reason the areas in the right-hand tail of the F density are tabulated and those in the left-hand tail are found as follows.

$$F_{N-1,\,M-1,\,\alpha} = \frac{1}{F_{M-1,\,N-1,\,1-\alpha}} \qquad (8.8.1)$$

Example 8.8.1 For $N = 10$, $N - 1 = 9$, $M = 16$, $M - 1 = 15$, then $F_{9,\,15,\,.95} = 2.59$. To find $F_{9,\,15,\,.05}$, use Equation (8.8.1), that is

$$F_{9,\,15,\,.05} = \frac{1}{F_{15,\,9,\,.95}} = \frac{1}{3.01} = .33.$$

When it is understood, we shall replace $F_{N-1,\,M-1,\,\alpha}$ by F_α.
To calculate a 95% confidence limit for σ_1^2/σ_2^2, write

$$P[F_{.025} < F < F_{.975}] = .95.$$

If for example $N = 10$, $M = 16$, then

$$P[.27 < F < 3.12] = .95$$

or

$$P\left[.27 < \frac{s_1^2/\sigma_1^2}{s_2^2/\sigma_2^2} < 3.12\right] = .95$$

The inequalities may be converted to give

$$P\left[\frac{s_1^2/s_2^2}{3.12} < \sigma_1^2/\sigma_2^2 < \frac{s_1^2/s_2^2}{.27}\right] = .95.$$

For confidence level $1 - \alpha$ we have the following definition.

Definition 8.8.3 If $X_{11}, X_{12}, \ldots, X_{1N}$ and $X_{21}, X_{22}, \ldots, X_{2M}$ are random samples of size N and M from normal populations with variances σ_1^2 and σ_2^2, respectively, then

$$P\left[\frac{s_1^2/s_2^2}{F_{1-\alpha/2}} < \sigma_1^2/\sigma_2^2 < \frac{s_1^2/s_2^2}{F_{\alpha/2}}\right] = 1 - \alpha \qquad (8.8.2)$$

defines a $1 - \alpha$ level confidence interval for σ_1^2/σ_2^2.

The comments in Section 8.6 on the construction of "equal-tail" confidence intervals and on the importance of the assumption of normality apply equally well to the work of this section.

Example 8.8.2 Use the observations of Example 1.1.4 to calculate a 95% confidence interval for σ_A^2/σ_B^2.

Solution

i) $s_A^2 = 117.12$, $s_B^2 = 135.46$ (see Example 8.7.2).
ii) $s_A^2/s_B^2 = .86$.
iii) $F_{1-\alpha/2} = F_{.975} = 6.68$ (with 9 and 5 degrees of freedom).

$F_{\alpha/2} = F_{.025} = \dfrac{1}{4.48} = .22.$

iv) The lower limit is $\dfrac{s_A^2/s_B^2}{F_{1-\alpha/2}} = .13$.

The upper limit is $\dfrac{s_A^2/s_B^2}{F_{\alpha/2}} = 3.91$.

v) A 95% confidence interval for σ_A^2/σ_B^2 is between .13 and 3.91.

8.9 INTERVAL ESTIMATION OF p IN A BINOMIAL PROBABILITY FUNCTION

In most sampling situations involving a random variable X which has a binomial probability function, we know n and desire to estimate p. Since X/n is an unbiased estimator of p (see Exercise 1) and since it can in fact be shown to be a minimum variance unbiased estimator of p, then it is a good starting point from which to calculate confidence intervals for p. For the case where n the sample size is small, the calculations are

186 Estimation

laborious, and graphical procedures are sufficient for most purposes. The upper and lower limits of 95% and 99% confidence intervals can be read in Appendix Tables 7a and 7b for various values of X/n.

Example 8.9.1 (Smoking Survey) 100 people are asked if they smoke, and 32 answer "yes." Find a 95% confidence interval for the proportion of the population that smokes.

Solution Reading from Table 7a with $X/n = .32$, we see that a 95% confidence interval for the true proportion is between .23 and .42.

If n is large and since X/n is the average number of "successes" in n trials, then from Theorem 8.2.4 X/n will be approximately normal. Further $E(X/n) = p$ and

$$\text{Var}(X/n) = \frac{1}{n^2}\text{Var}(X) = \frac{np(1-p)}{n^2} = \frac{p(1-p)}{n}.$$

Thus $Z = \dfrac{X/n - p}{\sqrt{\dfrac{p(1-p)}{n}}}$ has a unit normal density and

$$P\left[-Z_{1-\alpha/2} < \frac{X/n - p}{\sqrt{\dfrac{p(1-p)}{n}}} < Z_{1-\alpha/2}\right] = 1 - \alpha$$

so that

$$P\left[X/n - (Z_{1-\alpha/2})\sqrt{\frac{p(1-p)}{n}} < p < \frac{X}{n} + (Z_{1-\alpha/2})\sqrt{\frac{p(1-p)}{n}}\right] = 1 - \alpha \quad (8.9.1)$$

defines a $1 - \alpha$ level confidence interval for p. Although no hard and fast rules can be given concerning how large n must be in order for Equation (8.9.1) to hold, the approximation will usually be sufficient if $np(1-p) > 9$. For large n we may substitute X/n for p without seriously affecting the results.

Example 8.9.2 Use the normal approximation in Example 8.9.1 to find a 95% confidence interval for p, the true proportion of smokers.

Solution Since p is unknown we must approximate $\sqrt{p(1-p)/n}$, using $p = 0.32$, the point estimate of p. Thus

i) $\sqrt{\dfrac{p(1-p)}{n}} = \sqrt{\dfrac{(.32)(.68)}{100}} = \sqrt{.002176} = .0466.$

ii) $z_{1-\alpha/2} = z_{.975} = 1.96$.

iii) $(z_{1-\alpha/2})\sqrt{\dfrac{p(1-p)}{n}} = (1.96)(.0466) = .0913$.

iv) The upper limit is $.32 + .0913 = .411$.
The lower limit is $.32 - .0913 = .238$.

8.10 INTERVAL ESTIMATION OF THE DIFFERENCE OF TWO BINOMIAL PARAMETERS

If two binomial parameters p_1 and p_2 are estimated by X_1/n and X_2/m, respectively, then we will construct $1 - \alpha$ level confidence intervals in the case where n and m are reasonably large, so that the normal approximation can be used. Now

$$E\left(\frac{X_1}{n} - \frac{X_2}{m}\right) = p_1 - p_2,$$

$$\text{Var}\left(\frac{X_1}{n} - \frac{X_2}{m}\right) = \frac{p_1 q_1}{n} + \frac{p_2 q_2}{m}, \qquad q_i = 1 - p_i, \ i = 1, 2,$$

so

$$Z = \frac{(X_1/n - X_2/m) - (p_1 - p_2)}{\sqrt{p_1 q_1/n + p_2 q_2/m}}$$

has a unit normal density and

$$P\left[-Z_{1-\alpha/2} < \frac{(X_1/n - X_2/m) - (p_1 - p_2)}{\sqrt{p_1 q_1/n + p_2 q_2/m}} < Z_{1-\alpha/2}\right] = 1 - \alpha$$

yields as a $1 - \alpha$ level confidence interval

$$P\left[\left(\frac{X_1}{n} - \frac{X_2}{m}\right) - Z_{1-\alpha/2}\sqrt{\frac{p_1 q_1}{n} + \frac{p_2 q_2}{m}} < p_1 - p_2 \right.$$
$$\left. < \left(\frac{X_1}{n} - \frac{X_2}{m}\right) + Z_{1-\alpha/2}\sqrt{\frac{p_1 q_1}{n} + \frac{p_2 q_2}{m}}\right] = 1 - \alpha$$

where $q = 1 - p$.

For n and m large we may substitute X_1/n for p_1 and X_2/m for p_2 without seriously affecting the results.

Example 8.10.1 (Medicine) Two groups of subjects of sizes 100 and 150, were treated with analgesics (pain killers) of types A and B, respectively. In the first group 65 and in the second group 118 said that their pain level had been reduced. Calculate a 95% confidence interval for the true difference in response to the two drugs.

Solution

i) $n = 100$, $m = 150$, $X_1/n = .65$, $X_2/m = .79$,
$X_2/m - X_1/n = .14$.

ii) $\sqrt{\dfrac{p_1 q_1}{n} + \dfrac{p_2 q_2}{m}} = \sqrt{\dfrac{(.65)(.35)}{100} + \dfrac{(.79)(.21)}{150}} = \sqrt{.003381} = .0581.$

iii) $z_{1-\alpha/2} = z_{.975} = 1.96$.

iv) $z_{1-\alpha/2} \sqrt{\dfrac{p_1 q_1}{n} + \dfrac{p_2 q_2}{m}} = (1.96)(.0581) = .1139.$

v) The upper limit is $.14 + .1137 = .25$.
The lower limit is $.14 - .1137 = .03$.

Note that we have computed the difference $X_2/m - X_1/n$ in order to keep the difference positive.

8.11 INTERVAL ESTIMATION OF μ, THE POISSON PARAMETER

The Poisson density is defined in Section 4.8. Confidence intervals can be found for the parameter μ by using Appendix Table 8. The table is self-explanatory and the following example will clarify its use.

Example 8.11.1 Suppose that the number of typing errors on a page in a particular book is assumed to be Poisson with parameter λ. If the number of errors on a page chosen at random is four, what are the 95% and 99% confidence intervals for λ?

Solution From Appendix Table 8 the 95% confidence interval is between 1.09 and 10.24, while the 99% confidence interval is between .672 and 12.59.

8.12 SUMMARY

In the early part of the chapter properties of the mean and variance are discussed. We look at $E(\bar{X})$, $\text{Var}(\bar{X})$, and $E(s_{\bar{X}}^2)$ and at the density of $\bar{X}$; the central limit theorem is mentioned. Point estimation is considered briefly, as are the ideas of bias and minimum variance. The emphasis in the second part of the chapter is on interval estimation of the unknown parameters of a distribution by means of confidence intervals. The cases covered are as follows:

i) The mean of a normal population with known variance.
ii) The mean of a normal population with unknown variance.
iii) The variance and standard deviation of a normal population.

iv) The difference of two normal means in the cases where (a) the variances are known, (b) the variances are unknown but equal, (c) the variances are unknown and not equal, (d) the observations are paired.

v) The ratio of two normal variances.

vi) The parameter of a binomial distribution and the difference of two binomial parameters.

vii) The parameter of a Poisson distribution.

EXERCISES

1. For the binomial distribution, show that X/n is an unbiased estimator of p but that X^2/n^2 is a biased estimator of p^2.

2. Show that $\text{Var}(\overline{X}_1 - \overline{X}_2) = \text{Var}(\overline{X}_1) + \text{Var}(\overline{X}_2)$ if $\overline{X}_1$ and $\overline{X}_2$ are from independent samples.

3. Show that $\text{Var}(\overline{X}_1 + \overline{X}_2 - \overline{X}_3) = \text{Var}(\overline{X}_1) + \text{Var}(\overline{X}_2) + \text{Var}(\overline{X}_3)$ if $\overline{X}_1, \overline{X}_2, \overline{X}_3$ are from independent samples.

4. The amounts of ash in chemicals A and B are normally distributed with means 2 and 5 and variances 1 and 2, respectively. Samples of sizes 10 and 20 are taken from A and B and the average amount of ash for each computed. What is the distribution, expected value and variance of the sum of the two means?

5. Suggest an unbiased estimator of λ in the Poisson distribution and give a physical example.

6. Show that $\sum_{i=1}^{N} (X_i - \overline{X})^2 = \sum_{i=1}^{N} (X_i - \mu)^2 - N(\overline{X} - \mu)^2$ by writing $(X_i - \overline{X})$ as $[(X_i - \mu) + (\mu - \overline{X})]$.

7. Show that $E \sum_{i=1}^{N} (X_i - \mu)^2 = \sum_{i=1}^{N} E(X_i - \mu)^2$.

8. If X_i, $i = 1, 2, \ldots, N$ has expectation μ and variance σ^2, what is the value of $E(\overline{X} - \mu)^2$?

9. Suppose that $E(X_i) = \mu$, $\text{Var}(X_i) = \sigma^2$ for $i = 1, 2, \ldots, N$. Compare (a) $\overline{X}$ and (b) $(X_1 + X_2)/2$ for biasedness and variance.

10. Using bias and variance as criteria, is there any difference between $(X_1 + X_2)/2$ and $(X_1 + X_N)/2$?

11. The weights of sacks of grain are normally distributed with mean 100 lbs. and variance 64 lbs. What is the probability that the average weight of N sacks will be greater than 110 lbs. if $N = 1, 9, 16, 36$?

12. If the blood pressure of normal patients is normally distributed with mean 120 and standard deviation 10, what is the probability that the mean of a sample of 10 patients will be greater than 115?

13. If a bus is designed to carry a maximum load of 6000 lbs. and claims a capacity of 30 people, what is the probability that it will be overloaded, if people's weights have a normal density with mean 190 lbs. and standard deviation 25 lbs?

14. If $\bar{X}$, the mean of a sample of size N, is normally distributed with mean μ and variance σ^2/N, use Theorem 5.1.1 to show that ΣX_i is normally distributed. What are its mean and variance?

15. If the workers' wages in a certain factory are normally distributed with mean $100 and standard deviation $10, what is the probability that the total wages of 10 workers exceeds $1100?

16. In Exercise 15, how large a sample is required so that the probability of the mean lying between 95 and 105 is .90?

17. Suppose that the wages in factory A have mean 100 and variance 100 and those in factory B have mean 90 and variance 150. If a sample of size 10 is taken from factory A and 20 from factory B, what is the mean and variance of the difference of the means of the two samples?

18. Find the expectation of $\dfrac{1}{N} \sum_{i=1}^{N} (X_i - \bar{X})^2$ if $E(X_i) = \mu$, $\mathrm{Var}(X_i) = \sigma^2$.

19. The sample (1.1, 0.7, 2.3, 1.7, 1.0) was drawn from a normal population with mean μ and variance 10. Find 95% and 99% confidence intervals for μ.

20. In Exercise 19 calculate 95% and 99% confidence intervals if σ^2 is unknown.

21. The heights of males in a particular tribe are assumed to be normally distributed. A sample of size 25 yields a mean of 60 inches and a standard deviation of 10. Calculate a 90% confidence interval for the mean population height.

22. From a population with a proportion p of smokers, a sample of 100 yields 25 smokers. Calculate a 95% confidence interval for p by two methods.

23. From a population of trained mice, 200 are selected to perform a test (in psychology, for example). Suppose 70 pass the test and the others fail, calculate 95% confidence intervals for the true proportion of the population which can pass the test.

24. A radioactive sample emits a count of 10 during a randomly chosen counting period. Calculate a 95% confidence interval for the true number of emissions during a typical period.

25. The number of accidents at a busy intersection is Poisson distributed with mean λ/day, and during a ten day period 20 accidents occur. Calculate a 99% confidence interval for λ.

26. Suppose that two drugs are compared. Of 200 people using drug A, 70 have a negative reaction and of 50 using drug B, 10 have a negative reaction; calculate a 95% confidence interval for the difference in the effect of the two drugs.

27. The mean inside diameter of a sample of 200 washers produced by a machine is .50 inches. It is known from past experience that the standard deviation of the diameter of the washers produced by this machine is .05 inches. Find the (a) 95%, (b) 99% confidence limits for the mean inside diameter of all the washers produced by the machine.

28. The mean and the standard deviation of weights of a sample of 200 students from a university of 3,000 students are found to be 150 and 15, respectively. Find the (a) 90%, (b) 99% confidence limits for the mean weight of the 3,000 students.

29. Of two similar groups A and B consisting of 75 and 125 pigs each, the group A was given a new feed and group B was given a control feed. In group A the mean gain in weight in a certain period was found to be 60 lbs. with a standard deviation of 8 lbs. In group B the mean gain in weight in the same period was 50 lbs. with a standard deviation of 5 lbs. Find the (a) 95%, (b) 99% confidence limits for the difference in the mean gain in weight caused by the two kinds of feeds.

30. A sample of 100 nails manufactured by a company showed that 10 are defective. A sample of 150 nails of the same kind manufactured by another company showed 12 defectives. Find the (a) 90%, (b) 99% confidence limits for the difference in proportions of defective nails manufactured by the two companies.

31. A box contains an unknown proportion of black and white marbles. A random sample of 75 marbles selected with replacement from the box showed that 60% were black. Find the (a) 90%, (b) 95% confidence limits for the actual proportion of black marbles in the box.

32. A random sample of 150 from a large consignment of apples contained 20 bad ones. Find the 95% confidence interval for the actual proportion of bad apples in the consignment.

33. In a sample of 250 people from city A, 150 preferred coffee to tea after supper. In city B 190 out of 250 preferred coffee to tea after supper. Find the 99% confidence interval for the difference between the true proportions of persons in city A and B who prefer coffee to tea after supper.

34. The variance of the lifetimes of a sample of 20 electric bulbs was computed to be 2500 hours. Find the (a) 95%, (b) 99% confidence interval for the variance of all such electric bulbs.

35. How large a sample should be taken in order to be (a) 90% (b) 95% confident that a population standard deviation will not differ from a sample standard deviation by more than 5%?

36. The standard deviation of the voltage of a sample of 25 batteries from a large consignment is found to be .5 volts. Find the (a) 90%, (b) 98% confidence limits on the variance of the voltage of such batteries.

37. Twenty-five acres each of two kinds of corn are planted under uniform growing conditions. The first kind has a sample variance yield of 6.5 bushels and the second kind has a sample variance yield of 14.4 bushels. If σ_1^2 (unknown) denotes the population variance of the first kind and σ_2^2 (unknown) denotes that of the second kind, calculate the 99% confidence limits for σ_1^2/σ_2^2.

38. The intelligence scores of a random sample of 101 students at a certain university have a variance of 120. Construct the 95% confidence interval for the true variance of the intelligence scores of all the students at the university.

39. A random of four pages is selected from a set of pages typed by a typist. The errors per page are 1, 3, 2, 2. Find the 90% confidence limits for the average number of errors the typist makes per page.

40. If the standard deviation of the lifetimes of a certain valve is estimated as 75 hours, how large a sample must be taken in order to be 99% confident that the error in the estimated mean lifetime will not exceed 15 hours?

Chapter 9

HYPOTHESIS TESTING

9.1 INTRODUCTION

A statistical hypothesis is an assumption about the population which can be tested by using the observations in a sample from the population. Generally the hypothesis concerns the value of one or more parameters of the population.

Example 9.1.1 If the population is a group of bright children, we might hypothesize that μ, the average I.Q. of the population, was greater than 120.

Example 9.1.2 Suppose that in playing a dice game with one die, where you win if you get 1, 2, or 3 and lose otherwise, you suspect that the die is loaded against you. Normally the probability p of winning is $\frac{1}{2}$. You might hypothesize that in fact $p = \frac{1}{4}$.

Example 9.1.3 If a population consists of light bulbs whose lifetime μ in hours is of interest, you might hypothesize that $\mu > 100$.

Definition 9.1.1 If a statistical hypothesis completely specifies the form and parameters of the population density or probability function, it is called a simple hypothesis. If it is not simple it is composite.

Example 9.1.4 The hypothesis of Example 9.1.2 is simple because if X is the number of times that you win, then X has a binomial probability function with parameters n (usually known) and p (specified by the hypothesis).

Example 9.1.5 In Example 9.1.1 neither the density nor the value of the random variable is specified. Even if the density were known (say it were normal), then μ is still unspecified. Furthermore, the variance is not stated as being known. Thus the hypothesis is composite.

In order to discuss criteria for testing statistical hypotheses it is necessary to consider alternative hypotheses. For instance an alternative hypothesis in Example 9.1.1 is that μ is less than 120. In Example 9.1.2 an alternative hypothesis is that $p = \frac{1}{2}$.

Definition 9.1.2 The main hypothesis which we wish to test is called the null hypothesis and is denoted by H_0. Any other hypotheses are called alternative hypotheses and are denoted by H_A.

Example 9.1.6 If in Example 9.1.1 we hypothesize that $\mu > 120$ as the null hypothesis and that $\mu \leq 120$ as the alternative hypothesis, then we write
$$H_0 : \mu > 120,$$
$$H_A : \mu \leq 120.$$
We start by discussing the testing of simple null hypotheses against simple alternatives.

9.2 TESTING SIMPLE HYPOTHESES

In Example 9.1.2 where the null hypothesis is $p = \frac{1}{4}$ against the alternative that $p = \frac{1}{2}$ or
$$H_0 : p = \frac{1}{4},$$
$$H_A : p = \frac{1}{2}$$
both null and alternative hypotheses are simple. The parameter p is one of two values. Either $p = \frac{1}{4}$ and H_0 is true or $p = \frac{1}{2}$ and H_0 is false. Suppose that we test the null hypothesis by throwing the die 10 times and recording the number of wins and losses. On the basis of the observations we can accept H_0 (conclude that H_0 is true) or reject H_0 (conclude that H_0 is false). The following situations can occur.

a) H_0 is true and we accept it.
b) H_0 is true and we reject it.
c) H_0 is false and we accept it.
d) H_0 is false and we reject it.

Definition 9.2.1 A type I error is made when H_0 is wrongly rejected.

Definition 9.2.2 A type II error is made in wrongly accepting H_0.

Example 9.2.1 If in Example 9.1.2 the die is fair ($p = \frac{1}{2}$) and we conclude from the observations that H_0 is true, then we are making a type II error. If we reject H_0 when in fact $p = \frac{1}{4}$ we are making a type I error.

The two types of error are compared in Table 9.2.1.

TABLE 9.2.1

	Decision	
	accept H_0	reject H_0
H_0 true	correct decision	type I error
H_0 false	type II error	correct decision

Suppose that we decide beforehand that we will reject H_0 if we get four or more wins, then the numbers 4, 5, 6, 7, 8, 9, 10 form what is called the critical region.

Definition 9.2.3 The set of outcomes of the experiment for which H_0 is rejected is called the critical region. The set of outcomes for which H_0 is accepted may be called the acceptance region.

We would, of course, like to make correct decisions all the time. This is not possible, however, since random variables having unpredictable outcomes are involved. We must learn to live with probabilities of types I and II error.

Definition 9.2.4 The probability of making a type I error is called the level of significance of the experiment and is denoted by α. The probability of making a type II error is denoted by β

Definition 9.2.5 The power of the test is the probability of rejecting H_0. When H_0 is false, power $= 1 - \beta$.

Example 9.2.2 Find α, β, and power for testing

$$H_0 : p = \tfrac{1}{4},$$
$$H_A : p = \tfrac{1}{2}$$

where X has a binomial density with $n = 10$, and critical region defined by $X \geq 4$:

Solution α is the probability of rejecting H_0 when H_0 is true. We reject H_0 if $X \geq 4$, so

$$\alpha = P[X \geq 4 \mid p = \tfrac{1}{4}]$$
$$= \sum_{x=4}^{10} \binom{10}{x} \left(\tfrac{1}{4}\right)^x \left(\tfrac{3}{4}\right)^{10-x}$$
$$= .2241.$$

β is the probability of wrongly accepting H_0. We accept H_0 if $X < 4$, so
$$\beta = P[X < 4 \mid p = \tfrac{1}{2}]$$
$$= \sum_{x=0}^{3} \binom{10}{x} \left(\tfrac{1}{2}\right)^x \left(\tfrac{1}{2}\right)^{10-x}$$
$$= .1719,$$
$$\text{Power} = 1 - \beta = .8281.$$

Definition 9.2.6 The statistic on which the test is based is called the test statistic.

Example 9.2.3 In Example 9.2.2, X is the test statistic.

It is usual to select α in advance, and to calculate the critical region from it. Usual choices for α are .01, .05, and .10, although the consequences of a type I error may lead one to consider other values.

Example 9.2.4 Suppose $\alpha = .05$ in Example 9.2.2, then show that the critical region is $X \geq 6$.

Solution We intuitively reject H_0 if X is large, that is if $X \geq A$, where A must be calculated from a knowledge of α and the nature of the critical region. Thus
$$\alpha = .05 = P[X \geq A \mid p = \tfrac{1}{4}]$$
$$= \sum_{x=A}^{10} \binom{10}{x} \left(\tfrac{1}{4}\right)^x \left(\tfrac{3}{4}\right)^{10-x}.$$

In Example 9.2.2 if $A = 4$, $\alpha = .2281$. Similarly if $A = 5$, $\alpha = .0781$ and if $A = 6$, $\alpha = .0197$. Now since A must be an integer because X, the random variable, can only assume integer values, we see that no value of A gives $\alpha = .05$ exactly. If $A = 5$, α is too large and if $A = 6$, α is too small. If we pick $A = 6$, then α will be no larger than .05. When dealing with discrete random variables, it is common practice to pick α to be the upper bound for the probability of a type I error. Thus $\alpha = .05$ means that the probability of a type I error is no larger than .05. We now consider hypothesis testing involving continuous random variables.

Example 9.2.5 (Anthropology) Suppose that an anthropologist has observations on 9 skull measurements of members of a prehistoric tribe, which he knows belong to tribe A or tribe B. The skull dimensions for both tribes have normal densities, with means $\mu_A = 20$ and $\mu_B = 24$. The variance σ^2 is known to be 16, while the average of the sample is $\bar{x} = 22$. From which tribe did the sample come?

Solution The population from which the sample came has mean μ, which equals either μ_A or μ_B. If the null hypothesis is that the sample came from tribe A with $\mu_A = 20$ against the alternative that it came from tribe B with $\mu_B = 24$, then we write

$$H_0 : \mu = 20,$$
$$H_A : \mu = 24.$$

Let us choose a significance level $\alpha = .05$ and use as the test statistic $\overline{X}$, the minimum variance unbiased estimator of μ, then intuitively we will reject H_0 if $\overline{X}$ is large, that is reject H_0 if

$$\overline{X} > A \tag{9.2.1}$$

where A must be calculated from the definition of α. Now

$$\alpha = P[\text{Reject } H_0 \mid H_0 \text{ true}]$$
$$= P[\overline{X} > A \mid \mu = 20]. \tag{9.2.2}$$

But $Z = \dfrac{(\overline{X} - \mu_A)}{\sigma/\sqrt{N}}$ has a unit normal density if $\mu = \mu_A = 20$, so Equation (9.2.2) can be written

$$\alpha = P\left[\frac{\overline{X} - 20}{\sigma/\sqrt{N}} > \frac{A - 20}{\sigma/\sqrt{N}}\right]$$

where $\sigma = 4$, $\sqrt{N} = 3$, $\alpha = .05$, so

$$.05 = P\left[Z > \frac{A - 20}{\frac{4}{3}}\right].$$

But from Appendix Table 2, $P[Z > 1.645] = .05$ so

$$\frac{A - 20}{\frac{4}{3}} = 1.645$$

or

$$A = 22.193.$$

The test can be performed in two completely equivalent ways, namely:

a) Compute $Z = (\overline{X} - 20)/\frac{4}{3}$ and reject H_0 if $Z > 1.645$.
b) Compare $\overline{X}$ to $A = 22.193$ and reject H_0 if $\overline{X} > 22.193$.

Method (b) is perhaps easier to comprehend, since it is of the form of statement (9.2.1). The critical region consists of those values of $\overline{X}$ greater than 22.193. The power of the test when H_0 is false is

$$\text{Power} = P[\text{Reject } H_0 \mid H_0 \text{ false}]$$
$$= P[\overline{X} > 22.193 \mid \mu = \mu_B = 24].$$

But when $\mu = 24$, then $Z = \dfrac{\overline{X} - 24}{\sigma/\sqrt{N}}$ has a unit normal density, so

$$\text{Power} = P\left[\frac{\overline{X} - 24}{\sigma/\sqrt{N}} > \frac{22.193 - 24}{\frac{4}{3}}\right]$$
$$= P[Z > -1.355]$$
$$= 0.9131 \quad \text{from Appendix Table 2.}$$

In summary, if we reject H_0 when $\overline{X} > 22.193$, then there is a 5% probability of wrongly rejecting H_0 and a 91% probability of correctly rejecting it. The student should do Exercise 37 to see the effect of increasing the sample size.

Definition 9.2.7 A test whose critical region depends only on extreme values of the test statistics in one direction, for example only large or only small values, is called a one-sided test. A test whose critical region includes both small and large values is called a two-sided test.

Example 9.2.6 In Examples 9.2.2 and 9.2.5 both tests are one sided. Specifically, both reject H_0 for large values of the test statistic.

In reality few problems have simple hypotheses and simple alternatives. Usually one or both are composite. We shall discuss composite tests for the remainder of the chapter.

9.3 TESTING THE MEAN OF A NORMAL POPULATION WITH KNOWN VARIANCE

Given that $X_1, X_2, \ldots, X_N$ are N random variables from a normal population with unknown mean μ and known variance σ^2, consider the test

$$\begin{aligned} H_0 &: \mu = \mu_0, \\ H_A &: \mu \neq \mu_0, \end{aligned} \qquad (9.3.1)$$

where μ_0 is a specified number. The statement $\mu \neq \mu_0$ includes values of $\mu > \mu_0$ and $\mu < \mu_0$, so that H_A is a two-sided alternative and Equation (9.3.1) defines a two-sided test. As in Section 9.2 we base the test on $\overline{X}$. We know that if H_0 is true then

$$Z = \frac{\overline{X} - \mu_0}{\sigma/\sqrt{N}}$$

has a unit normal density. If the unknown value of μ is really larger than μ_0, then it is probable that $\overline{X}$ will also be larger than μ_0 and Z would therefore be larger than zero. If on the other hand μ were smaller than μ_0, then $\overline{X}$ would probably be smaller than μ_0 and Z would be negative.

Testing the mean of a normal population with known variance 199

Thus the critical region will be both small and large values of $\bar{X}$; that is, we will reject H_0 if $\bar{X} > A$ or $\bar{X} < B$ where A and B must be calculated, from the definition of α. If the significance level $\alpha = .05$, then

$$\alpha = .05 = P[\text{Reject } H_0 | H_0 \text{ true}]$$
$$= P[\bar{X} > A \text{ or } \bar{X} < B | \mu = \mu_0]$$

and from the normality of $Z = \dfrac{\bar{X} - \mu}{\sigma/\sqrt{N}}$,

$$.05 = P\left[Z > \frac{A - \mu_0}{\sigma/\sqrt{N}} \text{ or } Z < \frac{B - \mu_0}{\sigma/\sqrt{N}} \right].$$

But

$$Z > \frac{A - \mu_0}{\sigma/\sqrt{N}}$$

and

$$Z < \frac{B - \mu_0}{\sigma/\sqrt{N}}$$

are mutually exclusive events, so

$$.05 = P\left[Z > \frac{A - \mu_0}{\sigma/\sqrt{N}} \right] + P\left[Z < \frac{B - \mu_0}{\sigma/\sqrt{N}} \right].$$

From Appendix Table 2, we know that

$$.05 = P[Z > 1.96] + P[Z < -1.96] \tag{9.3.2}$$

so

$$\frac{A - \mu_0}{\sigma/\sqrt{N}} = 1.96$$

and $A = \mu_0 + 1.96\, \sigma/\sqrt{N}$.
similarly $B = \mu_0 - 1.96\, \sigma/\sqrt{N}$. There are values other than ± 1.96 which satisfy (9.3.2), but the symmetric one, that is the one with "equal tails" is generally regarded as most desirable in this case. To see what is happening graphically, consider Fig. 9.3.1.

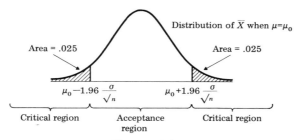

Figure 9.3.1

For general values of α, using the equal tail criterion, the Equation (9.3.2) can be written

$$\alpha = P[Z > Z_{1-\alpha/2}] + P[Z < Z_{\alpha/2}]$$

from which

$$A = \mu_0 + Z_{1-\alpha/2}\,\sigma/\sqrt{N},$$
$$B = \mu_0 + Z_{\alpha/2}\,\sigma/\sqrt{N}.$$

The critical region consists of values of $\bar{X} > A$ and $\bar{X} < B$. This discussion leads to the following theorem.

Theorem 9.3.1 To test

$$H_0 : \mu = \mu_0$$
$$H_A : \mu \neq \mu_0$$

on the basis of a sample of size N from a normal population with mean μ and known variance σ^2 at a significance level α, then reject H_0 if

$$\bar{X} > \mu_0 + Z_{1-\alpha/2}\,\sigma/\sqrt{N}$$

or

$$\bar{X} < \mu_0 + Z_{\alpha/2}\,\sigma/\sqrt{N}.$$

We shall next calculate the power of the test by first calculating β, the probability of a type II error.

$$\beta = P[\text{accept } H_0 \mid H_0 \text{ false}]$$
$$= P[\mu_0 + (\sigma/\sqrt{N})Z_{\alpha/2} < \bar{X} < \mu_0 + (\sigma/\sqrt{N})Z_{1-\alpha/2} \mid \mu \neq \mu_0] \quad (9.3.3)$$

Suppose that $\mu = \mu_1$ where $\mu_1 \neq \mu_0$, then

$$Z = \frac{\bar{X} - \mu_1}{\sigma/\sqrt{N}}$$

has a unit normal density, so that Equation (9.3.3) becomes

$$\beta = P\left[Z_{\alpha/2} + \left(\frac{\mu_0 - \mu_1}{\sigma/\sqrt{N}}\right) < \frac{\bar{X} - \mu_1}{\sigma/\sqrt{N}} < Z_{1-\alpha/2} + \left(\frac{\mu_0 - \mu_1}{\sigma/\sqrt{N}}\right)\right]. \quad (9.3.4)$$

Values of β can be calculated from Appendix Table 2, for varying μ_1. From Definition 9.2.5 power is calculated as $1 - \beta$.

Example 9.3.1 (Psychology) In a particular school district, the I.Q. is known to have a normal density with mean $\mu = 110$ and variance $\sigma^2 = 100$. From one school, a group of 25 students have an average I.Q. $\bar{x}$ of 115. Is the average in this school different from the district average? Use $\alpha = .10$ and plot the power of the test.

Testing the mean of a normal population with known variance

Solution The null hypothesis is that $\mu = 110$ against an alternative that $\mu \neq 110$.

$$H_0 : \mu = 110,$$
$$H_1 : \mu \neq 110.$$

Using Theorem 9.3.1 we reject H_0 if

$$\bar{X} > 110 + (1.65)\left(\tfrac{10}{5}\right) = 113.3$$

or

$$\bar{X} < 110 + (1.65)\left(\tfrac{10}{5}\right) = 106.7.$$

Since $\bar{x} = 115 > 113.3$, we reject H_0. Using Equation (9.3.4) the value of β is calculated as

$$\beta = P\left[-1.65 + \left(\frac{110 - \mu_1}{10/5}\right) < Z < 1.65 + \left(\frac{110 - \mu_1}{10/5}\right)\right].$$

For instance when $\mu_1 = 115$

$$\beta = P[-4.15 < Z < -.85] = .1977$$

from Appendix Table 2. This particular case is illustrated in Fig. 9.3.2. β and the power are shown in Table 9.3.1, for various values of μ_1, and the power is plotted in Fig. 9.3.3.

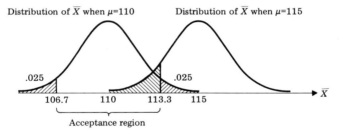

Figure 9.3.2

TABLE 9.3.1

μ_1	β	Power
117	.0322	.9678
115	.1977	.8023
113	.5588	.4412
111	.8591	.1409
109	.8591	.1409
107	.5588	.4412
105	.1977	.8023
103	.0322	.9678

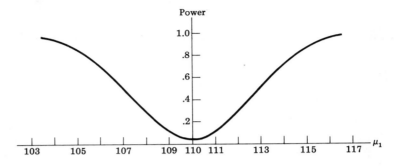

Figure 9.3.3
Comparing the Power of test for Example 9.3.1

Equations (9.3.1) define a two-sided test. For a one-sided test of the form

$$H_0 : \mu = \mu_0,$$
$$H_A : \mu > \mu_0 \qquad (9.3.5a)$$

or

$$H_0 : \mu = \mu_0,$$
$$H_A : \mu < \mu_0 \qquad (9.3.5b)$$

we reject H_0 for large values of $\bar{X}$ in Equation (9.3.5a) and small values of $\bar{X}$ in Equation (9.3.5b). For simplicity, consider first Equation (9.3.5a). We reject H_0 if $\bar{X} > C$.

$$\alpha = P[\text{reject } H_0 \mid H_0 \text{ true}]$$
$$= P[\bar{X} > C \mid \mu = \mu_0]$$
$$= P\left[Z > \frac{C - \mu_0}{\sigma/\sqrt{N}}\right]$$

so that

$$\frac{C - \mu_0}{\sigma/\sqrt{N}} = Z_{1-\alpha}$$

and

$$C = \mu_0 + \frac{\sigma}{\sqrt{N}} Z_{1-\alpha}.$$

The critical region for Equation (9.3.5a) is $\bar{X} > \mu_0 + (\sigma/\sqrt{N}) Z_{1-\alpha}$ and for Equation (9.3.5b) is $\bar{X} < \mu_0 + Z_\alpha \sigma/\sqrt{N}$. The calculation of the power follows as for the two-sided case.

Theorem 9.3.2 To test

$$H_0 : \mu = \mu_0,$$
$$H_A : \mu > \mu_0$$

on the basis of a sample of size N from a normal population with mean μ and known variance σ^2 at a significance level α, then reject H_0 if

$$\bar{X} > \mu_0 + Z_{1-\alpha} \sigma/\sqrt{N}.$$

The choice of whether to use a one-sided or two-sided test is not a statistical question, but depends on the problem being studied. If the problem logically admits only a one-sided alternative, then one is justified in using a one-sided test. The decision to use a one- or two-sided test should be made before the observations are collected. It is certainly wrong to run both one- and two-sided tests on the same observations and pick the result you like the best.

Example 9.3.2 Consider a population of typists with an average typing speed of μ (words per minute), and suppose that they take a course to improve their typing speed. If we wish to take a sample and test whether the course has improved their typing speed, the null hypothesis, is that μ has increased. The other alternative is excluded since it is unlikely that the course reduced their typing speed.

Example 9.3.3 Suppose in an industrial experiment, that we wish to test that the outputs from machines A and B are the same. The alternative hypothesis is two-sided, since if A and B are not the same, then either A is less than B or B is less than A; that is, two situations are possible.

An important consideration, applicable to all tests of hypotheses, is the following. When we reject H_0, we know the probability of our being wrong, namely α; whereas if we accept the null hypothesis, the probability of being wrong depends on the value of the parameter. For this reason one usually talks of "rejecting the null hypothesis" and "being unable to reject the null hypothesis." The expression "accepting the null hypothesis" is rarely used, and in fact if one is unable to reject the null hypothesis it is generally not because H_0 is true but because the experiment is not sensitive enough to reject it. If H_0 is not rejected, the power curve of the test should be shown if possible.

9.4 TESTING THE MEAN OF A NORMAL POPULATION WITH UNKNOWN VARIANCE

If in the discussion of Section 9.3 the value of σ^2 is unknown, then we can no longer use Z as the test statistic. In its place we use

$$t = \frac{\bar{X} - \mu_0}{s_X/\sqrt{N}} \tag{9.4.1}$$

which we know from Section 8.5 to have a Student's t density. To test $H_0 : \mu = \mu_0$ against the two-sided alternative

$$H_A : \mu \neq \mu_0 \tag{9.4.2}$$

we will reject H_0 if $t > A$ or $t < B$ where the values of A and B are calculated below.

$$\alpha = P[\text{Reject } H_0 \,|\, H_0 \text{ true}]$$
$$= P[t > A \text{ or } t < B \,|\, \mu = \mu_0].$$

But since t has a Student's t density, we know that

$$P[t > t_{1-\alpha/2} \text{ or } t < -t_{1-\alpha/2}] = \alpha. \tag{9.4.3}$$

Thus if t is calculated by Equation (9.4.1), then the critical region is

$$t > t_{1-\alpha/2} \text{ or } t < -t_{1-\alpha/2}.$$

This leads to the following theorem.

Theorem 9.4.1 To test
$$H_0 : \mu = \mu_0,$$
$$H_A : \mu \neq \mu_0$$

on the basis of a sample of size N from a normal population with mean μ and unknown variance σ^2, estimated by s_X^2, at a significant level α, then reject H_0 if

$$\frac{\bar{X} - \mu_0}{s_X/\sqrt{N}} > t_{1-\alpha/2}$$

or $\hspace{8cm}$ (9.4.4)

$$\frac{\bar{X} - \mu_0}{s_X/\sqrt{N}} < -t_{1-\alpha/2}.$$

To perform one-sided tests of the form

$$H_0 : \mu = \mu_0,$$
$$H_A : \mu > \mu_0$$

then reject H_0 if $t > t_{1-\alpha}$, that is,

$$\alpha = P[t > t_{1-\alpha} \,|\, \mu = \mu_0].$$

Example 9.4.1 (Industry) The average time required to perform a certain industrial task is known to be 12.5 minutes. Suppose 10 new employees are hired and trained. During a testing period their times for completion of the same task are as follows:

$$9.3, \; 12.1, \; 15.7, \; 10.3, \; 12.2, \; 14.8, \; 15.1, \; 13.2, \; 15.9, \; 14.5.$$

Test the hypothesis that these men are no different from the average, at the 5% level of significance.

Solution

i) $H_0 : \mu = 12.5 \; (= \mu_0),$
$\hspace{0.5cm} H_A : \mu \neq 12.5.$

ii) $N = 10$, $N - 1 = 9$.

iii) $\sum x_i^2 = 1818.67$, $\sum x_i = 133.1$, $s_x^2 = \dfrac{47.109}{9} = 5.23$,

$s_x = \sqrt{5.23} = 2.28$, $\bar{x} = 13.31$.

iv) $t = \dfrac{\bar{x} - \mu}{s_x/\sqrt{N}} = \dfrac{13.31 - 12.5}{2.28/\sqrt{10}} = 1.13$.

v) Using Equation (9.4.4) we reject H_0 if $t < -t_{1-\alpha/2}$. From Appendix Table 4, $t_{1-\alpha/2} = t_{.975} = 2.26$, thus we cannot reject H_0. There is no evidence to assume that the workers are any different from the average.

9.5 COMPARISON OF CONFIDENCE INTERVALS AND HYPOTHESIS TESTING

It should be clear by now that there is a relationship between the acceptance region of hypothesis testing and confidence intervals. For example to test

$$H_0 : \mu = \mu_0,$$
$$H_A : \mu \neq \mu_0$$

as in Section 9.4, the acceptance region is

$$-Z_{1-\alpha/2} < Z < Z_{1-\alpha/2} \qquad (9.5.1)$$

while the confidence interval for μ is

$$\bar{X} - Z_{1-\alpha/2}\, \sigma/\sqrt{N} < \mu < \bar{X} + Z_{1-\alpha/2}\, \sigma/\sqrt{N} \qquad (9.5.2)$$

from Definition 8.4.2. Rearranging Inequality (9.5.2) we obtain

$$-Z_{1-\alpha/2} < \dfrac{\bar{X} - \mu}{\sigma/\sqrt{N}} < Z_{1-\alpha/2}$$

But $(\bar{X} - \mu)/(\sigma/\sqrt{N})$ Z, and so expressions (9.5.1) and (9.5.2) are equivalent. This shows that the acceptance region for testing $H_0 : \mu = \mu_0$ against a two-sided alternative, when σ^2 is known, with a significance level α, is the same as a $1 - \alpha$ level confidence interval. In other words, accepting H_0 at an α level of significance is equivalent to μ_0 being contained in a $1 - \alpha$ level confidence interval. By similar reasoning, rejecting H_0 at the α level of significance is equivalent to μ_0 not being contained in a $1 - \alpha$ level confidence interval.

Similar results hold for all the hypotheses discussed in the remainder of this chapter. For this reason, we will present the results concerning hypothesis testing in an abbreviated form, the general ideas having already been met and discussed in Chapter 8.

9.6 TESTING THE VARIANCE AND STANDARD DEVIATION OF A NORMAL POPULATION

If we wish to test σ^2 and σ of a normal population, on the basis of N independent observations, then we know from Definition 8.6.1 that

$$\frac{(N-1)s_X^2}{\sigma^2}$$

has a χ^2 density with $N-1$ degrees of freedom. To test

$$H_0 : \sigma^2 = \sigma_0^2,$$
$$H_A : \sigma^2 \neq \sigma_0^2$$

the test statistic is

$$\frac{(N-1)s_X^2}{\sigma_0^2}.$$

We reject H_0 if

$$\frac{(N-1)s_X^2}{\sigma^2} \begin{cases} > \chi_{1-\alpha/2}^2 \\ < X_{\alpha/2}^2. \end{cases} \tag{9.6.1}$$

If the alternative hypothesis is $H_A : \sigma^2 \geq \sigma_0^2$, then reject H_0 if

$$\frac{(N-1)s_X^2}{\sigma_0^2} > \chi_{1-\alpha}^2 \tag{9.6.2}$$

and if we have $H_A : \sigma^2 \leq \sigma_0^2$, then reject H_0 if

$$\frac{(N-1)s_X^2}{\sigma_0^2} < \chi_\alpha^2 \tag{9.6.3}$$

Example 9.6.1 (Industry) Suppose that the iron ore melting points in region A of Example 1.1.4 are obtained by a new method. The old method had a variance of 110. Test the hypothesis that the new method is the same as the old versus the alternative that the new method is more variable. Use an α of .05.

Solution
i) $H_0 : \sigma^2 = 110 \; (= \sigma_0^2),$
$H_A : \sigma^2 > 110.$
ii) $N = 10$, degrees of freedom $= N - 1 = 9$.
iii) From Example 7.7.4, $s_x^2 = 117.12$.
iv) The test statistic is $\dfrac{(N-1)s_x^2}{\sigma_0^2} = \dfrac{(9)(117.12)}{110} = 9.58.$
v) Using Equation (9.6.2) we reject H_0 if

$$\frac{(N-1)s_x^2}{\sigma_0^2} > \chi_{1-\alpha}^2.$$ But from the Appendix Table 5, $\chi_{1-\alpha}^2 = \chi_{.95}^2 = 16.9$, so we cannot reject H_0.

The hypotheses concerning the standard deviation σ can be converted to one concerning σ^2, simply by squaring, and this is generally regarded as the preferable procedure.

9.7 TESTING THE DIFFERENCE OF TWO NORMAL MEANS

The discussion in this section runs parallel to that of Section 8.7. Suppose that we wish to compare the means μ_1 and μ_2 of two normal populations on the basis of two independent samples of sizes N and M and variances σ_1^2 and σ_2^2, respectively. Such tests would be of interest to an engineer in comparing two measuring techniques, or to a farmer comparing two types of feed.

We consider first the case where σ_1^2 and σ_2^2 are known. We know from Equation (8.7.1) that

$$Z = \frac{(\bar{X}_1 - \bar{X}_2) - (\mu_1 - \mu_2)}{\sqrt{\sigma_1^2/N + \sigma_2^2/M}} \quad (9.7.1)$$

follows the unit normal distribution.

In order to test

$$H_0 : \mu_1 = \mu_2,$$
$$H_A : \mu_1 \neq \mu_2$$

we know from the discussion of Section 9.5 and Equation (8.7.2) that we reject H_0 if

$$Z \begin{cases} > Z_{1-\alpha/2} \\ < -Z_{1-\alpha/2}. \end{cases} \quad (9.7.2)$$

To test

$$H_0 : \mu_1 = \mu_2,$$
$$H_A : \mu_1 > \mu_2$$

reject H_0 if $Z > Z_{1-\alpha}$, and to test

$$H_0 : \mu_1 = \mu_2,$$
$$H_A : \mu_1 < \mu_2$$

reject H_0 if $Z < -Z_{1-\alpha}$.

Example 9.7.1 Using the data of Example 1.1.4, test the hypothesis that there is no difference between the melting points of the two regions, A and B, if we assume that $\sigma_1^2 = \sigma_2^2 = 100$, and $\alpha = .05$.

Solution

i) $H_0: \mu_A = \mu_B$ or $\mu_A - \mu_B = 0$,
    $H_A: \mu_A \neq \mu_B$ or $\mu_A - \mu_B \neq 0$.

ii) $N = 10$, $M = 6$.

iii) From Example 8.7.1, $\bar{x}_A = 1506.7$, $\bar{x}_B = 1499.3$.

iv) $z = \dfrac{(1506.7 - 1499.3)}{\sqrt{\frac{100}{10} + \frac{100}{6}}} = 1.43$.

v) Using Equation (9.7.2) we reject H_0 if $Z > Z_{1-\alpha/2} = Z_{.975}$ or $Z < -Z_{.975}$. But from Appendix Table 2, $Z_{.975} = 1.96$, so we cannot reject H_0.

If σ_1^2 and σ_2^2 are unknown but assumed to be equal, and have estimators s_1^2 and s_2^2, then a pooled estimate of σ^2 is given by Equation (8.7.3) and we can then use Equation (8.7.4) that

$$t = \frac{(\bar{X}_1 - \bar{X}_2) - (\mu_1 - \mu_2)}{s_p \sqrt{1/N + 1/M}} \tag{9.7.3}$$

has a Student's t distribution with $N + M - 2$ degrees of freedom. From the discussion of Section 9.5 and Equation (8.7.5) we reject H_0: $\mu_1 = \mu_2$ against $H_A: \mu_1 \neq \mu_2$ if

$$\begin{aligned} t &> t_{1-\alpha/2}, \\ t &< -t_{1-\alpha/2} \end{aligned} \tag{9.7.4}$$

at the α level of significance. Similar tests can be made for one-sided alternatives.

Example 9.7.2   Test for no difference between the melting points of ore in regions A and B, assuming that $\sigma_A^2 = \sigma_B^2$, although both are unknown, with $\alpha = .05$.

Solution

i) $H_0: \mu_A = \mu_B$,
    $H_A: \mu_A \neq \mu_B$.

ii) From Example 9.7.2, $s_A^2 = 117.12$, $s_B^2 = 135.46$, $s_p^2 = 123.67$, $s_p = 11.12$, $\bar{x}_A - \bar{x}_B = 7.4$, $N = 10$, $= M = 6$.

iii) $t = \dfrac{(\bar{x}_A - \bar{x}_B) - (\mu_A - \mu_B)}{s_p \sqrt{1/N + 1/M}} = \dfrac{7.4}{5.74} = 1.29$.

iv) Using Equation (9.7.4) we reject H_0 if $t > t_{.975}$ or if $t < -t_{.975}$. But from Appendix Table 4, $t_{.975} = 2.15$ (with 14 degrees of freedom), so we cannot reject H_0.

If σ_1^2 and σ_2^2 are unknown and cannot be assumed to be equal, then we may use the approach of Welch (as discussed in Section 8.7), who approximates a t statistic.

In Equation (8.7.6)
$$t = \frac{(\bar{X}_1 - \bar{X}_2) - (\mu_1 - \mu_2)}{\sqrt{s_1^2/N + s_2^2/M}}$$

has an approximate Student's t distribution with ν degrees of freedom where ν is given by Equation (8.7.7). From Equation (8.7.8) we reject H_0 if
$$\begin{aligned} t &> t_{1-\alpha/2}, \\ t &< -t_{1-\alpha/2} \end{aligned} \qquad (9.7.5)$$

where $t_{1-\alpha/2}$ is read from Appendix Table 4 with ν degrees of freedom. Discussion of one-sided tests follow directly by similar reasoning.

Example 9.7.3 Using the data of Example 8.7.3, test the hypothesis that there is no difference between the melting points of regions A and C, if the unknown variances are not assumed to be equal. Use $\alpha = .05$.

Solution

i) $H_0: \mu_A = \mu_C,$
   $H_A: \mu_A \neq \mu_C.$

ii) From Example 8.7.3, $s_A^2 = 117.12$, $s_C^2 = 300.16$, $N = 10$, $M = 9$, $\bar{x}_A = 1506.7$, $\bar{x}_C = 1499.2$.

iii) $t = \dfrac{(\bar{x}_A - \bar{x}_C) - (\mu_A - \mu_C)}{\sqrt{s_A^2/N + s_C^2/M}} = \dfrac{(1506.7 - 1499.2)}{\sqrt{117.12/10 + 300.16/9}} = 1.12.$

iv) From Example 8.7.3, $\nu = 14.42$.

v) Using Equation (9.7.5) we reject H_0 if $t > t_{.975}$ or $t < -t_{.975}$. From Appendix Table 4, $t_{.975} = 2.14$, so we are unable to reject H_0.

The discussion in Section 8.7 concerning pairs of observations can also be applied to hypothesis testing. Suppose that the pairs of observations are (X_{1i}, X_{2i}) for $i = 1, 2, \ldots, N$ where X_1 and X_2 have normal distributions with means μ_1 and μ_2 and variances σ_1^2 and σ_2^2 respectively. To test that
$$\begin{aligned} H_0: \mu_1 &= \mu_2, \\ H_A: \mu_1 &\neq \mu_2, \end{aligned}$$
when σ_1^2 and σ_2^2 are known, use Equation (8.7.2) and Section 8.5 and reject H_0 if
$$\begin{aligned} Z_D &> Z_{1-\alpha/2}, \\ Z_D &< -Z_{1-\alpha/2}. \end{aligned} \qquad (9.7.6)$$

If σ_1^2 and σ_2^2 are unknown, then from Equation (8.7.5) reject H_0 if

$$t_D > t_{1-\alpha/2},$$
$$t_D < -t_{1-\alpha/2}. \qquad (9.7.7)$$

Discussion of one-sided alternatives presents no new difficulties and can be handled as discussed previously.

Example 9.7.4 (Education) The following figures are the typing speeds in words per minute for 10 students measured before and after taking a typing course. The figures are assumed to be normally distributed. Can we say that the course has been effective? Use a 5% significance level.

Student	1	2	3	4	5	6	7	8	9	10
Before	40	47	33	54	61	39	42	47	58	50
After	52	45	51	60	58	69	65	63	59	72

Solution

i) $H_0: \mu_1 = \mu_2$ or $\mu_1 - \mu_2 = 0$,
   $H_A: \mu_1 < \mu_2$ or $\mu_1 - \mu_2 < 0$.
ii) The differences d, (after-before) for each student are 12, -2, 18, 6, -3, 30, 23, 16, 1, 22.
iii) $\bar{d} = 12.3$, $s_d^2 = 130.45$, $s_d^2/10 = 13.045$, $s_d/\sqrt{10} = 3.61$.
iv) $t_d = \dfrac{\bar{d} - 0}{s_d/\sqrt{10}} = \dfrac{12.3}{3.61} = 3.41$.
v) We reject H_0 if $t > t_{.95}$. From Appendix Table 4, $t_{.95}$ (with 9 degrees of freedom) is 1.833. We therefore reject H_0 and conclude that $\mu_1 < \mu_2$.

9.8 TESTING THE RATIO OF TWO NORMAL VARIANCES

If $X_{11}, X_{12}, \ldots, X_{1N}$ and $X_{21}, X_{22}, \ldots, X_{2M}$ are N and M random variables from two normal populations with variances σ_1^2 and σ_2^2, respectively, then we know from Section 8.8 that if s_1^2 and s_2^2 are the unbiased estimates of σ_1^2 and σ_2^2 then $(s_1^2/\sigma_1^2)/(s_2^2/\sigma_2^2)$ has an F distribution with $N - 1$ and $M - 1$ degrees of freedom. In order to test the hypothesis

$$H_0: \sigma_1^2 = \sigma_2^2 \quad \text{or} \quad \sigma_1^2/\sigma_2^2 = 1,$$
$$H_A: \sigma_1^2 \neq \sigma_2^2 \quad \text{or} \quad \sigma_1^2/\sigma_2^2 \neq 1,$$

we can use Equation (8.8.2) and the discussion of Section 8.5 and reject H_0 if

$$s_1^2/s_2^2 > F_{1-\alpha/2},$$
$$s_1^2/s_2^2 < F_{\alpha/2}.$$

In order to test the one-sided alternatives

$$H_0: \sigma_1^2 = \sigma_2^2,$$
$$H_A: \sigma_1^2 > \sigma_2^2,$$

and

$$H_0: \sigma_1^2 = \sigma_2^2,$$
$$H_A: \sigma_1^2 < \sigma_2^2,$$

we reject H_0 if $s_1^2/s_2^2 > F_{1-\alpha}$ and if $s_1^2/s_2^2 < F_\alpha$, respectively.

Example 9.8.1 (Physics) In order to measure the carbon content of a chemical substance, two methods are available. The first is expensive but precise. The second is cheaper but suspected of being more variable or less precise. The two are compared by dividing a large amount of the chemical substance into 24 samples, and assigning 12 to each method of analysis. The results obtained were as follows:

	First method	Second method
No. of observations	$N = 12$	$M = 12$
Estimate of σ^2	$s_1^2 = .0413$	$s_2^2 = .0737$

Use a 5% level of significance to test that the second method is more variable than the first.

Solution

i) $H_0: \sigma_1^2 = \sigma_2^2,$
   $H_A: \sigma_1^2 < \sigma_2^2.$

ii) The test statistic is $s_1^2/s_2^2 = .5604$.

iii) We reject H_0 if $s_1^2/s_2^2 < F_\alpha$ where F_α is read from Appendix Table 6, with 11 degrees of freedom in the numerator and the denominator. From Equation (8.8.1) $F_\alpha = 1/F_{1-\alpha}$ or $F_{.05} = 1/F_{.95} = 1/2.82 = .355$. Since $s_1^2/s_2^2 > .355$, we cannot reject H_0. There is no evidence of greater variability in the second method of analysis.

9.9 TESTING THE PARAMETER OF A BINOMIAL DISTRIBUTION

We shall discuss the case of testing the parameter p of a binomial distribution when the sample size n is large enough for the normal approximation to be used. To test

$$H_0: p = p_0,$$
$$H_A: p \neq p_0$$

where X successes are obtained in n trials, we know that

$$Z = \frac{X/n - p_0}{\sqrt{p_0(1 - p_0)/n}}$$

has a unit normal distribution, so that an $\alpha\%$ significance level consists of rejecting H_0 if

$$Z > Z_{1 - \alpha/2}$$
$$Z < -Z_{1 - \alpha/2}. \qquad (9.9.1)$$

Example 9.9.1 (Medicine) A standard medication causes an improvement in 60% of the cases treated. A new medication gives an improvement in 150 out of 200 cases. Is the new medication different from the old? Use a 1% level test.

Solution

i) $H_0: p = .6,$
   $H_A: p \neq .6.$

ii) $n = 200$ $X = 150$, $p_0 = .6$, $\sqrt{p_0(1 - p_0)/n} = \sqrt{.6 \times .4/200} = .0346$.

iii) $Z = \dfrac{X/n - p_0}{\sqrt{p_0(1 - p_0)/n}} = \dfrac{.75 - .6}{.0346} = 4.34.$

iv) Using Equation (9.9.1) with $Z_{1 - \alpha/2} = Z_{.995} = 2.57$, we reject H_0. The new medication has a different effect.

9.10 TESTING THE DIFFERENCE OF TWO BINOMIAL PARAMETERS

If we are dealing with samples of sizes n and m from two binomial distributions, and X_1 and X_2 are the number of successes occurring with probabilities p_1 and p_2, respectively, we shall consider the test

$$H_0: p_1 = p_2 \quad \text{or} \quad p_1 - p_2 = 0,$$
$$H_A: p_1 \neq p_2 \quad \text{or} \quad p_1 - p_2 \neq 0,$$

in the case where n and m are large enough to use the normal approximation. Now

$$Z = \frac{(X_1/n - X_2/m) - (p_1 - p_2)}{\sqrt{p_1(1 - p_1)/n + p_2(1 - p_2)/m}}$$

has a unit normal distribution; so we reject H_0 at the α level of significance if

$$Z > Z_{1 - \alpha/2},$$
$$Z < -Z_{1 - \alpha/2}. \qquad (9.10.1)$$

As mentioned in Section 8.10, we may substitute X_1/n and X_2/m for p_1 and p_2 without seriously affecting the results.

Example 9.10.1 (Pollution) To measure the response to an anti-pollution program, 150 people in city A and 250 people in city B are asked if they would support such a program. In city A 100 people and in city B 150 people answer in the affirmative. Can we conclude in general that city B is less interested in the program than city A? Use $\alpha = .10$.

Solution

i) $H_0: p_A = p_B$ or $p_A - p_B = 0$,
   $H_A: p_A > p_B$.

ii) $n = 150, m = 250, X_A = 100, X_B = 150, X_A/n = \frac{2}{3}, X_B/m = \frac{3}{5}$.

iii) $\sqrt{\dfrac{p_A(1 - p_A)}{n} + \dfrac{p_B(1 - p_B)}{m}}$ is approximated by

$\sqrt{\dfrac{(\frac{2}{3})(\frac{1}{3})}{150} + \dfrac{(\frac{3}{5})(\frac{2}{5})}{250}} = \sqrt{.01429} = .12$.

iv) $z = \dfrac{(100/150 - 150/250)}{.12} = .56$.

v) Using Equation (9.10.1) we reject H_0 if $Z > Z_{1-\alpha/2} = Z_{.95}$ or if $Z < -Z_{1-\alpha/2}$. Since $z_{.95} = 1.28$, we cannot reject H_0. We cannot conclude that city B is less interested in the program than city A.

9.11 SUMMARY

The concept of a statistical hypothesis as an assumption about one or more of the parameters of a distribution is studied. One tests the truth of a null hypothesis as opposed to some alternative hypotheses by means of a test statistic based on the observations. Two types of error can result: type I error where the null hypothesis is wrongly rejected and type II error where the alternative hypothesis is wrongly rejected. The probability of a type I error is called the significance level of the test, and the probability of correctly rejecting the null hypothesis is called the power of the test. The idea of a critical region as those values of the test statistic for which the null hypothesis is rejected and an acceptance region are also studied. Both one-sided and two-sided tests are fully discussed, and the close relationship between tests of hypothesis and confidence intervals are brought out. Because of this close relationship, tests of hypothesis are discussed in an abbreviated form, the details having being discussed in Chapter 8.

EXERCISES

1. A transistor manufacturer claims that his product has 10% defectives. A sample of 15 transistors are examined and three are found to be defective. Would you reject his claim with $\alpha \leq .05$?

2. An experiment is set to test the hypothesis that a given coin is unbiased. The decision rule is the following. Accept the hypothesis if the number of heads in a sample of 200 tosses is between 90 and 110 inclusive. Reject the hypothesis otherwise.
   a) Find the probability of accepting the hypothesis when it is correct.
   b) Find the probability of rejecting the hypothesis when it is actually correct.

3. Set up a decision rule to test the hypothesis that a coin is unbiased if a sample of 100 tosses of the coin is taken and if the level of significance of .1 is used.

4. In an examination given to a large number of schools the mean grade is 75 with a standard deviation of 15. In a certain school of 100 students the mean grade is 72. Can one conclude at 5% level of significance that the standard of students in this school is inferior?

5. Suppose that the average height of American males is 71 inches. A sample of 100 male university students has an arithmetic mean $\bar{x} = 72.5$ inches and sample variance $s_x^2 = 25.0$. Can one claim that male university students are taller than average? Use $\alpha = .05$.

6. It is known in a pharmacological experiment that rats fed with a particular diet over a two month period gain an average of 40 grams in weight. A new diet is tried on a sample of 20 rats and their weight gain is 43 grams with an s^2 of 7 grams. Test the hypothesis at the 10% level of significance that the new diet is an improvement.

7. Suppose in a coin tossing game, you believe that the probability of a head is $\frac{2}{5}$ rather than $\frac{1}{2}$. Letting H_0 be that $p = \frac{1}{2}$, the coin is flipped 15 times, and X is the number of heads. If we reject H_0 for $X \leq 3$ find α. What is the smallest number of heads required in order to have a 10% level of significance? In this last case what is the power of the test?

8. A bushel is claimed to contain 10% of bad apples, but you believe that the actual figure is nearer 20%. You decide to test the claim by taking 20 apples and rejecting the 10% figure if X, the number of bad ones, is greater than 3. Find α and β for this experiment. Find α and β if you reject for greater than two bad ones out of 10 apples.

9. Consider students entering a typing course who have an average speed of 50 words per minute and known standard deviation $\sigma = 4$. At the end of the course the speeds for ten students are 57, 62, 48, 51, 63, 55, 44, 46, 59, 50. Use a 5% one-sided test that the course has been of value to them.

10. In Exercise 9, find the power of the test when μ in fact is equal to 55, 56, 57, 58, 60 words per minute. Plot the graph of the power.

11. If the grade of students in a statistics examination is known from experience to average 60 with $\sigma = 5$ and a new class of 30 students obtains an average of 64, can we say that the new class is brighter than average? Use $\alpha = .05$.

12. Do Exercise 9 without assuming that σ is known; that is, ignore the information that $\sigma = 4$.

13. A drug company claims that its medication takes effect within 10 minutes of administration. The following figures are the times that 12 patients took in order to feel the effects: 9.1, 13.2, 8.7, 8.1, 4.2, 9.5, 10.7, 15.3, 12.1, 14.0, 9.0, 12.7. Test the company's claim, with $\alpha = .01$.

14. If the pollution levels of a large city are known from past records to have an average of 50 (on an artificial scale); and if a new antipollution program is purported to have taken place giving 15 weekly pollution figures of 47, 56, 52, 51, 40, 44, 63, 48, 51, 59, 60, 55, 46, 53, and 55, then test the hypothesis that the program has indeed been effective. Use $\alpha = .05$.

15. Using the data of Exercise 9, test that σ is indeed equal to 4 as stated, using $\alpha = .05$.

16. In Exercise 13, test that the variance of the response is $\sigma^2 = 2$ against a one-sided alternative that $\sigma^2 > 2$, with $\alpha = .01$.

17. In Exercise 14, if the old figures had a variance of 50, are the new figures equally variable or not? Use $\alpha = .05$.

18. A machinist makes parts whose diameter variance is .01. A new machinist takes over the job and to test his performance, a sample of 20 parts have their diameters measured giving the following results: 1.3, 1.7, 1.4, 1.2, 1.3, 1.4, 1.1, 1.7, 1.3, 1.5, 1.4, 1.8, 1.6, 1.1, 1.6, 1.5, 1.4, 1.1, 1.6, 1.5. Is the new machinist more variable than the first?

19. If pulse rates of patients before and after surgery are assumed normal with variances 25 and 50, respectively, test that the mean pulse rate increases after surgery by considering the following 10 patients' results. Use $\alpha = .05$.

Patient	Before	After
1	69	72
2	75	73
3	68	78
4	71	81
5	73	70
6	77	75
7	70	83
8	65	74
9	60	75
10	74	70

20. In Exercise 19, do not assume that the variances are known, but assume that there is enough evidence to assume that they are equal. Test at the 5% level of significance that pulse rate increases after surgery.

21. Two methods A and B for teaching children how to read are compared by considering the comprehension, measured on a 1–20 scale, of 9 students using method A and 12 students using method B. Test that there is no difference between the two methods if the following results are obtained. Assume that the measurements are normal, that the variances are equal, and $\alpha = .05$.

Method A	15, 10, 13, 9, 6, 17, 14, 13, 10
Method B	9, 7, 10, 7, 15, 20, 13, 19, 7, 11, 14, 12

22. Suppose that we wish to compare two sets of normally distributed observations, obtained from two separate chemical analyses, even though we know that the variances of the two sets are unequal. Test the equality of the population means with the following data, and with $\alpha = .01$.

Method 1 (%)	11, 12, 14, 9, 7, 16, 14, 9, 10, 12
Method 2 (%)	17, 16, 24, 4, 8, 27, 31, 30, 27, 19, 6, 10, 20

23. In Exercise 21, use a 5% test to compare the variances of the two methods.

24. In Exercise 22, use a 1% test to compare the variances of the two methods.

25. An examination in mathematics was given to two sections of 100 and 120 students, respectively. The mean grade of the first section was 75 with a standard deviation of 15, whereas in the second section the mean grade was 80 with a standard deviation of 10. Is there a significant difference between the performance of the two sections at 1% level of significance if the unknown variances are assumed equal?

26. A machine producing ball bearings is considered to be properly set if it produces ball bearings of .50 inches in diameter. To check whether the machine is properly set, a sample of 17 bearings is chosen and the mean diameter is found to be .58 inches with a standard deviation .01 inches. Test the hypothesis that the machine is properly set.

27. On an examination in sociology a sample of 10 students in group A had a mean grade of 80 with a standard deviation of 7 while a sample of 15 students from group B was found to have a mean grade of 76 with a standard deviation of 5. Using a 5% level of significance, test the hypothesis that there is no significant difference between the two groups if the two unknown variances are assumed equal.

Exercises

28. The average earnings of samples from two separate industries (in dollars per hour) were as follows:

	Industry A	Industry B
Sample size	17	14
Mean ($\bar{x}$)	2.52	3.04
Variance (s^2)	1.00	1.83

Use 5% significance levels to test (a) that the variances are equal, (b) that the means are equal.

29. The I.Q.'s of two samples of mathematics majors at two separate universities were as follows:

University A: 124, 119, 150, 107, 127, 132, 117, 128, 160, 147, 133

University B: 146, 140, 112, 122, 131, 109, 127, 108, 144, 153, 105, 160, 151, 133, 130, 125

If the I.Q.'s are assumed to be normally distributed, test, at the 5% level of significance, that the variance of I.Q.'s at University B is greater than at University A.

30. The standard deviation of the lifetime of light bulbs produced by a company in the past was 125 hours. After changes were made in the machinery, a sample of 10 light bulbs showed a standard deviation of 90 hours. Investigate the significance of the apparent decrease in variability using a 5% level of significance.

31. A sample poll of 400 voters from Province A showed that 250 voted for a particular candidate, and that out of 350 from Province B, 170 favored the same candidate. At .01 level of significance, test the hypothesis that there is a difference in voting between the provinces.

32. The mean lifetime of a sample of 81 transistors produced by a company is computed to be 1535 days with a standard deviation of 60 days. If μ is the mean lifetime of all transistors produced by this company, test the null hypothesis $\mu = 1550$ days against $\mu \neq 1550$ days at .05 level of significance.

33. Suppose that 30% of the population smoke. It is suspected that the residents of a certain large city have a higher proportion of smokers among them. To test this, 1000 people are asked if they smoke to which 350 answer "yes." Test the suspicion using a one-sided 5% level test.

34. To compare the smoking habits of two areas, samples of 1000 and 1500 people, respectively, are asked if they smoke; 300 in the first and 500 in the second answer "yes." Test the hypothesis that there is no difference in the proportion of smokers for the two areas. Use $\alpha = .05$.

35. It is claimed that in a bushel of apples there are 10% defectives; 150 apples are examined and 30 are found to be bad. What would you conclude about the claim?

36. A poll taken with 1000 people before a candidate's campaign starts indicates that 600 people favor him. After his campaign another group of 2000 people shows that 1000 favor him. What would you say about his campaign efforts?

37. In Example 9.2.5, calculate the power of the test for $N = 15$ and $N = 25$.

Chapter 10

TESTS OF SIGNIFICANCE BASED ON CHI-SQUARE

10.1 INTRODUCTION

The chi-square distribution was introduced in Chapter 8, and the test for variance of a normal population was discussed in Chapter 9. In this chapter we turn to a different type of problem which deals with categorical, qualitative, or enumerative data. The difference is that we shall now assign the experimental units to specific categories and count the frequencies falling in specified categories. For example, in public opinion polls we count the number of favorable responses about a given issue; in medicine we count the number of patients who recover from a disease after being administered a certain therapeutic drug; in socio-economic studies we count the number of children in families at various income levels. There are many situations of the above type where the measurement data are changed to categorical data. The chi-square statistic, defined below, is a useful tool in determining whether a frequency with which a given event has occurred is significantly different from that which we expected.

Definition 10.1.1 A statistic which measures the discrepancy between K observed frequencies $o_1, o_2, \ldots, o_K$ and their corresponding expected frequencies $e_1, e_2, \ldots, e_K$, called the chi-square statistic, is defined in Table 10.1.1.

TABLE 10.1.1

Event	Observed frequency	Expected frequency
1	o_1	e_1
2	o_2	e_2
.	.	.
.	.	.
.	.	.
K	o_K	e_K

Tests of significance based on chi-square

$$\chi^2 = \frac{(o_1 - e_1)^2}{e_1} + \frac{(o_2 - e_2)^2}{e_2} + \cdots + \frac{(o_K - e_K)^2}{e_K} = \sum_{i=1}^{K} \frac{(o_i - e_i)^2}{e_i}.$$
(10.1.1)

If $\chi^2 = 0$, there is perfect agreement between observed and expected frequencies. The greater the discrepancy between observed and expected frequencies, the larger will be the value of chi-square.

The sampling distribution of the χ^2 statistic (10.1.1) is very closely approximated by the chi-square distribution defined in Chapter 8. In order to test the significance of χ^2 the calculated value of χ^2 is compared with the table value for the given degrees of freedom at a certain level of significance. If the calculated value of χ^2 is greater than the table value, the difference between the observed and expected frequencies is significant. On the other hand, if the calculated value of χ^2 is less than the table value, the difference between the observed and expected frequencies could have arisen due to chance fluctuation and is considered insignificant.

Conditions for the Application of the Chi-square Test

The following conditions must be satisfied in applying the chi-square test.

1. Each observation or frequency must be independent of all other observations.
2. The sample size N must be reasonably large in order that the difference between actual and expected frequencies be normally distributed. In practice we may say that N should be at least 50.
3. No expected cell frequency should be small. In practice 5 is regarded as the very minimum and 10 is considered better. In cases where the cell frequencies fall below these limits, they are amalgamated into a single cell.

Example 10.1.1 A die is rolled 120 times and the following observations are made:

Face	1	2	3	4	5	6
Observed frequency	25	15	12	26	15	27

Test the hypothesis that the die is honest using a level of significance of (a) .05, (b) .01.

Solution On the hypothesis that the die is honest, the probability of getting 1, 2, 3, 4, 5, or 6 is $\frac{1}{6}$. Hence the expected frequencies of getting

1, 2, 3, 4, 5, or 6 in 120 throws are 120 $(\frac{1}{6})$ = 20. The value of the χ^2 statistic is calculated in Table 10.1.2.

TABLE 10.1.2

Face	o	e	$o-e$	$(o-e)^2$	$(o-e)^2/e$
1	25	20	5	25	1.25
2	15	20	-5	25	1.25
3	12	20	-8	64	3.20
4	26	20	6	36	1.80
5	15	20	-5	25	1.25
6	27	20	7	49	2.45
				$\chi^2 =$	11.20

Since the number of events or categories is $K = 6$, the number of degrees of freedom $v = K - 1 = 6 - 1 = 5$.

a) Using a .05 level of significance, the value of χ^2 for 5 degrees of freedom from Appendix Table 5 is $\chi^2_{.95} = 11.07$. Since the calculated value of $\chi^2 = 11.20$ is greater than the table value, the hypothesis of an honest die is rejected.

b) Using a .01 level of significance, the value of χ^2 for 5 degrees of freedom from Appendix Table 5 is $\chi^2_{.99} = 15.09$. Since the calculated value of $\chi^2 = 11.20$ is less than the table value, the hypothesis of an honest die is accepted.

Example 10.1.2 (Sociology) In a survey of 400 families having 4 children, the frequency distribution of boys and girls is given as follows:

Number of boys	Number of girls	Number of families
4	0	23
3	1	98
2	2	165
1	3	87
0	4	27

Test whether the data are consistent with the hypothesis that the numbers of male and female births are equal.

Solution Let $p = P$ (a male birth) and $q = 1 - p = P$ (a female birth). Then the probabilities of 4 boys, 3 boys and 1 girl, 2 boys and 2 girls, 1 boy and 3 girls, and 4 girls are given by the successive terms of the binomial expansion

$$(p + q)^4 = p^4 + 4p^3q + 6p^2q^2 + 4pq^3 + q^4.$$

Since $p = q = \frac{1}{2}$, we have

$$P(4 \text{ boys}) = (\tfrac{1}{2})^4 = \tfrac{1}{16}.$$

Expected number of families with 4 boys $= 400 \times \tfrac{1}{16} = 25.$

$$P(3 \text{ boys and 1 girl}) = 4(\tfrac{1}{2})^3(\tfrac{1}{2}) = \tfrac{1}{4}.$$

Expected number of families with 3 boys and 1 girl $= 400 \times \tfrac{1}{4} = 100.$

$$P(2 \text{ boys and 2 girls}) = 6(\tfrac{1}{2})^2(\tfrac{1}{2})^2 = \tfrac{3}{8}.$$

Expected number of families with 2 boys and 2 girls $= 400 \times \tfrac{3}{8} = 150.$

$$P(1 \text{ boy and 3 girls}) = 4(\tfrac{1}{2})(\tfrac{1}{2})^3 = \tfrac{1}{4}.$$

Expected number of families with 1 boy and 3 girls $= 400 \times \tfrac{1}{4} = 100.$

$$P(4 \text{ girls}) = (\tfrac{1}{2})^4 = \tfrac{1}{16}.$$

Expected number of families with 4 girls $= 400 \times \tfrac{1}{16} = 25.$

The value of the χ^2 statistic is calculated in Table 10.1.3.

TABLE 10.1.3

o	e	$o - e$	$(o - e)^2$	$(o - e)^2 / e$
23	25	-2	4	.16
98	100	-2	4	.04
165	150	15	225	1.50
87	100	-13	169	1.69
27	25	2	4	.16
				$\chi^2 = 3.55$

Since the number of events or categories is $K = 5$, the degrees of freedom $v = K - 1 = 5 - 1 = 4$. Using a level of significance of .05, the value of χ^2 for 4 degrees of freedom from Appendix Table 5 is $\chi^2_{.95} = 9.49$. Since the calculated value of $\chi^2 = 3.55$ is less than the table value, the hypothesis that male and female births are equally likely is accepted.

10.2 GOODNESS OF FIT TEST

The chi-square statistic can be used to decide the goodness of fit. The term "goodness of fit" is used for the comparison of observed sample distributions or empirical distributions with expected or theoretical distributions such as the normal, binomial, Poisson etc. That is, the curve of the expected frequencies is superimposed on the curve of the observed frequencies, and the chi-square statistic determines whether the fit is good or not. We shall explain this procedure with the help of the following examples.

Example 10.2.1 (Biology) In 120 litters of 5 mice the number of litters which contained 0, 1, 2, 3, 4, 5 females were noted. The figures are given as follows:

Number of female mice (x)	0	1	2	3	4	5
Number of litters (f)	5	20	36	34	22	3

Fit a binomial distribution to this data and test the goodness of fit.

Solution

$$\bar{x} = \frac{\sum_{i=0}^{5} f_i x_i}{\sum_{i=0}^{5} f_i}$$

$$= \frac{0 \times 5 + 1 \times 20 + 2 \times 36 + 3 \times 34 + 4 \times 22 + 5 \times 3}{120}$$

$$= 2.475.$$

Therefore $5p = 2.475$, $p = .495$, and $q = 1 - p = .505$.
Hence the binomial distribution fitted to the above data is

$$f(x) = \binom{5}{x}(.505)^{5-x}(.495)^x, \quad x = 0, 1, 2, 3, 4, 5.$$

The expected frequencies and the value of the χ^2 statistic are calculated in Table 10.2.1.

Since the parameter p is estimated from the sample, the number of degrees of freedom is $v = K - r - 1 = 4 - 1 - 1 = 2$, where r is the number of parameters being estimated. Using a level of significance of .05, the value of χ^2 for 2 degrees of freedom from Appendix Table 5 is $\chi^2_{.95} = 5.99$. The calculated value of $\chi^2 = .9541$ is less than the table value $\chi^2_{.95} = 5.99$, hence our hypothesis is correct and the binomial fit is good.

TABLE 10.2.1

x	o		$e = 120\binom{5}{x}.505^{5-x}.495^x$		$o-e$	$(o-e)^2$	$(o-e)^2/e$
0	5 ⎫	25	3.9 ⎫	23.2	1.8	3.24	.1397
1	20 ⎭		19.3 ⎭				
2	36		37.8		−1.8	3.24	.0857
3	34		37.1		−3.1	9.61	.2590
4	22 ⎫	25	18.2 ⎫	21.8	3.2	10.24	.4697
5	3 ⎭		3.6 ⎭				
						$\chi^2 =$	.9541

Example 10.2.2 A typist commits the following number of mistakes per page in typing 100 pages. Fit a Poisson distribution and test the goodness of fit.

Mistakes per page (x)	0	1	2	3	4	5
Number of pages (f)	42	33	14	6	4	1

Solution The estimate of μ, the mean of the distribution

$$= \frac{\sum_{i=0}^{5} f_i x_i}{\sum_{i=0}^{5} f_i}$$

$$= \frac{0 \times 42 + 1 \times 33 + 2 \times 14 + 3 \times 6 + 4 \times 4 + 5 \times 1}{100}$$

$$= \tfrac{100}{100} = 1.$$

Hence the Poisson distribution fitted to the data is

$$f(x) = \frac{e^{-1}(1)^x}{x!}, \quad x = 0, 1, 2, 3, 4, 5.$$

The expected frequencies and the value of the χ^2 statistic are calculated in Table 10.2.2.

Since the parameter μ is estimated from the sample the number of degrees of freedom is $v = 4 - 1 - 1 = 2$. Using level of significance of .05, the value of $\chi^2_{.95} = 5.99$. The calculated value of $\chi^2 = 3.3958$ is less than the table value $\chi^2_{.95} = 5.99$. Hence our hypothesis is correct and the Poisson fit is good.

TABLE 10.2.2

x	o		$e = \dfrac{100 e^{-1}(1)^x}{x!}$		$o-e$	$(o-e)^2$	$(o-e)^2/e$
0	42		36.8		5.2	27.04	.7348
1	33		36.8		−3.8	14.44	.3924
2	14		18.4		−4.4	19.36	1.0521
3	6⎫		6.1⎫				
4	4⎬	11	1.5⎬	7.9	3.1	9.61	1.2165
5	1⎭		.3⎭				
						$\chi^2 =$	3.3958

Example 10.2.3 (Refer to Example 7.2.2) Examine the closeness of fit of the normal distribution to the data of 97 forty-day-old rats given in Table 7.2.2.

Solution Using formulas for grouped data from Chapter 7, the mean and the standard deviation of the distribution can be calculated as $\bar{x} = 150.0$ and $s = 15.92$. Hence the normal curve fitted to the data given in Table 7.2.2 is

$$f(x) = \frac{1}{\sqrt{2\pi (253.45)}} \exp\left[-\frac{(x-150.0)^2}{2(253.45)}\right].$$

Assuming normality, the probability that an individual will fall in the first class is equal to the area to the left of 123.5 under the normal curve. The probability that an individual will fall in the last class is equal to the area to the right of 186.5. The probability of an individual falling in a class other than the first and last can be found by computing the z-value for lower and upper class limits and then finding the area between the two z-values. Calculation of probability of an individual falling within a class and its corresponding expected frequency is shown in Table 10.2.3.

Hence

$$\chi^2 = \frac{(2 - 4.6)^2}{4.6} + \frac{(10 - 6.1)^2}{6.1} + \frac{(13 - 10.1)^2}{10.1} +$$

$$\frac{(11 - 14.4)^2}{14.4} + \frac{(17 - 16.8)^2}{16.8} + \frac{(15 - 16.1)^2}{16.1} +$$

$$\frac{(14 - 12.8)^2}{12.8} + \frac{(6 - 8.4)^2}{8.4} + \frac{(9 - 7.7)^2}{7.7}$$

$$= 1.470 + 2.493 + .833 + .803 + .002 + .075$$
$$\quad + .113 + .686 + .219$$
$$= 6.694.$$

TABLE 10.2.3

Class limits	Observed frequency	$z = \dfrac{x - \bar{x}}{s}$	Probability of an individual falling within class limits	Expected frequency
116.5–123.5	2	−1.67	.0475	4.6
123.5–130.5	10	−1.23	.0618	6.1
130.5–137.5	13	−.79	.1055	10.1
137.5–144.5	11	−.35	.1484	14.4
144.5–151.5	17	.09	.1727	16.8
151.5–158.5	15	.53	.1660	16.1
158.5–165.5	14	.97	.1321	12.8
165.5–172.5	6	1.41	.0867	8.4
172.5–179.5	4 ⎫	1.85	.0471	4.6 ⎫
179.5–186.5	4 ⎬ 9	2.29	.0212	2.1 ⎬ 7.7
186.5–193.5	1 ⎭	2.73	.0110	1.0 ⎭
Total	97		1.0000	97

Since the parameters μ and σ are estimated from the sample, the number of degrees of freedom is $v = 9 - 2 - 1 = 6$. Using a level of significance of .05, the value of χ^2 for 6 degrees of freedom from Appendix Table 5 is $\chi^2_{.95} = 12.95$. The calculated value $\chi^2 = 6.694$ is less than the table value $\chi^2_{.95} = 12.95$. Hence our hypothesis is correct and the normal fit is good.

10.3 TEST OF INDEPENDENCE: CONTINGENCY TABLES

The chi-square statistic defined in Section 10.1.1 can also be used to test the hypothesis of independence of two or more criteria of classification. More often we are interested to know whether or not two criteria may be considered to be independent of one another. For example, we might like to know the effect of immunization in controlling susceptibility to tuberculosis, we might study the attitude of male and female responses on a certain political issue. To study problems of this type we could take a random sample of size N and classify them according to two criteria. The observed frequencies can then be presented in the form of a table, defined below, known as a contingency table.

Definition 10.3.1 An $r \times c$ contingency table is an arrangement in which the data is classified into r classes $A_1, A_2, \ldots, A_r$ of attribute A and c classes $B_1, B_2, \ldots, B_c$ of attribute B. One attribute is entered in rows and the other in columns.

The $r \times c$ contingency table with r rows and c columns is presented in the Table 10.3.1.

TABLE 10.3.1

Rows	Columns				Row totals
	B_1	B_2	$\cdots$	B_c	
A_1	o_{11}	o_{12}	$\cdots$	o_{1c}	$N_{1.}$
A_2	o_{21}	o_{22}	$\cdots$	o_{2c}	$N_{2.}$
.	.	.		.	.
.	.	.		.	.
.	.	.		.	.
A_r	o_{r1}	o_{r2}	$\cdots$	o_{rc}	$N_{r.}$
Column totals	$N_{.1}$	$N_{.2}$	$\cdots$	$N_{.c}$	N

where
o_{ij} = observed frequency in the ith row and jth column,
$N_{i.}$ = total observed frequency in the ith row,
$N_{.j}$ = total observed frequency in the jth column,
N = the sample size.

Since entries in all cells except in one row and one column can be determined given the marginal totals, the number of degrees of freedom for an $r \times c$ contingency table is

$$v = (r-1)(c-1).$$

That is, $(r-1)(c-1)$ is the number of cells whose frequencies can be chosen arbitrarily.

The χ^2 statistic can be used to test whether there is any relationship between the two attributes. That is, the null hypothesis is that the two attributes are independent. We compute the statistic

$$\chi^2 = \sum_{i=1}^{r} \sum_{j=1}^{c} \frac{(o_{ij} - e_{ij})^2}{e_{ij}} \qquad (10.3.1)$$

where
o_{ij} = observed frequency in the ijth cell,
e_{ij} = expected frequency in the ijth cell,
$= \dfrac{N_{.i} \times N_{.j}}{N}$

and $\sum\sum o_{ij} = \sum\sum e_{ij}.$

228   Tests of significance based on chi-square

As before, the sampling distribution of the χ^2 statistic (10.3.1) is very closely approximated by the chi-square distribution defined in Chapter 8. In order to test the hypothesis of independence the calculated value of χ^2 is compared with the table value for given degrees of freedom at a certain level of significance. If the calculated value of χ^2 is greater than the table value, the hypothesis of independence is rejected.

We should remember that the distribution of χ^2 defined in Chapter 8 is a continuous distribution while the distribution of frequencies is discrete. In the past it was common to adjust the χ^2 statistic to take care of this discrepancy by the use of a correction factor called Yates' correction defined below.

Definition 10.3.2   The chi-square statistic (10.3.1) rewritten as

$$\chi^2 \text{ (adjusted)} = \sum_{i=1}^{r} \sum_{j=1}^{c} \frac{(|\,o_{ij} - e_{ij}\,| - .5)^2}{e_{ij}}$$

is defined as adjusted chi-square or corrected chi-square and is called Yates' correction for continuity.

Although the use of Yates' correction for continuity is very common, some recent work on this subject has cast doubt on the desirability of using the correction; that is, that it does more harm than good in terms of its effect on the correct probability statements. In the Example 10.3.1 we use Yates' correction for illustrative purposes, but we do not in general recommend its use.

Example 10.3.1   (Veterinary Medicine) (Refer to Example 1.1.5) An experiment to measure the effects of immunization of cattle against tuberculosis was performed by deliberately infecting both those cattle which had been inoculated and those which had not. The results are presented in the Table 10.3.2.

TABLE 10.3.2

	Died or seriously affected	Unaffected or slightly affected	Total
Inoculated	8	22	30
Not inoculated	10	4	14
Total	18	26	44

Examine the effect of immunization in controlling susceptibility to tuberculosis.

Test of independence: contingency tables

Solution On the hypothesis that vaccine has no effect in controlling susceptibility to tuberculosis, the expected frequencies may be calculated as follows:

$$e_{11} = \frac{30 \times 18}{44} = 12.3.$$

$$e_{12} = \frac{30 \times 26}{44} = 17.7.$$

$$e_{21} = \frac{14 \times 18}{44} = 5.7.$$

$$e_{22} = \frac{14 \times 26}{44} = 8.3.$$

Therefore

$$\begin{aligned}
\chi^2 &= \sum_{i=1}^{2} \sum_{j=1}^{2} \frac{(o_{ij} - e_{ij})^2}{e_{ij}} \\
&= \frac{(o_{11} - e_{11})^2}{e_{11}} + \frac{(o_{12} - e_{12})^2}{e_{12}} + \frac{(o_{21} - e_{21})^2}{e_{21}} + \frac{(o_{22} - e_{22})^2}{e_{22}} \\
&= \frac{(8 - 12.3)^2}{12.3} + \frac{(22 - 17.7)^2}{17.7} + \frac{(10 - 5.7)^2}{5.7} + \frac{(4 - 8.3)^2}{8.3} \\
&= \frac{(-4.3)^2}{12.3} + \frac{(4.3)^2}{17.7} + \frac{(4.3)^2}{5.7} + \frac{(-4.3)^2}{8.3} \\
&= \frac{18.49}{12.3} + \frac{18.49}{17.7} + \frac{18.49}{5.7} + \frac{18.49}{8.3} \\
&= 1.503 + 1.045 + 3.244 + 2.228 \\
&= 8.02.
\end{aligned}$$

The number of degrees of freedom for a 2×2 contingency table is

$$v = (r - 1)(c - 1) = (2 - 1)(2 - 1) = 1.$$

Using a level of significance of .05, the value of χ^2 for 1 degree of freedom from Appendix Table 5 is $\chi^2_{.95} = 3.84$. The calculated value $\chi^2 = 8.02$ is greater than the table value $\chi^2_{.95} = 3.84$. Hence our hypothesis of independence is rejected and the vaccine does have an effect in controlling susceptibility to tuberculosis.

Example 10.3.2 Work Example 10.3.1 using Yate's correction.

Solution Using Equation (10.3.2), we have

$$\chi^2 \text{ (adjusted)} = \sum_{i=1}^{2} \sum_{j=1}^{2} \frac{(|o_{ij} - e_{ij}| - .5)^2}{e_{ij}}$$

$$= \frac{(|8-12.3|-.5)^2}{12.3} + \frac{(|22-17.7|-.5)^2}{17.7}$$
$$+ \frac{(|10-5.7|-.5)^2}{5.7} + \frac{(|4-8.3|-.5)^2}{8.3}$$
$$= \frac{(3.8)^2}{12.3} + \frac{(3.8)^2}{17.7} + \frac{(3.8)^2}{5.7} + \frac{(3.8)^2}{8.3}$$
$$= \frac{14.44}{12.3} + \frac{14.44}{17.7} + \frac{14.44}{5.7} + \frac{14.44}{8.3}$$
$$= 1.1740 + .8158 + 2.5333 + 1.7398$$
$$= 6.2629.$$

The number of degrees of freedom for a 2×2 contingency table is
$$v = (r-1)(c-1) = (2-1)(2-1) = 1.$$
Using a level of significance of .05, the value of χ^2 for 1 degree of freedom from Appendix Table 5 is $\chi^2_{.95} = 3.84$. The calculated value $\chi^2 = 6.2629$ is greater than the table value $\chi^2_{.95} = 3.84$. Hence our hypothesis of independence is rejected and we arrive at the same conclusion as in Example 10.3.1.

Example 10.3.3 (Sociology) Two investigators A and B draw samples by different sampling techniques to estimate the number of persons falling in three income groups, poor, middle class, and well to do. The results are as follows:

Investigator	Poor	Middle class	Well to do	Total
A	66	40	14	120
B	35	28	7	70
Total	101	68	21	190

Test whether the sampling techniques are significantly different.

Solution On the hypothesis that the sampling techniques of both the investigators are the same, the expected frequencies may be calculated as
$$e_{11} = \frac{120 \times 101}{190} = 63.8, \quad e_{21} = \frac{70 \times 101}{190} = 37.2,$$
$$e_{12} = \frac{120 \times 68}{190} = 42.9, \quad e_{22} = \frac{70 \times 68}{190} = 25.1,$$

$$e_{13} = \frac{120 \times 21}{190} = 13.3, \quad e_{23} = \frac{70 \times 21}{190} = 7.7.$$

Therefore

$$\chi^2 = \sum_{i=1}^{2} \sum_{j=1}^{3} \frac{(o_{ij} - e_{ij})^2}{e_{ij}}$$

$$= \frac{(66 - 63.8)^2}{63.8} + \frac{(40 - 42.9)^2}{42.9} + \frac{(14 - 13.3)^2}{13.2}$$

$$+ \frac{(35 - 37.2)^2}{37.2} + \frac{(28 - 25.1)^2}{25.1} + \frac{(7 - 7.7)^2}{7.7}$$

$$= .0759 + .1960 + .0368 + .1301 + .3351 + .0636$$

$$= .8375.$$

The number of degrees of freedom for a 2 × 3 contingency table is

$$v = (r - 1)(c - 1) = (2 - 1)(3 - 1) = 2.$$

Using a level of significance of .05, the value of χ^2 for 2 degrees of freedom from Appendix Tables 5 is $\chi^2_{.95} = 5.991$. The calculated value $\chi^2 = .8375$ is less than the table value $\chi^2_{.95} = 5.991$. Hence our hypothesis that the sampling techniques of both the investigators is the same is accepted. That is, the present data does not show that the techniques used by investigators A and B are different.

10.4 SUMMARY

The sampling distribution of the chi-square statistic is very closely approximated by the chi-square distribution. The number of degrees of freedom associated with the chi-square distribution depends on the particular application. The chi-square statistic is useful in testing hypotheses of difference between the expected and observed frequencies. The chi-square statistic can be used to test the hypothesis of goodness of fit and the hypothesis of independence of two or more criteria of classification.

EXERCISES

1. In 100 tosses of a coin 65 heads and 35 tails were observed. Test the hypothesis that the coin is unbiased using a level of significance of .01.
2. In an experiment with peas one observed 280 round and yellow, 100 round and green, 92 wrinkled and yellow, and 25 wrinkled and green. According to the Mendelian theory of heredity the numbers should be in the ratio 9 : 3 : 3 : 1. Is there any evidence of difference from the theory at the .05 level of significance?
3. A survey of 200 families with three children showed the following distribution:

Number of boys and girls	3 boys 0 girls	2 boys 1 girl	1 boy 2 girls	0 boys 3 girls
Number of families	15	75	100	10

Is this result consistent with the hypothesis that male and female births are equally likely at .05 level of significance?

4. Six cents are tossed 64 times and the following number of heads were observed:

Number of heads	0	1	2	3	4	5	6
Frequency	1	9	10	25	12	5	2

Fit a binomial distribution to this data and test the goodness of the data at .05 level of significance.

5. The distribution of the number of deaths due to a rare disease in a certain year among the cities of a certain country is as follows:

Number of deaths	0	1	2	3	4	5	6
Number of townships	93	70	26	8	2	0	1

Test the goodness of fit of a Poisson distribution to this data.

6. A gambler suspects that a normal-looking die is biased. He throws the die 108 times and gets the following results:

Number of spots	1	2	3	4	5	6
Frequency	25	18	15	20	22	8

What decision would he make, using level of α of .05?

7. A forestry expert claims that birch, oak, and pine trees occur with equal frequency in a particular forest. He gives as evidence, the following data:

Type of tree	birch	oak	pine
Frequency	54	72	75

Test his claim at the 5% level of significance.

8. The number of mosquitoes caught in a mosquito-attracting device on six successive days were 26, 32, 9, 17, 41, 19. Test the hypothesis that the number of mosquitoes attracted to the device is the same from day to day.

9. Cities are categorized with regard to their air pollution problems as extremely bad, bad, average, good, extremely good. If of 200 autopsies in each city, the number of deaths from lung cancer is as follows: 32, 41, 17, 9, 2 test the hypothesis that air pollution does not affect the chance of a person dying from lung cancer.

10. In order to determine the relationship between the I.Q.'s of children and the mothers' ages at the time of their births, 144 children with I.Q.'s over 120 were selected, and yielded the following results:

Age of mother	less than 20	20–30	30–40	greater than 40
Number of children	45	41	37	21

Test the hypothesis that the mother's age does not affect the child's I.Q.

11. The following table gives the observed results when a new serum against a disease was tested in some animals.

	Affected	Not affected
Vaccinated	7	43
Not vaccinated	15	25

Test the hypothesis that the serum has no effect on the disease. Use $\alpha = .05$.

12. In a certain election the voters were divided into two classes, Rural and Urban.

	Rural	Urban
Candidate A	52	18
Candidate B	20	45

Test the hypothesis that voting for candidates A and B is independent of the classification of voters.

13. From the following data test the hypothesis that the division according to sex is independent of the results of the examination in mathematics.

234   Tests of significance based on chi-square

	Boys	Girls
Passed in mathematics	80	20
Failed in mathematics	40	30

14. Show that for a 2 × 2 contingency table where the frequencies are

a	b
c	d

χ^2 calculated from independent frequencies is

$$\chi^2 = \frac{(a+b+c+d)(ad-bc)^2}{(a+b)(c+d)(b+d)(a+c)}.$$

15. Show that for a 2 × 2 contingency table where the frequencies are

a	b
c	d

the value of adjusted χ^2 calculated from independent frequencies is

$$\chi^2 \text{ (adjusted)} = \frac{N(|ad-bc|-N/2)^2}{(a+b)(c+d)(b+d)(a+c)}$$

where $N = a+b+c+d$.

16. Five hundred individuals are classified according to sex and according to whether they are color blind.

	Male	Female
Color blind	20	5
Not color blind	240	235

Do these data indicate that color blindness depends on sex? Use $\alpha = .05$.

17. A group of 25 pigs suffering from a particular disease was used to test a serum. The serum was administered to 12 of the pigs and the other 13 were kept as a control. The following results were observed:

	Recovered	Died
Inoculated	3	10
Not inoculated	8	4

Test at 5% level of significance whether the serum has any effect on the disease.

18. The following table provides the effect of a kind of fumigation on the spoiling of fruits:

	Unspoiled	Spoiled
Fumigated	45	5
Unfumigated	40	10

Does the spoiled fruit depend on whether it has been fumigated or not? Use $\alpha = .05$.

19. A firm produces two kinds of water heaters. A year record shows that the following number of heaters were sold by two salesmen:

	Salesman 1	Salesman 2
Type I	240	80
Type II	180	100

Does this record support the claim that the salesman's ability to sell depends on the type he is selling? Use $\alpha = .05$.

20. The alumni association of the ABC university sent a set of questionaires to graduates of the university. The returns were classified as follows:

	Bachelors	Masters	Ph.D.
Returned	105	35	10
Not returned	33	5	2

Do these data indicate that the rate of return of the questionaires is independent of the degree earned at the ABC university? Use $\alpha = .05$.

Chapter 11

REGRESSION AND CORRELATION

11.1 INTRODUCTION

In this chapter we shall study the joint behavior of two or more variables. Very often an experimenter or a researcher is interested to find if there is a relationship between two or more variables and how this relationship can be expressed in the form of an equation. For example, an agricultural biologist might like to know the relationship between yield of wheat and the amount of fertilizer or between yield of wheat, the amount of fertilizer, and the number of inches of rainfall. An engineer might like to know the relationship between pressure and temperature. An economist might like to know the relationship between various levels of income and consumption expenditures in a certain community. Problems of the above type in predicting behavior of one variable knowing the other variable are known as regression problems. The known variable is defined as an independent variable and the other as the dependent variable.

To study the joint behaviour of two variables x and y, one must first collect data from a bivariate population.

Definition 11.1.1 Let $x_1, x_2, \ldots, x_N$ be the values of a variable x and let $y_1, y_2, \ldots, y_N$ be the corresponding values of a second variable y. Then a set of N pairs $(x_1, y_1), (x_2, y_2), \ldots, (x_N, y_N)$ is called bivariate data.

Example 11.1.1 Let $x_1, x_2, \ldots, x_{50}$ be the heights of 50 fathers in a certain city and let $y_1, y_2, \ldots, y_{50}$ be the heights of their oldest sons. Then the set of measurements $(x_1, y_1), (x_2, y_2), \ldots, (x_{50}, y_{50})$ is bivariate data.

Example 11.1.2 Let $x_1, x_2, \ldots, x_{20}$ be the achievement test score and let $y_1, y_2, \ldots, y_{20}$ be the final score of 20 students in a freshman course. Then the set of scores $(x_1, y_1), (x_2, y_2), \ldots, (x_{20}, y_{20})$ is bivariate data.

Example 11.1.3 Let $x_1, x_2, \ldots, x_{10}$ be the various levels of temperatures and let $y_1, y_2, \ldots, y_{10}$ be the corresponding pressures in a fixed volume of

Introduction

gas. Then the set of pairs $(x_1, y_1), (x_2, y_2), \ldots, (x_{10}, y_{10})$ is bivariate data.

To study a possible relationship between the pairs of values of x and y, we plot the values of x along the horizontal axis and the values of y along the vertical axis in a coordinate system. The resulting diagram known as a scatter diagram or dot diagram may reveal a linear relationship or a nonlinear relationship between the variables.

Definition 11.1.2 The graph of a set of N pairs $(x_1, y_1), (x_2, y_2), \ldots, (x_N, y_N)$ in a coordinate system is called a scatter diagram.

Example 11.1.4 The heights in inches of eight fathers and their oldest sons are given in the Table 11.1.1. Draw the scatter diagram.

TABLE 11.1.1

Height of father (x)	Height of oldest son (y)
63	65
64	67
70	69
72	70
65	64
67	68
68	71
66	63

Solution The scatter diagram is shown in Fig. 11.1.1.

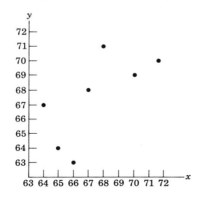

Figure 11.1.1

11.2 LINEAR REGRESSION

You should recall that every straight line has an equation of the form
$$y = a + bx \qquad (11.2.1)$$
where a is the y-intercept and b is the slope of the line. The y-intercept a and the slope of the line b completely determine the straight line.

We now return to the problem of finding a "best line" or a "best fit" to a given set of N points $(x_1, y_1), (x_2, y_2), \ldots, (x_N, y_N)$. In other words we want to determine the values of a and b in such a manner that the N points lie as close to the line as possible. There will be a difference between the actual y-values $y_1, y_2, \ldots, y_N$ and the corresponding values determined from the line. Such a difference is called an error, a residual, or a deviation. To distinguish the two y-values, we make the following definition.

Definition 11.2.1 Let the estimated or the predicted value of y obtained from the line of best fit be $\hat{y}$, then the equation
$$\hat{y} = a + bx \qquad (11.2.2)$$
is called the prediction equation or the regression equation of y on x and is an estimate of the population regression equation
$$y = \alpha + \beta x.$$

Definition 11.2.2 Let y_i be the actual y-value of the ith point and $\hat{y}_i$ be the predicted or the estimated value of the ith point, then the difference, denoted by e_i,
$$e_i = y_i - \hat{y}_i$$
is defined as the deviation or the error in the estimated and actual y-value.

The errors e_i which may be positive, negative, or zero are shown in Fig. 11.2.1.

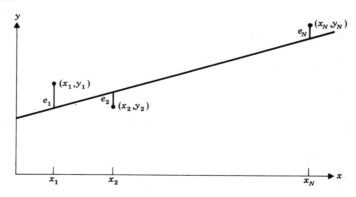

Figure 11.2.1

In finding the best line to the given set of points it seems intuitively reasonable to minimize the sum of squares of deviations or errors between the actual y-values and the estimated y-values. A method which accomplishes this criterion of "goodness of fit" is called the principle of least squares.

Definition 11.2.3 The method of least squares is the method of estimating the "best line" to a given set of N points in such a way that the sum of squares of the errors between the actual y-values and the estimated y-values is minimized.

Symbolically, we wish to minimize

$$\sum_{i=1}^{N} (y_i - \hat{y}_i)^2 = \sum_{i=1}^{N} e_i^2. \qquad (11.2.3)$$

The line $\hat{y} = a + bx$ is called the least squares line.

The method of least squares may be applied in fitting parabola, cubic, exponential, or any other curve to a given set of N points.

The next problem is to find the values of a and b in such a way that $\sum_{i=1}^{N} (y_i - \hat{y}_i)^2$ is minimized. The values of a and b can be determined using calculus, but are beyond the scope of this book. We state the following theorem without proof.

Theorem 11.2.1 Given N pairs of values of x and y, then the constants a and b of the equation

$$\hat{y} = a + bx$$

are solutions of two linear equations, called normal equations

$$aN + b \sum_{i=1}^{N} x_i = \sum_{i=1}^{N} y_i \qquad (11.2.4)$$

and

$$a \sum_{i=1}^{N} x_i + b \sum_{i=1}^{N} x_i^2 = \sum_{i=1}^{N} x_i y_i \qquad (11.2.5)$$

Theorem 11.2.2 The constants a and b of the equation $\hat{y} = a + bx$ are given by

$$b = \frac{N \sum_{i=1}^{N} x_i y_i - \sum_{i=1}^{N} x_i \sum_{i=1}^{N} y_i}{N \sum_{i=1}^{N} x_i^2 - (\sum_{i=1}^{N} x_i)^2} \qquad (11.2.6)$$

and

$$a = \bar{y} - b\bar{x} \qquad (11.2.7)$$

where $\bar{x}$ is the mean of the x-values and $\bar{y}$ is the mean of the y-values.

Proof. From Theorem 11.2.1 the constants a and b satisfy the two normal equations

$$aN + b \sum_{i=1}^{N} x_i = \sum_{i=1}^{N} y_i \qquad (11.2.8)$$

and

$$a \sum_{i=1}^{N} x_i + b \sum_{i=1}^{N} x_i^2 = \sum_{i=1}^{N} x_i y_i. \qquad (11.2.9)$$

Equation (11.2.8) and Equation (11.2.9) is a system of two linear equations in two unknowns. Multiplying Equation (11.2.8) by $\sum_{i=1}^{N} x_i$ and Equation (11.2.9) by N and then subtracting the first equation from the second, we obtain

$$Nb \sum_{i=1}^{N} x_i^2 - b(\sum_{i=1}^{N} x_i)^2 = N \sum_{i=1}^{N} x_i y_i - \sum_{i=1}^{N} x_i \sum_{i=1}^{N} y_i. \qquad (11.2.10)$$

Thus

$$b = \frac{N \sum_{i=1}^{N} x_i y_i - \sum_{i=1}^{N} x_i \sum_{i=1}^{N} y_i}{N \sum_{i=1}^{N} x_i^2 - (\sum_{i=1}^{N} x_i)^2}. \qquad (11.2.11)$$

Dividing numerator and denominator by N and replacing $\sum_{i=1}^{N} x_i/N$ by $\bar{x}$ and $\sum_{i=1}^{N} y_i/N$ by $\bar{y}$, Equation (11.2.11) can be written as

$$b = \frac{\sum_{i=1}^{N} x_i y_i - \sum_{i=1}^{N} x_i \sum_{i=1}^{N} y_i/N}{\sum_{i=1}^{N} x_i^2 - (\sum x_i)^2/N} = \frac{\sum_{i=1}^{N} x_i y_i - N\bar{x}\bar{y}}{\sum_{i=1}^{N} x_i^2 - N\bar{x}^2}. \qquad (11.2.12)$$

Alternatively, Equation (11.2.12) can be written as

$$b = \frac{\sum_{i=1}^{N} (x_i - \bar{x})(y_i - \bar{y})}{\sum_{i=1}^{N} (x_i - \bar{x})^2}. \qquad (11.2.13)$$

From Equation (11.2.8), a is given by

$$a = \frac{\sum_{i=1}^{N} y_i}{N} - b \frac{\sum_{i=1}^{N} x_i}{N} = \bar{y} - b\bar{x}$$

where b is given in Equation (11.2.12).

Example 11.2.1

a) Find the regression line of y on x for the data of Table 11.1.1.
b) From the data of Table 11.1.1 estimate the height of the oldest son for a father's height of 69 inches.
c) Graph the line obtained in (a).

Solution

a) Calculations of the constants a and b of the regression line

$$\hat{y} = a + bx$$

and other related quantities are shown in Table 11.2.1.

TABLE 11.2.1

	x	y	xy	x^2	y^2	$\hat{y}$	$y - \hat{y}$	$(y - \hat{y})^2$
	63	65	4095	3969	4225	64.55	.45	.2025
	64	67	4288	4096	4489	65.21	1.79	3.2041
	70	69	4830	4900	4761	69.20	− .20	.0400
	72	70	5040	5184	4900	70.53	− .53	.2809
	65	64	4160	4225	4096	65.88	− 1.88	3.5344
	67	68	4556	4489	4624	67.21	.79	.6241
	68	71	4828	4624	5041	67.87	3.13	9.7969
	66	63	4158	4356	3969	66.54	− 3.54	12.5316
Total	535	537	35955	35843	36105			30.2145

Since $\bar{x} = \frac{535}{8} = 66.875$ and $\bar{y} = \frac{537}{8} = 67.125$, from Equation (11.2.12), we obtain

$$b = \frac{\sum_{i=1}^{N} x_i y_i - N\bar{x}\bar{y}}{\sum_{i=1}^{N} x_i^2 - N\bar{x}^2} = \frac{35955 - 8(66.875)(67.125)}{35843 - 8(66.875)^2} = \frac{43.125}{64.875} = .6647$$

and from Equation (11.2.7)

$$a = \bar{y} - b\bar{x} = 67.125 - (.6647)(66.875)$$
$$= 67.1250 - 44.4518 = 22.6732.$$

Hence the equation of the regression line of y on x is

$$\hat{y} = 22.6732 + .6647x. \qquad (11.2.14)$$

b) Using the regression Equation (11.2.14) for $x = 69$, we find
$$\hat{y} = 22.6732 + .6647x$$
$$= 22.6732 + (.6647)(69)$$
$$= 68.5375.$$

c) In order to graph the regression line we need only two points fairly far apart. Using Equation (11.2.14) we find

$$\hat{y} = 22.6732 + (.6647)(62) = 63.8846, \qquad \text{when } x = 62.$$
$$\hat{y} = 22.6732 + (.6647)(69) = 68.5375, \qquad \text{when } x = 69.$$

The graph of the regression line
$$\hat{y} = 22.6732 + .6647x$$
is obtained by connecting two points (62, 63.8846) and (69, 68.5375) by a straight line. The scatter diagram of data given in Table 11.2.2 and the graph of the regression line (11.2.14) are shown in Fig. 11.2.2.

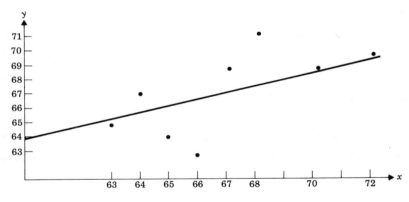

Figure 11.2.2

Theorem 11.2.3 The regression line $\hat{y} = a + bx$ passes through the point $(\bar{x}, \bar{y})$. In other words the point $(\bar{x}, \bar{y})$ always lies on the regression line.

Proof. The equation of the regression line of y on x is
$$\hat{y} = a + bx \qquad (11.2.15)$$
where $a = \bar{y} - b\bar{x}$.
Equation (11.2.15) can be rewritten as
$$\hat{y} = \bar{y} - b\bar{x} + bx.$$

Thus for $x = \bar{x}$ we get $\hat{y} = \bar{y}$. That is, the regression line $\hat{y} = a + bx$ passes through the point $(\bar{x}, \bar{y})$.

Example 11.2.2 Theorem 11.2.3 can be verified numerically for the data of Example 11.2.1. The regression line of y on x obtained in Example 11.2.1 is

$$\hat{y} = 22.6732 + .6647x.$$

For $x = \bar{x} = 66.875$, we find

$$\hat{y} = 22.6732 + (.6647)(66.875)$$
$$= 67.125 = \bar{y}.$$

So far we have considered the regression line of y on x. The regression line of y on x predicts the y-values given the x-values. In other words x is the independent variable and y is the dependent variable. There are many occasions when it is of interest to predict the x-values given the y-values. The regression line of x on y is different from the regression line of y on x. That is, y is the independent variable and x is the dependent variable. For example, from the data of Example 11.2.1 we might wish to determine a regression line to predict the height of a father given the height of an oldest son. We might also wish to compare the two regression lines. The regression line of x on y can be written as

$$\hat{x} = c + dy. \qquad (11.2.16)$$

To find the values of the constants c and d we apply the principle of least squares by minimizing $\sum_{i=1}^{N}(x_i - \hat{x})^2$. We note that the roles of x and y have been interchanged and the following theorems follow from Theorem 11.2.1 and Theorem 11.2.2.

Theorem 11.2.4 Given N pairs of values of x and y, the constants c and d of the equation $\hat{x} = c + dy$ are solutions of two linear equations, called normal equations

$$cN + d\sum_{i=1}^{N} y_i = \sum_{i=1}^{N} x_i \qquad (11.2.17)$$

$$c\sum_{i=1}^{N} y_i + d\sum_{i=1}^{N} y_i^2 = \sum_{i=1}^{N} x_i y_i. \qquad (11.2.18)$$

Theorem 11.2.5 The constants c and d of the equation $\hat{x} = c + dy$ are given by

$$d = \frac{N\sum_{i=1}^{N} x_i y_i - \sum_{i=1}^{N} x_i \sum_{i=1}^{N} y_i}{N\sum_{i=1}^{N} y_i^2 - (\sum_{i=1}^{N} y_i)^2} = \frac{\sum_{i=1}^{N} x_i y_i - N\bar{x}\bar{y}}{\sum_{i=1}^{N} y_i^2 - N\bar{y}^2} \qquad (11.2.19)$$

and

$$c = \bar{x} - d\bar{y} \qquad (11.2.20)$$

where $\bar{x}$ is the mean of the x-values and $\bar{y}$ is the mean of the y-values.

Example 11.2.3 Find the regression line of x on y for the data of Example 11.2.1. Use this equation to predict the height of the father when the height of the son is 66 inches.

Solution From Table 11.2.1 and using Equation (11.2.8) we obtain

$$d = \frac{\sum_{i=1}^{N} x_i y_i - N\bar{x}\bar{y}}{\sum_{i=1}^{N} y_i^2 - N\bar{y}^2} = \frac{43.125}{36105 - 8(67.125)^2} = \frac{43.125}{58.875} = .7325$$

and from Equation (11.2.20)

$$c = \bar{x} - d\bar{y}$$
$$= 66.875 - (.7325)(67.125)$$
$$= 17.7059.$$

Hence the equation of the regression line of x on y is

$$\hat{x} = 17.7059 + .7325y. \qquad (11.2.21)$$

For $y = 66$ we find

$$\hat{x} = 17.7059 + (.7325)(66)$$
$$= 66.0509.$$

11.3 PARTITIONING THE SUM OF SQUARES AND ESTIMATE OF VARIANCE

The estimated regression line

$$\hat{y} = a + bx \qquad (11.3.1)$$

is an estimate of the population regression line

$$y = \alpha + \beta x. \qquad (11.3.2)$$

Equation (11.3.1) is called the prediction equation and its main use is to predict a value of y, given a value of x. Since a and b are only estimates of α and β and these are estimated from a sample of N pairs of observations, they are subject to sampling variability. If we are to use $\hat{y}$ as an estimate of y we would like to know the degree of precision attached to $\hat{y}$, the estimated value of y. To answer this question we make the following definitions.

Definition 11.3.1 The sum of squares of the deviations of the actual y-values from their mean $\bar{y}$, denoted by S.S., is defined as

$$\text{S.S.} = \sum_{i=1}^{N} (y_i - \bar{y})^2. \qquad (11.3.3)$$

Definition 11.3.2 The sum of squares of the deviations of the predicted y-values from their mean $\bar{y}$, called the sum of squares for regression, denoted by S.S.R., is defined as

$$\text{S.S.R.} = \sum_{i=1}^{N} (\hat{y}_i - \bar{y})^2. \tag{11.3.4}$$

Definition 11.3.3 The sum of squares of the deviations of actual y-values from their corresponding predicted values on the regression line, denoted by S.S.E., is defined as

$$\text{S.S.E.} = \sum_{i=1}^{N} (y_i - \hat{y}_i)^2. \tag{11.3.5}$$

Now it can be shown that the total variation of y as defined in Equation (11.3.3) is equal to the sum of the variation of the predicted y-values $\hat{y}$ from their mean $\bar{y}$ and the variation of the actual y-values from their corresponding predicted values $\hat{y}$. We state this fact in the following theorem.

Theorem 11.3.1 If S.S., S.S.R., and S.S.E. are defined as in Definitions 11.3.1, 11.3.2, and 11.3.3, respectively, then the following relationship holds.

$$\text{S.S.} = \text{S.S.R.} + \text{S.S.E.} \tag{11.3.6}$$

Proof.

$$\begin{aligned}
\text{S.S.} &= \sum_{i=1}^{N} (y_i - \bar{y})^2 \\
&= \sum_{i=1}^{N} (y_i - \hat{y}_i + \hat{y}_i - \bar{y})^2 \\
&= \sum_{i=1}^{N} (y_i - \hat{y}_i)^2 + \sum_{i=1}^{N} (\hat{y}_i - \bar{y})^2 + 2 \sum_{i=1}^{N} (y_i - \hat{y}_i)(\hat{y}_i - \bar{y}).
\end{aligned} \tag{11.3.7}$$

Since from Equation (11.3.1)

$$\hat{y} = a + bx, \text{ where } a = \bar{y} - b\bar{x}, \text{ then } \hat{y} = \bar{y} - b\bar{x} + bx.$$

The last term of Equation (11.3.7) can be written as

$$\begin{aligned}
\sum_{i=1}^{N} (y_i - \hat{y}_i)(\hat{y}_i - \bar{y}) &= \sum_{i=1}^{N} (y_i - \bar{y} + b\bar{x} - bx_i)(\bar{y} - b\bar{x} + bx_i - \bar{y}) \\
&= b[\sum_{i=1}^{N} (x_i - \bar{x})(y_i - \bar{y}) - b \sum_{i=1}^{N} (x_i - \bar{x})^2].
\end{aligned} \tag{11.3.8}$$

Now using the value of b given in Equation (11.2.13) in the bracketed expression of Equation (11.3.8), we find

$$\sum_{i=1}^{N}(y_i-\hat{y}_i)(\hat{y}_i-\bar{y})=0.$$

Thus from Equation (11.3.8), we get

$$\text{S.S.} = \text{S.S.R.} + \text{S.S.E.}$$

The partitioning of the sum of squares discussed in Theorem 11.3.1 can be summarized in Table 11.3.1.

TABLE 11.3.1

Source of variation	Degrees of freedom	Sum of squares
Due to regression (S.S.R.)	1	$\sum_{i=1}^{N}(\hat{y}_i-\bar{y})^2$
Deviations from regression (S.S.E.)	$N-2$	$\sum_{i=1}^{N}(y_i-\hat{y}_i)^2$
Total (S.S.)	$N-1$	$\sum_{i=1}^{N}(y_i-\bar{y})^2$

We note from Table 11.3.1 the degrees of freedom as well as sum of squares, are partitioned.

Example 11.3.1 For the data given in Example 11.2.1 compute S.S., S.S.R. and S.S.E.

Solution Using calculations given in Table 11.2.1, we find

$$\text{S.S.} = \sum_{i=1}^{N}(y_i-\bar{y})^2 = \sum_{i=1}^{N} y_i^2 - N(\bar{y})^2$$
$$= 36105 - 8(67.125)^2 = 58.875$$

and

$$\text{S.S.R.} = \sum_{i=1}^{N}(\hat{y}_i-\bar{y})^2$$
$$= (-2.575)^2 + (-1.915)^2 + (2.075)^2 + (3.405)^2$$
$$+ (-1.245)^2 + (.085)^2 + (.745)^2 + (.585)^2$$
$$= 28.6518$$

and

$$\text{S.S.E.} = \sum_{i=1}^{N}(y_i-\hat{y})^2 = 30.2145.$$

To estimate σ^2, the variance of the random error for a regression line fitted to a sample of N pairs of x, y-values, we find that S.S.E. given in Definition 11.3.3 provides a measure of variation of actual y-values from the predicted y-values, $\hat{y}$. We use S.S.E. to make the following definitions.

Definition 11.3.4 The unbiased estimate of the variance of the regression line of y on x, denoted by $\hat{\sigma}^2$, is given by the quantity

$$\hat{\sigma}^2 = \frac{\text{S.S.E.}}{N-2} = \frac{\sum_{i=1}^{N}(y_i - \hat{y}_i)^2}{N-2}. \tag{11.3.9}$$

Definition 11.3.5 The standard error of estimate for a regression line of y on x is defined as

$$\hat{\sigma} = \sqrt{\frac{\sum(y_i - \hat{y}_i)^2}{N-2}}. \tag{11.3.10}$$

We note from Equation (11.3.9) that the nearer the points lie to a regression line, the smaller will be the standard error of estimate.

Example 11.3.2 Find the standard error of estimate for the data of Example 11.2.1.

Solution Using calculations from Table 11.2.2, we find

$$\hat{\sigma} = \sqrt{\frac{\sum(y_i - \hat{y}_i)^2}{N-2}} = \sqrt{\frac{30.2145}{6}} = \sqrt{5.0357} = 2.2439.$$

11.4 CONFIDENCE INTERVALS AND TESTS OF HYPOTHESES

Since a and b are estimates of α and β, estimated from sample data, they are subject to sampling variability. Let the different estimates of α and β based on different sample sizes be denoted by the random variables A and B. In order to test the hypotheses and for calculating confidence intervals for α and β we need to assume that the random variables y_1, y_2, ..., y_N are independent and normally distributed. Now it can be shown that A and B are unbiased estimates of α and β and the sampling distributions of A and B are normal with variances

$$\sigma_A^2 = \frac{\sigma^2 \left(\sum_{i=1}^{N} x_i^2\right)}{N \sum_{i=1}^{N}(x_i - \bar{x})^2} \tag{11.4.1}$$

and

$$\sigma_B^2 = \frac{\sigma^2}{\sum_{i=1}^{N}(x_i - \bar{x})^2}. \qquad (11.4.2)$$

Since the variance σ^2 is usually unknown, the estimator $\hat{\sigma}^2$ is used. Then

$$\hat{\sigma}_A^2 = \frac{\hat{\sigma}^2 \left(\sum_{i=1}^{N} x_i^2\right)}{N \sum_{i=1}^{N}(x_i - \bar{x})^2} \qquad (11.4.3)$$

and

$$\hat{\sigma}_B^2 = \frac{\hat{\sigma}^2}{\sum_{i=1}^{N}(x_i - \bar{x})^2}. \qquad (11.4.4)$$

To test the hypothesis $H_0: \alpha = \alpha_0$ against $H_A: \alpha \neq \alpha_0$, we use the statistic

$$t = \frac{a - \alpha_0}{\hat{\sigma}\sqrt{\sum x_i^2}/\sqrt{N \sum(x_i - \bar{x})^2}} \qquad (11.4.5)$$

which we know from Section 8.5 to have a t-distribution with $N-2$ degrees of freedom. The above discussion leads to us the following theorem.

Theorem 11.4.1 To test

$$H_0: \alpha = \alpha_0,$$
$$H_A: \alpha \neq \alpha_0$$

using a level of significance α, reject H_0 if

$$t = \frac{a - \alpha_0}{\hat{\sigma}\sqrt{\sum x_i^2}/\sqrt{N \sum(x_i - \bar{x})^2}} > t_{1-\alpha/2} \qquad (11.4.6)$$

or

$$t = \frac{a - \alpha_0}{\hat{\sigma}\sqrt{\sum x_i^2}/\sqrt{N \sum(x_i - \bar{x})^2}} < -t_{1-\alpha/2} \qquad (11.4.7)$$

Example 11.4.1 Using the estimated value $a = 22.673$ for the data of Example 11.2.1, test the hypothesis that $\alpha = 25$ at the .05 level of significance.

Solution $H_0: \alpha = 25,$
$H_A: \alpha \neq 25.$

Using Theorem 11.4.1 we get

Confidence intervals and tests of hypotheses 249

$$t = \frac{a - \alpha_0}{\hat{\sigma}\sqrt{\sum_{i=1}^{N} x_i^2} / \sqrt{N \sum_{i=1}^{N} (x_i - \bar{x})^2}}$$

$$= \frac{22.673 - 25}{\sqrt{5.035}\sqrt{35843}/\sqrt{8(64.875)}} = \frac{(-2.327)\sqrt{519}}{\sqrt{180469.505}}$$

$$= \frac{-53.0128}{424.8170} = -.1248.$$

From Appendix Table 4 we find $t_{.025} = -2.447$ and $t_{.975} = 2.447$ for $v = 6$ degrees of freedom. Since the computed value of $t = -.1248$ lies between $t_{.025} = -2.447$ and $t_{.975} = 2.447$, we are unable to reject the null hypothesis that $\alpha = 25$.

We can construct a confidence interval for the parameter α using the procedures of Chapter 8 for setting confidence intervals. A $(1 - \alpha)\ 100\%$ confidence interval for the parameter α is given by

$$P\left[-t_{1-\alpha/2} < (A - \alpha)\frac{\sqrt{N \sum_{i=1}^{N}(x_i - \bar{x})^2}}{\hat{\sigma}\sqrt{\sum_{i=1}^{N} x_i^2}} < t_{1-\alpha/2}\right] = 1 - \alpha. \quad (11.4.8)$$

After some algebraic manipulations inequality (11.4.8) can be written as

$$P\left[A - t_{1-\alpha/2}\frac{\hat{\sigma}\sqrt{\sum_{i=1}^{N} x_i^2}}{\sqrt{N \sum_{i=1}^{N}(x_i - \bar{x})^2}} < \alpha < A + t_{1-\alpha/2}\frac{\hat{\sigma}\sqrt{\sum_{i=1}^{N} x_i^2}}{\sqrt{N \sum_{i=1}^{N}(x_i - \bar{x})^2}}\right] = 1 - \alpha. \quad (11.4.9)$$

Thus for a given sample of size N, a confidence interval for the parameter α is

$$\left(a - t_{1-\alpha/2}\frac{\hat{\sigma}\sqrt{\sum_{i=1}^{N} x_i^2}}{\sqrt{N \sum_{i=1}^{N}(x_i - \bar{x})^2}},\ a + t_{1-\alpha/2}\frac{\hat{\sigma}\sqrt{\sum_{i=1}^{N} x_i^2}}{\sqrt{N \sum_{i=1}^{N}(x_i - \bar{x})^2}}\right)$$

where a and $\hat{\sigma}$ are the values of A and σ, respectively, for a given sample.

Example 11.4.2 Find the 95% confidence interval for α for the data of Example 11.2.1.

Solution Using the data of Table 11.2.2 and the estimated value $a = 22.673$ from Example 11.2.1, the 95% confidence interval for α is given by

$$22.673 - \frac{(2.447)(2.2439)\sqrt{35843}}{\sqrt{8[35843 - 8(66.875)^2]}} < \alpha < 22.673$$
$$+ \frac{(2.447)(2.2439)\sqrt{35843}}{\sqrt{8[35843 - 8(66.875)^2]}}$$

or
$$22.673 - 45.6304 < \alpha < 22.673 + 45.6304$$

or
$$-22.9574 < \alpha < 68.3034.$$

To test the hypothesis $H_0: \beta = \beta_0$ against $H_A: \beta \neq \beta_0$, we use the statistic

$$t = \frac{b - \beta_0}{\hat{\sigma}/\sqrt{\sum_{i=1}^{N}(x_i - \bar{x})^2}} \tag{11.4.10}$$

which we know from Section 8.5. to have a t-distribution with $N - 2$ degrees of freedom. The above discussion leads us to this theorem.

Theorem 11.4.2 To test
$$H_0: \beta = \beta_0,$$
$$H_A: \beta \neq \beta_0$$

using a level of significance α, reject H_0 if

$$t = \frac{b - \beta_0}{\hat{\sigma}/\sqrt{\sum_{i=1}^{N}(x_i - \bar{x})^2}} > t_{1-\alpha/2} \tag{11.4.11}$$

or
$$t = \frac{b - \beta_0}{\hat{\sigma}/\sqrt{\sum_{i=1}^{N}(x_i - \bar{x})^2}} < -t_{1-\alpha/2}. \tag{11.4.12}$$

Example 11.4.3 Using the estimated value of $b = .6647$ for the data of Example 11.2.1, test the hypothesis that $\beta = 0$ against $\beta > 0$ at the .05 level of significance.

Solution $H_0: \beta = 0$,
$H_A: \beta > 0$.

Using Theorem 11.4.2 we get

Confidence intervals and tests of hypotheses

$$t = \frac{(b - \beta_0)\sqrt{\sum_{i=1}^{N}(x_i - \bar{x})^2}}{\hat{\sigma}}$$

$$= \frac{(.6647 - 0)\sqrt{\sum_{i=1}^{N} x_i^2 - N\bar{x}^2}}{2.2439}$$

$$= \frac{.6647\sqrt{35843 - 8(66.875)^2}}{2.2439}$$

$$= \frac{(.6647)(8.0545)}{2.2439} = 2.3859.$$

From Appendix Table 4 we find $t_{.95} = 1.943$ for $v = 6$ degrees of freedom. Since the computed value of $t = 2.3859$ is greater than the table value $t_{.95} = 1.943$ we reject the null hypothesis that $\beta = 0$.

We can construct a confidence interval for the parameter β using the procedures of Chapter 8 for setting confidence intervals. A $100(1 - \alpha)\%$ confidence interval for the parameter β is given by

$$P\left[-t_{1-\alpha/2} < \frac{(B - \beta)\sqrt{\sum_{i=1}^{N}(x_i - \bar{x})^2}}{\hat{\sigma}} < t_{1-\alpha/2}\right] = 1 - \alpha \quad (11.4.13)$$

After some algebraic manipulations inequality (11.4.13) can be written as

$$P\left[B - t_{1-\alpha/2}\frac{\hat{\sigma}}{\sqrt{\sum_{i=1}^{N}(x_i - \bar{x})^2}} < \beta \right.$$

$$\left. < B + t_{1-\alpha/2}\frac{\hat{\sigma}}{\sqrt{\sum_{i=1}^{N}(x_i - \bar{x})^2}}\right] = 1 - \alpha. \quad (11.4.14)$$

Thus for a given sample of size N a confidence interval for the parameter β is

$$\left(b - t_{1-\alpha/2}\frac{\hat{\sigma}}{\sqrt{\sum_{i=1}^{N}(x_i - \bar{x})^2}},\ b + t_{1-\alpha/2}\frac{\hat{\sigma}}{\sqrt{\sum_{i=1}^{N}(x_i - \bar{x})^2}}\right)$$

where b and $\hat{\sigma}$ are the values of B and σ, respectively, for a given sample.

Example 11.4.4 Find the 95% confidence interval for β for the data of Example 11.2.1.

Solution Using the data of Table 11.2.2 and the estimated value $b = .6647$ from Example 11.2.1, the 95% confidence interval for β is given by

or
$$.6647 - \frac{(2.447)(2.2439)}{\sqrt{64.875}} < \beta < .6647 + \frac{(2.447)(2.2439)}{\sqrt{64.875}}$$

or
$$.6647 - .6817 < \beta < .6647 + .6817$$

$$-.0170 < \beta < 1.3464.$$

11.5 MULTIPLE LINEAR REGRESSION

A simple linear regression equation with one independent variable and one dependent variable can be easily extended to several independent variables. For example, let us assume that a variable y is a function of two independent variables x_1 and x_2 and we might wish to fit the relationship

$$y = a + bx_1 + cx_2. \tag{11.5.1}$$

Linear Equation (11.5.1) in three variables y, x_1, and x_2 represents a plane in three-dimensional space.

We now return to the problem of finding a "best plane" or "best fit" to a given set of N points, which are listed in Table 11.5.1 where (y_i, x_{1i}, x_{2i}) are the coordinates of the ith point. In other words we want to determine the values of a, b, and c in such a manner that the N points lie as close to the plane as possible. To distinguish between the actual y-values and the predicted y-values, we make the following definition.

TABLE 11.5.1

y	x_1	x_2
y_1	x_{11}	x_{21}
y_2	x_{12}	x_{22}
.	.	.
.	.	.
.	.	.
y_i	x_{1i}	x_{2i}
.	.	.
.	.	.
.	.	.
y_N	x_{1N}	x_{2N}

Definition 11.5.1 Let the estimated or the predicted value of y obtained from the plane of best fit be $\hat{y}$, then the equation

$$\hat{y} = a + bx_1 + cx_2 \tag{11.5.2}$$

is called the prediction equation or the regression equation of y on x_1 and x_2 and is an estimate of the population regression equation

$$y = \alpha + \beta x_1 + \gamma x_2. \quad (11.5.3)$$

The method of least squares can now be applied in finding the values of a, b, and c of Equation (11.5.2). The values of a, b, and c are determined, by minimizing $\sum_{i=1}^{N} (y_i - \hat{y}_i)^2$, in the same manner as for simple linear regression. This can be done using calculus, but is beyond the scope of this book so we state the following theorem without proof.

Theorem 11.5.1 Given N values of x_1, x_2, and y, the constants a, b and c of the equation

$$\hat{y} = a + bx_1 + cx_2 \quad (11.5.4)$$

are solutions of three linear equations, called normal equations

$$aN + b \sum_{i=1}^{N} x_{1i} + c \sum_{i=1}^{N} x_{2i} = \sum_{i=1}^{N} y_i. \quad (11.5.5)$$

$$a \sum_{i=1}^{N} x_{1i} + b \sum_{i=1}^{N} x_{1i}^2 + c \sum_{i=1}^{N} x_{1i} x_{2i} = \sum_{i=1}^{N} x_{1i} y_i. \quad (11.5.6)$$

$$a \sum_{i=1}^{N} x_{2i} + b \sum_{i=1}^{N} x_{1i} x_{2i} + c \sum_{i=1}^{N} x_{2i}^2 = \sum_{i=1}^{N} x_{2i} y_i. \quad (11.5.7)$$

The values of a, b, and c are obtained by solving the above set of equations.

Regression problems when a variable is dependent on two or more independent variables in a linear relationship are called problems of multiple linear regression. Theorem 11.5.1 can be easily generalized for N independent variables.

11.6 CORRELATION

In this section we consider another measure of association between the two variables called correlation. Regression analysis estimates the value of the dependent variable when the independent variable is known. Correlation measures the joint variation of two variables when both are random and neither is restricted. Correlation may be either positive or negative. If an increase in the value of one variable is associated with an increase in the value of the other variable, the two variables are said to be positively correlated. If an increase in the value of one variable is associated with a decrease in the value of the other variable, the variables are said to be negatively correlated. If there is no change in the value of

254   Regression and correlation

one variable as the other changes, we say that there is no correlation between them, that is they are uncorrelated. Note that "no correlation" does not mean the two variables are statistically independent. When we say the two variables have no correlation we mean the two variables have no linear correlation and the variables may have a non-linear relationship. A measure of linear correlation between the two variables is called the coefficient of correlation and is defined as follows.

Definition 11.6.1 Given N pairs of values of x and y, the sample correlation coefficient, denoted by r, is defined as

$$r = \frac{\sum_{i=1}^{N}(x_i - \bar{x})(y_i - \bar{y})}{\sqrt{\sum_{i=1}^{N}(x_i - \bar{x})^2}\sqrt{\sum_{i=1}^{N}(y_i - \bar{y})^2}} \quad (11.6.1)$$

Alternatively, the coefficient of correlation denoted by r, defined above, is given by

$$r = \frac{\sum_{i=1}^{N}(x_i - \bar{x})(y_i - \bar{y})}{N s_x s_y} \quad (11.6.2)$$

where

$$s_x = \sqrt{\frac{\sum(x_i - \bar{x})^2}{N-1}}, \quad s_y = \sqrt{\frac{\sum(y_i - \bar{y})^2}{N-1}}.$$

For purposes of computation Equation (11.6.1) may be rewritten as

$$r = \frac{\sum_{i=1}^{N} x_i y_i - N\bar{x}\bar{y}}{\sqrt{(\sum x_i^2 - N\bar{x}^2)(\sum y_i^2 - N\bar{y}^2)}} \quad (11.6.3)$$

Example 11.6.1 Compute r, the coefficient of correlation, for the data given in Example 11.2.1.

Solution Using calculations from Table 11.2.2, we get

$$r = \frac{\sum_{i=1}^{N} x_i y_i - N\bar{x}\bar{y}}{\sqrt{(\sum_{i=1}^{N} x_i^2 - N\bar{x}^2)(\sum_{i=1}^{N} y_i^2 - N\bar{y}^2)}}$$

$$= \frac{35955 - 8(66.875)(67.125)}{\sqrt{[35843 - 8(66.875)^2][36105 - 8(67.125)^2]}}$$

$$= \frac{43.125}{\sqrt{(64.875)(58.875)}} = \frac{43.125}{\sqrt{3819.5156}}$$

$$= \frac{43.125}{61.8022} = .6978.$$

In order to interpret the meaning of the coefficient of correlation we prove the following theorem.

Theorem 11.6.1 Given N pairs of values of x and y, the following relationship holds between the coefficient of correlation r, the sum of squares for regression and the total sum of squares

$$r^2 = \frac{\text{S.S.R.}}{\text{S.S.}} = \frac{\sum_{i=1}^{N}(\hat{y}_i - \bar{y})^2}{\sum_{i=1}^{N}(y_i - \bar{y})^2}. \tag{11.6.4}$$

Proof.

$$\frac{\text{S.S.R.}}{\text{S.S.}} = \frac{\sum_{i=1}^{N}(\hat{y}_i - \bar{y})^2}{\sum_{i=1}^{N}(y_i - \bar{y})^2} = \frac{\sum_{i=1}^{N}[\bar{y} + b(x_i - \bar{x}) - \bar{y}]^2}{\sum_{i=1}^{N}(y_i - \bar{y})^2}$$

$$= \frac{b^2 \sum_{i=1}^{N}(x_i - \bar{x})^2}{\sum_{i=1}^{N}(y_i - \bar{y})^2}$$

$$= \left[\frac{\sum_{i=1}^{N}(x_i - \bar{x})(y_i - \bar{y})}{\sum_{i=1}^{N}(x_i - \bar{x})^2}\right]^2 \frac{\sum_{i=1}^{N}(x_i - \bar{x})^2}{\sum_{i=1}^{N}(y_i - \bar{y})^2}$$

$$= \frac{[\sum_{i=1}^{N}(x_i - \bar{x})(y_i - \bar{y})]^2}{\sum_{i=1}^{N}(x_i - \bar{x})^2 \sum_{i=1}^{N}(y_i - \bar{y})^2} = r^2.$$

It can be seen from Equation (11.6.4) that r^2 is the proportion of total variance in y which is due to linear association between x and y.

Definition 11.6.2 The square of the coefficient of correlation is defined as the coefficient of determination.

Example 11.6.2 The coefficient of determination r^2 for the data of Example 11.2.1 is

$$r^2 = (.6978)^2 = .4869.$$

That is 48.69% of the variation in y is due to the linear regression of y on x.

Theorem 11.6.2 The coefficient of correlation r lies between -1 and 1.

Proof. From Theorem 11.6.1 we have

$$r^2 = \frac{\sum_{i=1}^{N} (\hat{y}_i - \bar{y})^2}{\sum_{i=1}^{N} (y_i - \bar{y})^2} = \frac{\text{S.S.R.}}{\text{S.S.}}. \qquad (11.6.5)$$

If all points of the scatter diagram lie on a line, then S.S.E. $= 0$ and from Theorem 11.3.1, S.S. $=$ S.S.R. Therefore using Equation (11.6.5) we get

$$r^2 = 1. \qquad (11.6.6)$$

If no linear relationship exists between the x and y values, then S.S.R. $= 0$. Therefore using Equation (11.6.5) we get

$$r^2 = 0. \qquad (11.6.7)$$

From Equations (11.6.6) and (11.6.7) it follows

$$0 \leq r^2 \leq 1. \qquad (11.6.8)$$

Thus the coefficient of correlation r satisfies the inequality

$$-1 \leq r \leq 1. \qquad (11.6.9)$$

Examples of $r = +1$, r approximately equal to -1 and r approximately equal to zero are shown in Figs. 11.6.1, 11.6.2, and 11.6.3, respectively.

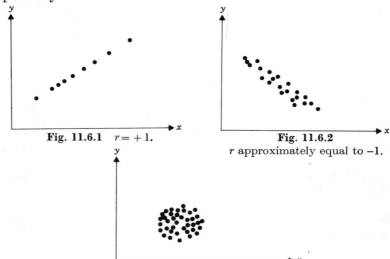

Fig. 11.6.1 $r = +1$.

Fig. 11.6.2 r approximately equal to -1.

Fig. 11.6.3 r approximately equal to 0.

11.7 SIGNIFICANCE OF THE COEFFICIENT OF CORRELATION

The sample coefficient of correlation r is an estimate of the population coefficient of correlation ρ. Like other statistics, the estimate r is subject to sampling variations. To test the significance of r we need the sampling distribution of r. Assuming that the samples are drawn from a bivariate normal distribution, it can be shown that the distribution of r is normal as the size of the sample increases. For $\rho = 0$, the distribution of r is symmetric and r is normally distributed even for smaller N. However, when $\rho \neq 0$, the distribution of r is skewed when N is small and r will be normally distributed only when N is very large.

To test the hypothesis $\rho = 0$, that is, to determine if r is significantly different from zero, we state the following theorem.

Theorem 11.7.1 If all samples of size N are drawn from a bivariate normal distribution with population coefficient of correlation $\rho = 0$ and if their sample coefficients of correlation are denoted by r, then

$$t = \frac{r}{\sqrt{(1 - r^2)/(N - 2)}} \tag{11.7.1}$$

has t-distribution with $N-2$ degrees of freedom.

Example 11.7.1 In a sample of 27 pairs the coefficient of correlation is found to be .25. Using a .05 level of significance, test the hypothesis that $\rho = 0$.

Solution $H_0: \rho = 0,$
$H_A: \rho > 0.$

Using Theorem 11.7.1 we find

$$t = \frac{r}{\sqrt{(1 - r^2)/(N - 2)}} = \frac{.25}{\sqrt{(1 - (.25)^2)/(27 - 2)}} = 1.2911.$$

From Appendix Table 4 we find $t_{.95} = 1.708$ for $v = 25$ degrees of freedom. Since the computed value of $t = 1.2911$ is less than the table value, we are unable to reject the null hypothesis that $\rho = 0$.

To test the hypothesis $\rho = \rho_0$, we need the distribution of r when $\rho \neq 0$. For $\rho \neq 0$ the distribution of r is skewed even for larger values of N. In such a case, R. A. Fisher suggested the following transformation, called Fisher's Z transformation.

$$Z = \tfrac{1}{2} \log_e \frac{1 + r}{1 - r}.$$

It can be shown that the distribution of Z is approximately normal. Thus we state the following theorem.

Theorem 11.7.2 If all samples of size N are drawn from a bivariate normal distribution with coefficient of correlation ρ, then the statistic

$$Z = \tfrac{1}{2} \log_e \frac{1+r}{1-r} \tag{11.7.2}$$

is approximately normally distributed with mean

$$\mu = \tfrac{1}{2} \log_e \frac{1+\rho}{1-\rho} \tag{11.7.3}$$

and standard deviation

$$\sigma = 1/\sqrt{N-3}. \tag{11.7.4}$$

Example 11.7.2 In a sample of 25 pairs the coefficient of correlation is found to be .75. Using a .05 level of significance, test the hypothesis that $\rho = .85$.

Solution $H_0: \rho = .85,$
$H_A: \rho \neq .85.$

From Appendix Table 9 we find $Z = .973$ for $r = .75$. Using the same table we find $\mu = 1.256$ for $\rho = .85$. Using Equation (11.7.4) we find

$$\sigma = \frac{1}{\sqrt{N-3}} = \frac{1}{\sqrt{22}} = .213.$$

Thus

$$z = \frac{.973 - 1.256}{.213} = 1.332.$$

Using a .05 level of significance and a two-sided test, we would reject H_0 if $z > 1.96$ or $z < -1.96$. Since the computed value of z is -1.332, we cannot reject H_0 that the population coefficient of correlation is $\rho = .85$.

Example 11.7.3 In a sample of 28 pairs, the coefficient of correlation is found to be .80. Find the 95% confidence limits for the population coefficient of correlation.

Solution From Appendix Table 9 we find $Z = 1.099$ for $r = .80$. Using Equation (11.7.4) we find

$$\sigma = \frac{1}{\sqrt{N-3}} = \tfrac{1}{5} = .2.$$

95% confidence limits for μ are

$$Z \pm 1.96\sigma = 1.099 \pm (1.96)(.2)$$
$$= 1.099 \pm .392.$$

Therefore μ lies between .707 and 1.491. From Appendix Table 9 we find $\rho = .61$ for $\mu = .707$ and $\rho = .90$ for $\mu = 1.491$. Thus 95% confidence limits for ρ are .61 and .90.

Example 11.7.4 Two coefficients of correlation, $r_1 = .60$, $r_2 = .50$ are obtained from samples of size $N_1 = 28$, $N_2 = 39$, respectively. Is there a significant difference between the population coefficients of correlation at a .05 level of significance?

Solution $H_0: \rho_1 = \rho_2,$
$H_A: \rho_1 \neq \rho_2.$

From Appendix Table 9 we find $Z_1 = .693$ for $r_1 = .60$ and $Z_2 = .549$ for $r_2 = .50$. The standard deviation of the difference $Z_1 - Z_2$ is given by

$$\sigma_{Z_1 - Z_2} = \sqrt{\frac{1}{N_1 - 3} + \frac{1}{N_2 - 3}} = \sqrt{\frac{1}{25} + \frac{1}{36}}$$
$$= \frac{\sqrt{61}}{30} = .260.$$

Thus
$$z = \frac{(Z_1 - Z_2) - (\mu_{Z_1} - \mu_{Z_2})}{\sigma_{Z_1 - Z_2}}$$
$$= .553.$$

Using a .05 level of significance and a two-sided test, we would reject H_0 if $z > 1.96$ or $z < -1.96$. Since the computed value of $z = .553$, we cannot reject H_0 that the population coefficients of correlation are equal.

11.8 SUMMARY

Problems concerning the prediction of a dependent random variable knowing the independent random variable are known as regression problems. To study a possible relationship between the pairs of x and y, we plot the values of x along the horizontal axis and the values of y along the vertical axis in a coordinate system. The resulting dot diagram is known as a scatter diagram. The method of least squares may be applied in fitting a straight line, cubic, exponential, or any other curve to a given set of N points.

The regression line of y on x predicts the y-values given the x-values. There are many occasions when it is of interest to predict the x-values given the y-values. The regression line of x on y is different from the regression line of y on x. The variance of the random error for a regression line fitted to a sample of N pairs of x, y values is given. In order to test the hypotheses and to calculate confidence intervals, we assume that the variables $y_1, y_2, \ldots, y_N$ are independent and normally distributed. A simple linear regression equation with one independent variable can be easily extended to several independent variables, and this is called multiple linear regression.

Regression analysis estimates the value of the dependent variable when the independent variable is known. Correlation measures the joint variation of two variables when both are random and neither is restricted. Correlation may be positive, negative, or zero. Note that "no correlation" does not mean the two variables are statistically independent. When we say the two variables have no correlation we mean the two variables have no linear correlation and the variables may have a nonlinear relationship. A linear correlation between the two variables is called the coefficient of correlation. The square of the coefficient of correlation is defined as the coefficient of determination.

EXERCISES

1. The following table gives the weights of a chemical compound y, dissolved in 100 grams of water at temperature $x°$ centigrade.

x	0	10	20	30	40	50	60
y	55	59	65	70	75	81	86

   a) Find the equation of the regression line of y on x.
   b) Use this equation to estimate y when $x = 55$.
   c) Draw the scatter diagram and graph the line.

2. The following table gives the ages in years, x, of a group of 10 persons and their blood pressures, y.

Age (x)	40	55	35	70	45	60	46	65	50	38
Blood pressure (y)	125	152	120	158	142	150	125	158	145	122

   a) Find the equation of the regression line of y on x.
   b) Use this equation to estimate the blood pressure of a person whose age is 56.

3. The data in the table show y, the stopping distance in feet, of a car travelling at x miles per hour on a particular surface.

x	20	25	30	40	50	60	70
y	16	27	42	55	72	98	140

a) Calculate the regression of y on x.
b) Test the hypothesis that $\alpha = -25$.
c) Test the hypothesis that $\beta = 2.5$.

4. The complexity of 8 mathematical problems were graded on a (linear) scale from 1 to 8 and the time taken by a particular student to do the problem (presented in random order) were as follows:

Problem (complexity) (y)	1	2	3	4	5	6	7	8
Time to solve (x)	19	28	25	38	41	49	55	62

Draw a scatter diagram, calculate the regression equation y on x and test that $\beta = 0$.

5. In a certain city, age and income were found to be related as follows:

Age (x)	31	47	25	60	65	37	34	40
Income (y)	5280	6300	5700	7425	7470	6200	6350	7000

a) Plot a scatter diagram.
b) Find the regression of y on x.
c) Estimate the income for age 50.
d) Does the line pass through $(\bar{x}, \bar{y})$?
e) Test that $\beta = 50$.
f) Calculate the correlation coefficient.
g) Find a 90% confidence interval for the population coefficient of correlation.

6. The height in centimeters, y, of a certain plant on age in weeks, x, after germination is given below:

x	1	2	3	4	5	6	7
y	4	15	22	30	37	44	50

Find the equation of the regression line of y on x.

7. In Exercise 4,
a) Find the equation of the regression line of x on y.
b) Calculate the coefficient of correlation.
c) Find the 95% confidence interval for the population coefficient of correlation.

8. The following table gives the height in inches, x, and the weight in pounds, y, of 13 individuals

$$\sum_{i=1}^{N} x_i = 78.0 \qquad \sum_{i=1}^{N} y_i = 1.196$$

$$\sum_{i=1}^{N} x_i^2 = 650.0 \qquad \sum_{i=1}^{N} y_i^2 = .156338$$
$$\sum_{i=1}^{N} x_i y_i = 9.911$$

a) Find the coefficient of correlation between height and weight.
b) Write down the regression equations.

9. In Exercise 1, test the hypothesis that $\beta = 0$, using a .05 level of significance.
10. In Exercise 1, test the hypothesis that $\alpha = 60$, using a .05 level of significance.
11. In Exercise 2, find 99% confidence intervals for α and β.
12. In Exercise 6, find 95% confidence intervals for α and β.
13. In Exercise 6, test the hypothesis that $\beta = 10$, using a .01 level of significance.
14. In Exercise 6, test the hypothesis that $\alpha = -3$, using a .01 level of significance.
15. A coefficient of correlation of .60 is found in a sample of 18 pairs. Can this be regarded as significantly different from zero on the basis of a 5% level of significance?
16. In a sample of 50, the coefficient of correlation is found to be .42. Find the 95% confidence interval for the population coefficient of correlation.
17. If in a sample of 53 pairs the coefficient of correlation is found to be .95, test the hypothesis that $\rho = .70$, using a .05 level of significance.
18. A sample is found to have a coefficient of correlation of .63. If the hypothesis that this sample was taken from a population with a coefficient of correlation of .49 is rejected at .05 level, find the smallest possible size of the sample.
19. What should be the maximum size of a sample for which a coefficient of correlation of .35 would not be considered significant at the 5% level?
20. The 99% confidence limits for the population coefficient of correlation are found to be .12 and .7. What should be the 95% confidence limits of the coefficient of correlation of this population?
21. If r is the coefficient of correlation between x and y, b is the slope of the regression line of y on x, and d the slope of the regression line of x on y, prove that $bd = r^2$.
22. In a sample the regression line of y on x is found to be
$$\hat{y} = 15 - \tfrac{8}{5}x$$
and of x on y is
$$\hat{x} = 18 - \tfrac{2}{5}y$$
Find the coefficient of correlation, $\bar{x}$, and $\bar{y}$ for this sample.

23. Two coefficients of correlation from samples of sizes 27 and 22 are found to be .85 and .75, respectively. Can we conclude that there is a significant difference between the two coefficients of correlation? Use $\alpha = .05$.

24. From a sample of size 32 the coefficient of correlation is found to be .65. Can we reject the hypothesis that the population coefficient of correlation is as large as $\rho = .80$, using a .01 level of significance?

Chapter 12
THE ANALYSIS OF VARIANCE

12.1 INTRODUCTION

In Section 9.7 the problem of testing the difference between the means of two populations was considered. In particular we considered testing $H_0: \mu_1 = \mu_2$ against $H_A: \mu_1 \neq \mu_2$. We now generalize this procedure to testing the equality of the means of several populations. Such studies could be important to the industrial engineer in comparing the output of several machines, to the agricultural scientist in comparing several feeds, or to the psychologist in comparing various testing procedures.

The technique used to make the comparison of the population means is called the analysis of variance and basically it compares the variation between population means with the variation within the populations. The reason for this is that even if there were no difference between the population means, all of the observations would not be the same because of the random nature of the measurements. Hence we think of the variation within a population, which after all has only one mean, as the inherent variation in the experiment and try to estimate its variance. We also obtain a similar estimate for the variability between the sample means, allowing for the fact that the variance between means is smaller than that between the original observations. If the variabilities within populations and between means are of the same order of magnitude, we conclude that there is no difference between the population means. If the variation between means is significantly larger than the within population variation, then we conclude that there is a difference between population means.

The analysis of variance is performed by considering, let us say, K populations with unknown means $\mu_1, \mu_2, \ldots, \mu_K$, and from each taking a random sample. The null hypothesis is

$$H_0: \mu_1 = \mu_2 = \cdots = \mu_K$$

and the alternative hypothesis is

$$H_A: \text{at least one of the equalities is violated.}$$

Definition 12.1.1 The analysis of variance is a technique for comparing the unknown means $\mu_1, \mu_2, \ldots, \mu_K$ of K populations by taking samples from each population and comparing the variation between populations with the variation within populations.

Example 12.1.1 (Education) In order to compare the mathematical abilities of children in four different school districts, samples of sizes 4, 3, 4, and 5 were chosen from districts 1, 2, 3, and 4 and given the same mathematics test. Table 12.1.1 gives scores (in %) that were obtained.

TABLE 12.1.1

	District			
	1	2	3	4
	47	51	68	63
	32	74	46	85
	63	70	49	80
	54		53	95
				82
Sums	196	195	216	405
Means	49	65	54	81
N_i	4	3	4	5

In due course we shall test the hypothesis that the children in the different districts have the same mathematical ability. In Example 12.1.1 where the populations are distinguished on the basis of a single criterion (e.g. districts), we talk of a one factor analysis of variance.

12.2 NOTATION

Before proceeding it will be necessary to define some notation for use in this chapter, concerning a one factor analysis of variance.

Definition 12.2.1

i) K = the number of populations.
ii) μ_i = the (unknown) mean of population i.
iii) N_i = the number of observations in sample i.
iv) N = the total number of observations.

v) x_{ij} = the jth observation in sample i.
vi) $T_{i.}$ = the sum of observations in sample i.
vii) $\bar{x}_{i.}$ = the mean of observations in sample i.
viii) $T_{..}$ = the total of all observations.
ix) $\bar{x}_{..}$ = the mean of all observations.
x) σ^2 = the common variance of all populations.

Example 12.2.1 In Example 12.1.1 we have

$N_1 = 4$, $N_2 = 3$, $N_3 = 4$, $N_4 = 5$, $N = 16$.
$x_{11} = 47$, $x_{12} = 32$, etc., $x_{45} = 82$.
$T_{1.} = 196$, $T_{2.} = 195$, $T_{3.} = 216$, $T_{4.} = 405$, $T_{..} = 1012$.
$\bar{x}_{1.} = 49$, $\bar{x}_{2.} = 65$, $\bar{x}_{3.} = 54$, $\bar{x}_{4.} = 81$, $\bar{x}_{..} = 63.25$.

Pictorially this may be shown as follows:

	Population 1	2	$\cdots$	K	
	x_{11}	x_{21}		x_{K1}	
	x_{12}	x_{22}		x_{K2}	
	.	.		.	
	.	.		.	
	.	.		.	
	x_{1N_1}	x_{2N_2}		x_{KN_K}	
Totals	$T_{1.}$	$T_{2.}$	$\cdots$	$T_{K.}$	$T_{..}$
Means	$\bar{x}_{1.}$	$\bar{x}_{2.}$	$\cdots$	$\bar{x}_{K.}$	$\bar{x}_{..}$

12.3 TESTING EQUALITY OF POPULATION MEANS

The total variation, the variation between population means, and the variation within population means are measured by three quantities, called sums of squares, defined as follows.

Definition 12.3.1 The total variability of the observations is measured by an expression called the "total sum of squares", denoted by S.S.T., whose formula is as follows:

$$\text{S.S.T.} = \sum_{i=1}^{K} \sum_{j=1}^{N_i} (x_{ij} - \bar{x}_{..})^2 \qquad (12.3.1)$$

Definition 12.3.2 The variability between population means is measured by an expression called the "between sum of squares," denoted by S.S.B., whose formula is as follows:

$$\text{S.S.B.} = \sum_{i=1}^{K} \sum_{j=1}^{N_i} (\bar{x}_{i.} - \bar{x}_{..})^2. \qquad (12.3.2)$$

Definition 12.3.3 The variability within populations is measured by an expression called the "within sum of squares," denoted by S.S.W., whose formula is as follows:

$$\text{S.S.W.} = \sum_{i=1}^{K} \sum_{j=1}^{N_i} (x_{ij} - \bar{x}_{i.})^2. \qquad (12.3.3)$$

Note that S.S.W. is the sum of squares of deviations of each sample from its own mean $\bar{x}_{i.}$, and then these squared deviations are added for all samples. If there were only one population and therefore only one sample, then S.S.W. would be the numerator of the equation used to calculate the estimate of the variance; or S.S.W./$(N_i - 1)$ would be s^2. However, since there are K populations and samples, we must divide by $(N_1 - 1) + (N_2 - 1) + \cdots + (N_K - 1) = N - K$. The resulting quantity is called the "within mean square" denoted by M.S.W. and is an unbiased estimate of σ^2, the variance of the observations.

Similarly, looking at Equation (12.3.2), we may write

$$\text{S.S.B.} = \sum_{i=1}^{K} N_i (\bar{x}_{i.} - \bar{x}_{..})^2$$

which is the sum of squares of deviations of the sample mean $\bar{x}_{i.}$ from the overall mean $\bar{x}_{..}$ weighted by N_i. This weighting is to compensate for the fact that the variance of the mean is $1/N_i$ as large as the variance of the original observations. Since there are K means, we know by appealing to Equation (7.7.1) that if there is no difference between the means, then the estimate of σ^2 is found by dividing S.S.B. by $K - 1$. The resulting quantity is called the "between mean square" and denoted by M.S.B.

Definition 12.3.4 In an analysis of variance with K populations based on N observations where S.S.B. and S.S.W. are defined in Definitions (12.3.2) and (12.3.3) the quantity called "between mean square" denoted by M.S.B. and the quantity "within mean square" denoted by M.S.W. are defined as follows:

$$\text{M.S.B.} = \text{S.S.B.}/(K - 1),$$
$$\text{M.S.W.} = \text{S.S.W.}/(N - K).$$

Below we state without proof an important theorem concerning M.S.B. and M.S.W.

Theorem 12.3.1

i) M.S.W. is an unbiased estimator of σ^2.
ii) When H_0 is true, M.S.B. is an unbiased estimator of σ^2.
iii) M.S.B. and M.S.W. are statistically independent.

Theorem 12.3.2 S.S.T. = S.S.B. + S.S.W.

Proof

$$\text{S.S.T.} = \sum_{i=1}^{K} \sum_{j=1}^{N_i} (x_{ij} - \bar{x}_{..})^2$$

$$= \sum_{i=1}^{K} \sum_{j=1}^{N_i} [(x_{ij} - \bar{x}_{i.}) + (\bar{x}_{i.} - \bar{x}_{..})]^2$$

$$= \sum_{i=1}^{K} \sum_{j=1}^{N_i} (x_{ij} - \bar{x}_{i.})^2 + \sum_{i=1}^{K} \sum_{j=1}^{N_i} (\bar{x}_{i.} - \bar{x}_{..})^2$$

$$+ 2 \sum_{i=1}^{K} \sum_{j=1}^{N_i} (x_{ij} - \bar{x}_{i.})(\bar{x}_{i.} - \bar{x}_{..}). \qquad (12.3.4)$$

The first term is S.S.W., the second is S.S.B. We now show that the third term is zero.

$$\sum_{i=1}^{K} \sum_{j=1}^{N_i} (x_{ij} - \bar{x}_{i.})(\bar{x}_{i.} - \bar{x}_{..}) = \sum_{i=1}^{K} (\bar{x}_{i.} - \bar{x}_{..}) \sum_{j=1}^{N_i} (x_{ij} - \bar{x}_{i.})$$

and

$$\sum_{j=1}^{N_i} (x_{ij} - \bar{x}_{i.}) = \sum_{j=1}^{N_i} x_{ij} - N_i \bar{x}_{i.} = T_{i.} - N_i \bar{x}_{i.}.$$

But

$$\bar{x}_{i.} = T_{i.}/N_i \quad \text{or} \quad T_{i.} = N_i \bar{x}_{i.}.$$

Thus the third term in Equation (12.3.4) is zero and the proof is complete.

In Section 7.7 we derived Equation (7.7.3) as the computational form of Equation (7.7.1). We may treat Equations (12.3.2) and (12.3.3) in a similar way.

Theorem 12.3.3

i) $\text{S.S.B.} = \sum_{i=1}^{K} \left(\frac{T_{i.}^2}{N_i} \right) - \frac{T_{..}^2}{N}$

ii) $\text{S.S.W.} = \sum_{i=1}^{K} \sum_{j=1}^{N_i} x_{ij}^2 - \sum_{i=1}^{K} \left(\frac{T_{i.}^2}{N_i} \right)$

Proof. We prove only part (i). By Equation (12.3.2)

$$S.S.B. = \sum_{i=1}^{K} \sum_{j=1}^{N_i} (\bar{x}_{i.} - \bar{x}_{..})^2$$

$$= \sum_{i=1}^{K} N_i(\bar{x}_{i.}^2 - 2\bar{x}_{i.}\bar{x}_{..} + \bar{x}_{..}^2)$$

$$= \sum_{i=1}^{K} N_i \bar{x}_{i.}^2 - 2\bar{x}_{..} \sum_{i=1}^{K} N_i \bar{x}_{i.} + N\bar{x}_{..}^2.$$

But

i) $\bar{x}_{i.} = T_{i.}/N_i$

ii) $\sum_{i=1}^{K} N_i \bar{x}_{i.} = T_{..}$

iii) $\bar{x}_{..} = T_{..}/N$

Putting (i), (ii), and (iii) into S.S.B. yields the desired result.

Example 12.3.1 For the data of Example 12.1.1:

$T_{1.}^2/N_1 = (196)^2/4 = 9604.$
$T_{2.}^2/N_2 = 12675.$
$T_{3.}^2/N_3 = 11664.$
$T_{4.}^2/N_4 = 32805.$
$T_{..}^2/N = (1012)^2/16 = 64009.$
S.S.B. $= 9604 + 12675 + 11664 + 32805 - 64009$
$\qquad = 66648 - 64009 = 2639.$
S.S.W. $= (47)^2 + (32)^2 + \ldots + (82)^2 - 66648$
$\qquad = 68388 - 66648 = 1740.$

If we wish to test $H_0: \mu_1 = \mu_2 = \cdots = \mu_K$, and if the null hypothesis is true, then we would expect $\bar{x}_{1.}, \bar{x}_{2.}, \ldots, \bar{x}_{K.}$ to be similar to each other and not very different from $\bar{x}_{..}$. This in turn would imply that S.S.B. and M.S.B. would be "small." By "small" we mean not very different from M.S.W. If on the other hand H_0 is not true, and at least one value of μ is different from the others, then from Equation (12.3.2), S.S.B. and M.S.B. will be large in comparison with S.S.W. and M.S.W. A fundamental question is "what do we mean by small and large?"

It can be shown that if the K populations are normal with common variance and if the observations are mutually independent, then the ratio M.S.B./M.S.W. follows an F-distribution with $K - 1$ degrees of freedom in the numerator and $N - K$ in the denominator. Further one can show that a desirable critical region is defined by the upper tail of the F-distribution. If we choose a level of significance α, then we reject H_0 if the ratio M.S.B./M.S.W. is larger than $F_{1-\alpha}$.

Theorem 12.3.4 To test that the means of K normal populations, each with common variance σ^2, are equal on the basis of independent samples from each population, using a $100\alpha\%$ level of significance, we reject the null hypothesis if the ratio M.S.B./M.S.W. is greater than $F_{1-\alpha}$, where $F_{1-\alpha}$ is read from the F-tables with $K-1$ degrees of freedom in the numerator and $N-K$ degrees of freedom in the denominator.

The results may be summarized in an analysis of variance table similar to Table 12.3.1.

TABLE 12.3.1

Source of variation	Sum of squares	Degrees of freedom	Mean squares	Ratio
Between population	S.S.B. $= \sum_{i=1}^{K} \left(\dfrac{T_{i.}^2}{N_i}\right) - \dfrac{T_{..}^2}{N}$	$K-1$	$\dfrac{\text{S.S.B.}}{K-1} = \text{M.S.B.}$	$\dfrac{\text{M.S.B.}}{\text{M.S.W.}}$
Within population	S.S.W. $= \sum_{i=1}^{K}\sum_{j=1}^{N_i} x_{ij}^2 - \sum_{i=1}^{K}\left(\dfrac{T_{i.}^2}{N_i}\right)$	$N-K$	$\dfrac{\text{S.S.W.}}{N-K} = \text{M.S.W.}$	
Total	$\sum_{i=1}^{K}\sum_{j=1}^{N_i}(x_{ij}-\bar{x}_{..})^2$	$N-1$		

Example 12.3.2 The analysis of variance table for the data of Example 12.1.1 is shown in Table 12.3.2.

TABLE 12.3.2

Source of variation	Sum of squares	Degrees of freedom	Mean squares	Ratio
Between districts	2639	3	879.67	6.07
Within districts	1740	12	145.00	
Total	4379	15		

If we wish to test the null hypothesis that the children in the different districts have the same mathematical ability using $\alpha = .05$, then we read from Appendix Table 6, that $F_{1-\alpha} = F_{.95} = 3.49$. Since M.S.B./M.S.W. $> F_{.95}$, reject H_0, and conclude that there are differences between children in different school districts.

Example 12.3.3 (Pharmacology) Three new drugs are put forward as comparable analgesics, and it is desired to test them. The variable measured is the number of hours before relief begins to take place. In an initial study the following results are obtained:

	Drug 1	Drug 2	Drug 3
	.6	1.2	1.5
	1.1	1.0	1.2
	1.0	.9	1.3
	.7	.9	1.2
Sum	3.4	4.0	5.2

We wish to know if the effects of the drugs are significantly different at the 5% level of significance.

Solution

i) $H_0: \mu_1 = \mu_2 = \mu_3$,
H_A: at least one μ_i is different from the others.

ii) $N_1 = 4$, $N_2 = 4$, $N_3 = 4$, $N = 12$, $K = 3$.

iii) $T_{1.} = 3.4$, $T_{2.} = 4.0$, $T_{3.} = 5.2$, $T_{..} = 12.6$.

iv) S.S.B. $= \dfrac{(3.4)^2}{4} + \dfrac{(4.0)^2}{4} + \dfrac{(5.2)^2}{4} - \dfrac{(12.6)^2}{12} = 13.65 - 13.23 = .42$,

S.S.W. $= (.6)^2 + (1.1)^2 + \cdots + (1.2)^2 - 13.65 = 13.94 - 13.65 = .29$.

v) M.S.B. $= .42/2 = .21$,
M.S.W. $= .29/9 = .032$.

vi) The analysis of variance table

Source of variation	Sum of squares	Degrees of freedom	Mean squares	Ratio
Between drugs	.42	2	.21	6.56
Within drugs	.29	9	.032	
Total	.71	11		

vii) We reject H_0 if the ratio is greater than $F_{1-\alpha} = F_{.95}$. Since $F_{.95} =$ 4.26, reject H_0. The drugs do not have the same effects.

12.4 MULTIPLE COMPARISONS

If H_0 is rejected, all we can conclude is that some of the K population means differ from others. In practice, the experimenter wishes to know which population means are different, and even to rank them according to their magnitude. A few procedures with this aim will now be considered.

In the past if H_0 was rejected, then every pair of means was compared by means of a t-test. It can be shown that if each t-test is made at the 5% level of significance, then the overall level of significance is considerably higher than 5%. Hence the multiple t-test approach may be regarded as completely unsatisfactory. To overcome this difficulty, Newman put forward a method, later modified by Keuls, known as the Newman-Keuls multiple range test which, while making several comparisons at once, kept the significance level fixed at α. Duncan regarded the Newman-Keuls method as too strict in the sense that true differences between the means will be missed too often, and he developed a modified multiple range test, in which, if we compare any subset of p means, $2 \leq p \leq K$, then the significance level of wrongly rejecting H_0 is $\gamma_p = 1 - (1-\alpha)^{p-1}$. If $\alpha = .05$, $p = 5$, then the probability of wrongly rejecting H_0 is .1855. Duncan's method, when each mean is computed from the same number of observations M is as follows.

i) Arrange the K means in ascending order and denote them by $\bar{x}_{(1)}$, $\bar{x}_{(2)}, \ldots, \bar{x}_{(K)}$ so that $\bar{x}_{(1)} < \bar{x}_{(2)} < \cdots < \bar{x}_{(K)}$.

ii) Read the value "critical difference multiplier" $r_{p,\,v'}$, for $p = 2, 3, 4, \ldots, K$ from Appendix Table 10 where $v = N - K = K(M-1)$.

iii) Calculate the least significant ranges
$$R_p = r_p \sqrt{\text{M.S.W.}/M}, \qquad p = 2, 3, \ldots, K.$$

iv) Compute the differences of all pairs of adjacent means and compare them to R_2. If such differences are less than R_2, draw lines under the corresponding means.

v) Consider all sets of three successive means, and compute the difference of the two extreme means in each set. If any differences are less than R_3, draw lines under the three means in question.

vi) Continue this process until the set of K means is compared to R_K.

Any means which have a common line under them are regarded as not significantly different.

Example 12.4.1 Apply the Duncan Multiple Range Test to Example 12.3.3 using $\alpha = .05$.

Solution

i) The ordered means are as follows:
   Drug 1   Drug 2   Drug 3
   .85   1.0   1.3

ii) $p = 2$ 3,
    $r_p = 3.20$ 3.34.

iii) $p = 2$ 3,
     $R_p = .287$ .299.

iv) The differences of adjacent means are .15, .3. The differences of the extremes in the set of three is .45.

v) Drawing lines under the appropriate means gives
   Drug 1   Drug 2   Drug 3
   .85   1.0   1.3

vi) We conclude that drugs 1 and 2 are not significantly different from each other but both are different from drug 3.

12.5 THE ANALYSIS OF VARIANCE WITH TWO FACTORS

The analysis of variance can be used to measure the effects of two or more variables on a single response in one experiment. More specifically, by a judicious choice of levels of the variables, we can measure the effect of each factor separately, and further we can measure the effects of combinations of two or more variables acting together.

In this text we shall only consider experiments consisting of two factors, which for convenience we shall call rows and columns. We shall test the hypotheses (a) that the row effects are equal, and (b) that the column effects are equal, and (c) that the row and column effects do not interact. By saying that rows and columns do not interact we mean that the effect of rows will be the same for each column and that the effect of columns will be the same for each row.

Example 12.5.1 (Industry) Suppose that in comparing two factors for instance, for the amount of carbon added (at levels .01% and .005%) and temperature (at temperatures A and B) on the quality of steel (measured on an artificial scale) we get the following results.

	Temperature A	Temperature B
.005% Carbon	17	24
.01% Carbon	32	39

We see that changing from temperature A to temperature B gives an increase in quality of 7 units, at both levels of carbon added. Similarly, changing from .005% to .01% carbon added increases the quality by 15 units, no matter at which temperature it is done. In this case we would say that the two factors, temperature and carbon added, do not interact in their effects on quality. If the effect of an increase in temperature depended on the level of carbon added, we would say that the two factors interact. Because the notation must be expanded slightly from Section 12.2, we will use the following definition for the analysis of two factors.

Definition 12.5.1

R = the number of rows.
C = the number of columns.
N_{ij} = the number of observations in the cell defined by row i and column j.
N = the total number of observations.
x_{ijk} = the kth observation in cell ij.
$\bar{x}_{ij.}$ = the mean of observations in cell ij.
$\bar{x}_{i..}$ = the mean of observations in row i.
$\bar{x}_{.j.}$ = the mean of observations in column j.
$\bar{x}_{...}$ = the mean of all observations.
R_i = the sum of observations in row i.
C_j = the sum of observations in column j.
T_{ij} = the sum of observations in cell ij.
$T_{...}$ = the total of all observations.

Pictorially, Definition 12.5.1 may be shown as follows.

	Column 1	Column 2		Column C	Total	Mean
Row 1	x_{111} $\vdots$ $x_{11 N_{11}}$	x_{121} $\vdots$ $x_{12 N_{12}}$	$\cdots$	x_{1C1} $\vdots$ $x_{1C N_{1C}}$	R_1	$\bar{x}_{1..}$
Row 2	x_{211} $\vdots$ $x_{21 N_{21}}$	x_{221} $\vdots$ $x_{22 N_{22}}$			R_2	$\bar{x}_{2..}$
$\vdots$					$\vdots$	$\vdots$
Row r	x_{r11} $\vdots$ $x_{r1 N_{r1}}$	x_{r21} $\vdots$ $x_{r2 N_{r2}}$		x_{rC1} $\vdots$ $x_{rC N_{rC}}$	R_r	$\bar{x}_{r..}$
Total	C_1	C_2	$\cdots$	C_C	$T_{...}$	
Mean	$\bar{x}_{.1.}$	$\bar{x}_{.2.}$	$\cdots$	$\bar{x}_{.C.}$		$\bar{x}_{...}$

We shall consider only the case that N_{ij} is the same for each cell, that is, for $N_{ij} = M$ for all i and j. In analogy with Section 12.3 we have the following definitions.

Definition 12.5.2 The total variability of the observations is measured by an expression called the "total sum of squares" denoted by S.S.T., whose formula is as follows:

$$\text{S.S.T.} = \sum_{i=1}^{R} \sum_{j=1}^{C} \sum_{k=1}^{M} (x_{ijk} - \bar{x}_{...})^2 \quad (12.5.1)$$

Definition 12.5.3 The variabilities of rows and columns are measured by expressions called the "row sum of squares" and "column sum of squares,": denoted by S.S.R. and S.S.C., respectively, whose formulas are as follows:

$$\text{S.S.R.} = \sum_{i=1}^{R} \sum_{j=1}^{C} \sum_{k=1}^{M} (\bar{x}_{i..} - \bar{x}_{...})^2 = CM \sum_{i=1}^{R} (\bar{x}_{i..} - \bar{x}_{...})^2, \quad (12.5.2)$$

$$\text{S.S.C.} = \sum_{i=1}^{R} \sum_{j=1}^{C} \sum_{k=1}^{M} (\bar{x}_{\cdot j \cdot} - \bar{x}_{...})^2 = RM \sum_{j=1}^{C} (\bar{x}_{\cdot j \cdot} - \bar{x}_{...})^2. \quad (12.5.3)$$

Definition 12.5.4 The interaction is measured by an expression called the "interaction sum of squares" denoted by S.S.I., whose formula is as follows:

$$\text{S.S.I.} = \sum_{i=1}^{R} \sum_{j=1}^{C} \sum_{k=1}^{M} (\bar{x}_{ij\cdot} - \bar{x}_{i..} - \bar{x}_{\cdot j \cdot} + \bar{x}_{...})^2. \quad (12.5.4)$$

Note from Definition 12.5.4 that the expression inside the parentheses can be written as $(\bar{x}_{ij\cdot} - \bar{x}_{i..}) - (\bar{x}_{\cdot j \cdot} - \bar{x}_{...})$ which, from the earlier discussion on the meaning of interaction, will on the average be zero if there is no interaction.

Definition 12.5.5 The variability within the cells is measured by an expression called the "within sum of squares", denoted by S.S.W. whose formula is as follows:

$$\text{S.S.W.} = \sum_{i=1}^{R} \sum_{j=1}^{C} \sum_{k=1}^{M} (x_{ijk} - \bar{x}_{ij\cdot})^2. \quad (12.5.5)$$

Analogously with Definition 12.3.4, we obtain estimates of σ^2, called mean squares, from the sums of squares as follows.

Definition 12.5.6 In a two-factor analysis of variance with R rows, C columns and M observations per cell, where S.S.R., S.S.C., S.S.I. and S.S.W. are defined in Definitions 12.5.3, 12.5.4, and 12.5.5, we define quantities called mean squares as follows:

i) The "row mean square" denoted by M.S.R. is M.S.R. = S.S.R.$/(R-1)$. (12.5.6)

ii) The "column mean square" denoted by M.S.C. is M.S.C. = S.S.C.$/(C-1)$. (12.5.7)

iii) The "interaction mean square" denoted by M.S.I. is M.S.I. = S.S.I.$/(R-1)(C-1)$. (12.5.8)

iv) The "within mean square" denoted by M.S.W. is M.S.W. = S.S.W.$/CR(M-1)$. (12.5.9)

The reasoning for the use of Expressions (12.5.6) through (12.5.9) is similar to that of Section 12.3. We shall trace it through just for the variability of rows. If H_0 is true and all the true row means are equal then, since each row mean $\bar{x}_{i..}$ is composed of CM observations, $\sum_{i=1}^{R}(\bar{x}_{i..}-\bar{x})^2/(R-1)$ is an estimate of σ^2/CM. Hence Expression (12.5.6) is an estimate of σ^2. It is compared to another estimate of σ^2 to see if it is significantly large. The computational forms of Equations (12.5.1) through (12.5.5) are given by the following theorem.

Theorem 12.5.1

i) S.S.T. = $\sum_{i=1}^{R}\sum_{j=1}^{C}\sum_{k=1}^{M} x_{ijk}^2 - T_{...}^2/N.$

ii) S.S.R. = $\sum_{i=1}^{R} R_i^2/CM - T_{...}^2/N.$

iii) S.S.C. = $\sum_{j=1}^{C} C_j^2/RM - T_{...}^2/N.$

iv) S.S.I. = $\dfrac{\sum_{i=1}^{R}\sum_{j=1}^{C} T_{ij}^2}{M}$ - S.S.R. - S.S.C. - $T_{...}^2/N.$

v) S.S.W. = $\sum_{i=1}^{R}\sum_{j=1}^{C}\sum_{k=1}^{M} x_{ijk}^2 - \sum_{i=1}^{R}\sum_{j=1}^{C} T_{ij}^2/M.$

Proof. (We prove only part (ii)).
By Equation (12.5.2)

$$\text{S.S.R.} = CM \sum_{i=1}^{R} (\bar{x}_{i..} - \bar{x}_{...})^2$$

$$= CM \sum_{i=1}^{R} (\bar{x}_{i..}^2 - 2\bar{x}_{i..}\bar{x}_{...} + \bar{x}_{...}^2)$$

$$= CM \sum_{i=1}^{R} \bar{x}_{i..}^2 - 2CM\bar{x}_{...} \sum_{i=1}^{R} \bar{x}_{i..} + RCM\bar{x}_{...}^2$$

But

a) $\bar{x}_{i..} = R_i/CM$.

b) $\bar{x}_{...} = T_{...}/RCM$.

c) $\sum_{i=1}^{R} \bar{x}_{i..} = R\bar{x}_{...}$.

Putting (a), (b), and (c) into S.S.R. yields the desired result.

12.6 FIXED AND RANDOM EFFECTS

In Section 12.3, we tested H_0, that the population means were equal, by calculating the ratio of M.S.B. and M.S.W. In the analysis of variance involving two factors, the mean square with which one compares M.S.R., M.S.C., and M.S.I. depends on the nature of the factors. The calculation depends on whether the levels of the two factors are regarded as samples from two larger populations, or whether the levels are of interest in their own right.

Definition 12.6.1 If the levels of a factor are regarded as randomly chosen from a larger population, the factor is called random. A factor which is not random is called fixed.

Example 12.6.1 If the temperature at which an experiment is performed is a factor of interest, then generally a few levels of temperature, covering a practical range of investigation, will be used for the experiment. Temperature is then a random factor.

Example 12.6.2 If in an experiment to determine the effect of increasing blood pressure, we wish to determine the effects on men and women separately, then the factor "sex of patient" is a fixed factor, since all levels of that factor are included.

Factors must be identified as either fixed or random before tests of hypothesis are performed. In most cases the identification will follow from the nature of the experiment.

We shall now consider the three tests of hypothesis:

a) $H_{0(R)}$: all row means are equal.

b) $H_{0(C)}$: all column means are equal.

c) $H_{0(I)}$: the interaction between rows and columns is zero.

Three different situations may arise, namely:

i) Both factors are random.
ii) Both factors are fixed.
iii) One factor is fixed and one is random (for convenience assume that rows are fixed and columns are random).

Theorem 12.6.1 For situation (i) where both factors are random, in order to test hypotheses (a), (b), and (c), one uses as test statistics, the ratios F_a = M.S.R./M.S.I., F_b = M.S.C./M.S.I. and F_c = M.S.I./M.S.W., respectively. If the null hypothesis (a) is true then F_a follows an F-distribution with $R - 1$ degrees of freedom in the numerator and $(R - 1) \times (C - 1)$ degrees of freedom in the denominator. Similarly if hypothesis (b) is true F_b follows an F-distribution with $C - 1$ degrees of freedom in the numerator and $(R - 1)(C - 1)$ degrees of freedom in the denominator. If (c) is true, F_c follows an F-distribution with $(R - 1)(C - 1)$ degrees of freedom in the numerator and $CR(M - 1)$ degrees of freedom in the denominator.

For situation (ii) where both factors are fixed, the hypotheses (a), (b), and (c) are tested using F_a = M.S.R./M.S.W., F_b = M.S.C./M.S.W., and F_c = M.S.I./M.S.W. Each test statistic follows an F-distribution with the appropriate degrees of freedom.

For situation (iii) where rows are fixed and columns are random, the hypotheses (a), (b), and (c) are tested using F_a = M.S.R./M.S.I., F_b = M.S.C./M.S.W., and F_c = M.S.I./M.S.W. Each test statistic follows an F-distribution with the appropriate degrees of freedom. In every case the test statistic is compared to the tabulated F-value of Appendix Table 6, and the appropriate null hypothesis is rejected if the test ratio is larger than $F_{1-\alpha}$.

If $M = 1$ and there is only one observation per cell, then the denominator of Expression (12.5.9) is zero and S.S.W. cannot be calculated. In this case one uses M.S.I. as the denominator for all tests. In those cases where the test statistic would normally use M.S.I. in the denominator the results are unaffected, but in those cases where M.S.W. is normally used, the results are only approximate. It is of course impossible to test hypothesis (c) if $M = 1$. The results may be summarized in an analysis of variance table, at the top of page 279.

Example 12.6.3 (Industry) The purity of an industrial product depends on two factors. Factor A is length of time for which the product is boiled and factor B represents the solvent used. There are three levels of A and two of B, and each cell has two observations. It is reasonable to consider

Fixed and random effects

Source of variation	Sums of squares	Degrees of freedom	Mean squares	Ratio
Between rows	S.S.R. = $\dfrac{\Sigma R_i^2}{CM} - \dfrac{T_{...}^2}{N}$	$R-1$	M.S.R. = $\dfrac{\text{S.S.R.}}{R-1}$	
Between columns	S.S.C. = $\dfrac{\Sigma C_j^2}{RM} - \dfrac{T_{...}^2}{N}$	$C-1$	M.S.C. = $\dfrac{\text{S.S.C.}}{C-1}$	See Section 12.6
Interaction	S.S.I. = $\dfrac{\Sigma T_{ij}^2}{M} - \text{S.S.R.}$ $- \text{S.S.C.} - \dfrac{T_{...}^2}{N}$	$(R-1)(C-1)$	M.S.I. = $\dfrac{\text{S.S.I.}}{(R-1)(C-1)}$	
Within cells	S.S.W. = $\Sigma x_{ijk}^2 - \dfrac{\Sigma T_{ij}^2}{M}$	$RC(M-1)$	M.S.W. = $\dfrac{\text{S.S.W.}}{RC(M-1)}$	
Total	S.S.T. = $\Sigma x_{ijk}^2 - \dfrac{T_{...}^2}{N}$	$N-1$		

A to be random and B to be fixed. Run the analysis of variance with the following data.

	A_1	A_2	A_3
B_1	3.1	2.0	1.3
	2.1	1.6	1.9
B_2	4.3	2.8	4.1
	6.1	3.9	2.6

Solution $R = 2$, $C = 3$, $N_{ij} = M = 2$, $N = 12$, $x_{111} = 3.1$, $x_{112} = 2.1$, etc., $x_{232} = 2.6$, $T_{11} = 5.2$, $T_{12} = 3.6$, etc., $T_{23} = 6.7$, $R_1 = 12.0$, $R_2 = 23.8$, $C_1 = 15.6$, $C_2 = 10.3$, $C_3 = 9.9$. $T_{...} = 35.8$. Using Theorem 12.5.1 we obtain the sums of squares as follows.

$$\text{S.S.T.} = 3.1^2 + 2.1^2 + 2.0^2 + \cdots + 2.6^2 - 35.8^2/12$$
$$= 128.20 - 106.80 = 21.40.$$

$$\text{S.S.R.} = (12.0^2 + 23.8^2)/6 - 35.8^2/12 = 118.41$$
$$- 106.80 = 11.61.$$

$$\text{S.S.C.} = (15.6^2 + 10.3^2 + 9.9^2)/4 - 35.8^2/12 = 111.87 -$$
$$106.80 = 5.07.$$

$$S.S.I. = (5.2^2 + 3.6^2 + \cdots + 6.7^2)/2 - 11.61 - 5.07 - 106.80 = .61.$$

$$S.S.W. = 3.1^2 + 2.1^2 + \cdots + 2.6^2 - (5.2^2 + 3.6^2 + \cdots + 6.7^2)/2 = 4.11.$$

Putting the results in an analysis of variance table we have the following.

Source of variation	Sum of squares	Degrees of freedom	Mean squares	Ratio
Between rows	S.S.R. = 11.61	$R - 1 = 1$	$\dfrac{S.S.R.}{R-1} = 11.61$ = M.S.R.	37.45
Between columns	S.S.C. = 5.07	$C - 1 = 2$	$\dfrac{S.S.C.}{C-1} = 2.54$ = M.S.C.	3.68
Interaction	S.S.I. = .61	$(R-1)(C-1) = 2$	$\dfrac{S.S.I.}{(R-1)(C-1)} = .31$ = M.S.I.	.45
Within cells	S.S.W. = 4.11	$RC(M-1) = 6$	$\dfrac{S.S.W.}{RC(M-1)} = .69$ = M.S.W.	
Total	S.S.T. = 21.40	$N - 1 = 11$		

The ratios are calculated as M.S.R./M.S.I., M.S.C./M.S.W., and M.S.I./M.S.W., respectively. Comparing M.S.R./M.S.I. to Appendix Table 6, with 1 and 2 degrees of freedom, we see that the ratio is significant for $\alpha = .05$. Similarly M.S.C./M.S.W. and M.S.I./M.S.W. are not significant for $\alpha = .05$. We therefore conclude that only the rows (i.e. factor B) have an effect on the purity of the industrial product.

12.7 TESTING THE EQUALITY OF THE VARIANCES

We stated in Section 12.3 that the ratio M.S.B./M.S.W. followed an F-distribution if the K populations were normal with common variance. Although it is not recommended that this assumption be tested before applying the analysis of variance techniques, there is sometimes interest in comparing the variances of several populations. We shall discuss one such method, called Bartlett's test, for testing

$$H_0: \sigma_1^2 = \sigma_2^2 = \cdots = \sigma_i^2,$$
$$H_A: \text{at least one } \sigma_i^2 \text{ is unequal.}$$

Theorem 12.7.1 In order to test the null hypothesis that the variances of K normal populations are equal, on the basis of K independent estimates $s_1^2, s_2^2, \ldots, s_K^2$ of this variance, where

$$s_i^2 = \frac{1}{N_i - 1} \sum_{j=1}^{N_i} (x_{ij} - \bar{x}_{i.})^2, \quad i = 1, 2, \ldots, K,$$

$$s^2 = \frac{\sum_{i=1}^{N} (N_i - 1) s_i^2}{\sum_{i=1}^{K} (N_i - 1)},$$

$$C = 1 + \frac{1}{3(K-1)} \left[\sum_{i=1}^{K} \left(\frac{1}{N_i - 1} \right) - \frac{1}{\sum_{i=1}^{K} (N_i - 1)} \right],$$

then the test statistic is

$$B = \frac{1}{C} \left[\sum_{i=1}^{K} (N_i - 1) \log_e s^2 - \sum_{i=1}^{K} (N_i - 1) \log_e s_i^2 \right]$$

and B has approximately a chi-square distribution with $K - 1$ degrees of freedom. We reject H_0 for large values of B.

Example 12.7.1 Compare the variances of the populations in Example 12.3.4. Use $\alpha = .001$.

Solution

i) $H_0: \sigma_1^2 = \sigma_2^2 = \sigma_3^2,$
   $H_A:$ at least one σ_i^2 is different.

ii) $s_1^2 = .5667$, $s_2^2 = .02$, $s_3^2 = .02$.

iii) $s^2 = \dfrac{(3)(.5667) + (3)(.02) + (3)(.02)}{9} = .2022.$

iv) $C = 1 + \dfrac{1}{(3)(2)} [\tfrac{1}{3} + \tfrac{1}{3} + \tfrac{1}{3} - \tfrac{1}{9}] = 1.1481.$

v) $B = \dfrac{1}{1.1481} [(9)(-1.5985) - (3)(-.5679) - (3)(-3.9120)$
   $\quad - (3)(3.9120)] = 9.397.$

vi) Using the upper tail of the chi-square distribution, we compare B to $\chi^2_{.999}$ with 2 degrees of freedom. Since $\chi^2_{.999} = 13.80$ we cannot reject H_0.

If we reject the null hypothesis that the population variances are equal, then it is sometimes possible to transform the data, using an appropriate transformation, to stabilize the variances. The most common situation arising is where the variance is related to the mean, so that as the mean increases, the variance increases (or decreases). This is particularly common in biological experiments. We shall consider a random variable X with expected value μ_X and variance σ_X^2.

Theorem 12.7.2 If for a constant k, $\sigma_X^2 = k\mu_X$, and $Y = \sqrt{X}$, then the variance of Y will be approximately constant.

Example 12.7.2 If X is a Poisson random variable then $k = 1$. Thus the square root transformation will stabilize the variance of Poisson variables.

Theorem 12.7.3 For k a constant and $\sigma_X^2 = k\mu_X^4$ and $Y = \log_e X$, then the variance of Y will be approximately constant.

Theorem 12.7.4 For k a constant, $\sigma_X^2 = k\mu_X^4$ and $Y = 1/X$, then the variance of Y will be approximately constant.

Theorem 12.7.5 If X is a binomial proportion with mean p and variance $p(1 - p)/n$ and $Y = 2 \arcsin \sqrt{X}$ (Y is measured in radians), then the variance of Y will be approximately constant. Values of Y may be obtained from Appendix Table 11.

Example 12.7.3 Consider the following summary of data describing an experiment comparing diuretics by measuring the sodium content of urine in control and treated rats.

Means and variance of raw data

	Control	Treated (1)	Treated (2)
Mean	106	370	542
Variance	7,410	24,530	33,340

There appears to be a linear relationship between mean and variance, so that we can say that
$$\text{Variance} = k \text{ mean}.$$

Applying the square root transformation to the data, we have the following summary.

Means and variance of transformed data

	Control	Treated (1)	Treated (2)
Mean	9.5	18.5	29.0
Variance	13.1	10.9	11.7

Although the means of the transformed observations still differ, the variances are approximately equal.

12.8 SUMMARY

The technique for comparing several population means is called the analysis of variance. In its simplest case which may be considered as an extension of the ideas of the t-test, one calculates the values S.S.T., S.S.B., and S.S.W., the last two of which measure the effect of the treatments and the inherent variability, respectively. Both the technical calculations and some intuitive reasoning are discussed as to how the technique works. Once a significant difference between treatments is detected, the actual ranking of the treatments is performed by means of the method of Duncan. The simple analysis of variance is extended to the case of two factors with the appropriate calculations and reasoning. A distinction between fixed and random effects is discussed. Random effects are those whose levels are considered to be a sample from a population. The levels of a fixed factor are of interest in themselves. Tests of hypotheses in a two factor situation depend on whether the factors are fixed, random, or mixed. Finally, the equality of population variances is tested using Bartlett's test.

EXERCISES

1. Three varieties of wheat planted in 12 plots of land at random gave the following yields in bushels per acre.

Variety A	Variety B	Variety C
35	55	47
38	48	43
45	52	53
40	54	50

Perform the analysis of variance and decide whether there is a significant difference between the varieties at the .05 level of significance.

2. Sixteen pigs are divided into four lots of four pigs each at random. Each lot is given a different feed. The weight gain in pounds by each of the pigs for a particular length of time is given in the following table.

Feeds (treatments)			
1	2	3	4
130	155	200	185
135	160	205	178
128	145	190	175
140	158	210	190

Construct an analysis of variance table and see whether there is a significant difference between the treatments at the .05 level of significance.

3. Consider the following table.

Group 1	Group 2	Group 3
25	22	8
15	20	16
10	8	14
12	15	
18		

a) Construct the analysis of variance table.
b) Estimate the group means.
c) Are the group means significantly different?

4. The pollution levels of three major rivers are sampled five times each, yielding, on an artificial scale, the following figures.

River A	River B	River C
2.7	2.9	.6
1.4	2.4	1.2
2.1	3.7	1.5
2.0	1.6	1.7
1.2	2.4	2.1

Test the equality of the mean pollution level of the three rivers. All observations are assumed to be from independent normal populations. Use $\alpha = .05$.

5. The weight gains of pigs (in lbs.) on four different types of feed assumed to be normal with equal variances, are as follows.

Feed 1	Feed 2	Feed 3	Feed 4
42	36	45	23
37	22	28	17
29	17	52	20
53	31	67	
	30	54	

Use the analysis of variance to compare the four feed types. Use $\alpha = .01$.

6. The I.Q.'s of students in three different school districts are as follows.

District A	District B	District C
110	104	117
103	137	122
97	125	99
120	102	124
135	109	142
	113	

Test the hypothesis that the mean I.Q. in each district is the same.

7. In a psychological experiment three types of rats, well trained, partially trained, and untrained, are required to pass through a maze. The times (in minutes) are as follows. Does training seem to have any effect? Use $\alpha = .05$.

Well trained	Partially trained	Untrained
.7	1.9	1.3
1.2	1.4	1.8
1.3	1.8	2.4
.9	2.7	1.9
1.6	1.3	2.8
.4	2.1	1.3
.8	2.7	1.9
	2.6	1.6
	1.8	

8. Prove part (b) of Theorem 12.3.3.
9. Suppose in Example 12.3.4 that in fact there were five observations on drug 3, the fifth observation having been mislaid. Let the value of the fifth observation be 1.0. Go through the analysis of variance again and state your conclusions. Use $\alpha = .05$.
10. In Exercise 3, use Bartlett's test to compare the variances within groups.
11. In Section 12.5, show that S.S.T. = S.S.R. + S.S.C. + S.S.I. + S.S.W.
12. Prove parts (a), (c), (d) and (e) of Theorem 12.5.1.
13. In Exercise 4 discuss whether the factor is fixed or random.
14. In Exercise 5 discuss whether the factor is fixed or random.
15. In Exercise 7 discuss whether the factor is fixed or random.
16. Consider the following data summary for the type of experiment considered in Example 12.7.3.

	Control	Treated (1)	Treated (2)
Mean	100	300	1000
Variance	5000	45000	405000

What transformation of the raw data could be expected to stabilize the variance?

17. Use Bartlett's test to test the equality of the variances in Exercise 6, given that the number 113 in district B is known to be an aberration and is discarded.
18. With the number 113 in district B discarded, rerun the analysis of variance.
19. Using the results of Exercise 1, apply Duncan's test with $\alpha = .05$.
20. Apply Duncan's test to the data of Exercise 4.
21. Use Bartlett's test to test the equality of variety variances in Exercise 1.
22. Use Bartlett's test to test the equality of treatment variances in Exercise 2.
23. Cable wire is manufactured by three firms. A test is performed to determine whether the mean breaking strength is the same. Five pieces of wire are picked up at random from each firm and put under tension. The results are given in the following table.

Firm A	Firm B	Firm C
10	12	5
8	7	8
12	15	9
11	14	7
9	10	10

Construct the analysis of variance table. Using a .05 level of significance, test whether there is a significant difference between mean breaking strength of cables manufactured by the three firms.

24. It is decided to test the yield of three types of apples. The yields in bushels are given in the following table.

Apple 1	Apple 2	Apple 3
10	8	15
14	9	11
15	12	14
18	10	16
20	11	10
16	7	12
15	14	9

Perform the analysis of variance and conclude whether there is a significant difference, at the .01 level of significance, between the mean yields of different types of apples.

25. Four varieties of wheat are being compared for yield. Five locations each containing four plots of one acre each are available. One variety is assigned at random to a plot. The results after harvesting are as follows.

Locations	Variety of wheat			
	1	2	3	4
1	30	35	25	15
2	40	36	30	20
3	55	45	32	25
4	60	40	28	30
5	50	46	40	32

a) Test whether there is a significant difference, at the .05 level of significance, between the varieties.

b) Test whether there is a significant difference between the locations with respect to mean production of wheat.

26. An experiment is conducted to determine the merits of three feeds in regard to gain in weight of pigs. Twelve pigs, three each from four litters are given the feeds at random. The results are as follows.

Litters	Feeds		
	A	B	C
1	140	170	200
2	130	140	180
3	145	160	185
4	135	155	190

a) Are the feed differences significant at the .01 level of significance?

b) Can we conclude that the litters have some effect on the gain in weight?

27. The following table gives the number of units of production per day turned out by four different employees using four different types of machines.

Employee	Type of machine			
	M_1	M_2	M_3	M_4
E_1	40	36	45	30
E_2	38	42	50	41
E_3	36	30	48	35
E_4	46	47	52	44

a) Test the hypothesis that the mean production is the same for the four machines.

b) Test the hypothesis that the four employees do not differ with respect to mean productivity.

Chapter 13
NONPARAMETRIC STATISTICS

13.1 INTRODUCTION

Many of the statistical techniques discussed so far have an underlying assumption that the population from which the sample is drawn has a normal distribution. In some cases this assumption is not too restrictive, since some techniques, most notably those based on the t-test, are robust; that is, they do not depend too critically on this assumption. Other techniques may be made applicable by an appropriate transformation of the data. When the samples are large, the assumption of normality is not too critical since in many cases the test statistic will be approximately normal due to the application of the central limit theorem.

It is of practical interest to discuss techniques which do not depend on the underlying assumption of population normality, even for small samples. We shall call such techniques nonparametric techniques, although some authors refer to them as ranking methods or ordering techniques. Of the large number of available nonparametric tests which could be discussed, we have chosen six for inclusion in this text. These six cover many of the main types of problems which a statistician is apt to meet. The names of the tests, usually named after their inventors, and the areas in which they can be applied, are as follows:

Test	Area of Application
Runs test	Test of randomness
Wilcoxon's test for paired data	Test of location for paired samples
Wilcoxon's Rank Sum test	Test of location for two independent samples
Kruskal-Wallis's test	Test of location for K independent samples
Friedman's test	Test of location for K related samples
Spearman's test	Rank correlation of two samples

13.2 TEST OF RANDOMNESS: THE RUNS TEST

The basic assumption in all statistical methods for estimation and hypothesis testing is that the samples obtained are random. The researcher may wish to test the randomness of a given sample, that is, he may wish to test whether or not the sample observations are independent and identically distributed. For example, a biologist may like to know whether or not the number of cases of malaria reported from a certain district fluctuate randomly from one year to another. An economist may like to know the randomness of stock-market prices of a certain stock over a period of time, a manufacturer may like to know the quality of an article produced by machine and checked several times during the day. There are several tests available to judge the randomness of a given sample. We present in this section one such test known as the runs test based on the number of runs.

Definition 13.2.1 A sequence of elements of one kind followed and preceded by elements of another kind is defined as a run. The number of like elements of a particular kind in a run is defined as the length of a run.

Example 13.2.1 Consider the following sequence of defective items, D, and nondefective items, N, produced by a machine in the following order.

$$\underline{NNN}\ \underline{DD}\ \underline{NNNNN}\ \underline{DD}\ \underline{NNN}$$

According to Definition 13.2.1, we find five runs of varying length. The first run consists of three nondefective items, the second run consists of two defective items, the third run consists of five nondefective items, the fourth run consists of two defective items, and the fifth run consists of three nondefective items.

Example 13.2.2 Suppose a coin is tossed 10 times and the following sequence of heads, H, and tails, T, is observed.

$$\underline{HHHHH}\ \underline{TTTTT}$$

According to Definition 13.2.1, we find only two runs and this would seem too few for a fair coin.

Example 13.2.3 (Industry) Consider the diameter in inches of nine bolts made by a machine.

$$1.3,\ 1.6,\ 1.5,\ 1.2,\ 1.4,\ 1.1,\ 1.7,\ 1.8,\ 1.9.$$

Let each measurement be designated as low, L, if smaller than the median and designated as high, H, if larger than the median. Since the median of the above set of measurements is 1.5, the designation of L's and H's to the above set of measurements is

$$L\ H\ LLL\ HHH.$$

Let N_1 be the number of L's, N_2 be the number of H's and, r be the total number of runs. From the above sequence we find

$$N_1 = 4,\ N_2 = 4,\quad \text{and}\quad r = 4.$$

The sampling distribution of the total number of runs, R, can be obtained by mathematical methods, but is beyond the scope of this book. The distribution of R depends on N_1 and N_2, where N_1 is the number of L's smaller than the median or some other convenient reference point and N_2 is the number of H's larger than the median or some other reference point. Appendix Table 12 gives critical values of r for a 5 percent two-sided test. The smaller integer is the left-tail critical value and the larger integer is the right-tail critical value.

The Runs test for testing the hypothesis of randomness consists in counting the number of runs. Too few or too many runs in a sequence would indicate nonrandomness. To test the null hypothesis of randomness at $\alpha = .05$, we count the number of runs in a sample and compare it with table values given in Appendix Table 12. If the number of runs, r, in a given sample falls between the two values given in Appendix Table 12, we accept the null hypothesis of randomness.

Example 13.2.4 (Pollution) The following figures (on an artificial scale) give 15 weekly pollution levels of a large city.

47, 56, 52, 51, 40, 44, 63, 48, 51, 59, 60, 55, 46, 53, 55.

Use the method of runs above and below the median to test whether the fluctuation from one week to another is random or whether a trend is present. Use $\alpha = .05$.

Solution H_0: the L's and H's occur in random order,

H_A: the order of L's and H's deviates from randomness.

The assignment of L's and H's below and above the median to the above data is shown in Table 13.2.1.

From the table we find the number of L's is 7, that is $N_1 = 7$, the number of H's is 7, that is $N_2 = 7$, and the number of runs is 8, that is, $r = 8$. From Appendix Table 12 we find 3 and 13 as the critical values for $N_1 = 7$, $N_2 = 7$, and $\alpha = .05$. Since $r = 8$ falls in the acceptance region, 3 and 13, we accept the null hypothesis that the pollution levels occurred in random order.

TABLE 13.2.1

Week	Pollution level	Position of pollution level with respect to median
1	47	L
2	56	H
3	52	Median
4	51	L
5	40	L
6	44	L
7	63	H
8	48	L
9	51	L
10	59	H
11	60	H
12	55	H
13	46	L
14	53	H
15	55	H

If either N_1 or N_2 is larger than 20, Appendix Table 12 cannot be used. The random variable R is normally distributed as N_1 and N_2 become large. Therefore, for large sample sizes, the hypothesis of randomness may be tested by the statistic

$$Z = \frac{R - E(R)}{\sqrt{V(R)}} \quad (13.2.1)$$

where

$$E(R) = \frac{2N_1 N_2}{N_1 + N_2} + 1 \quad (13.2.2)$$

and

$$V(R) = \frac{2N_1 N_2 (2N_1 N_2 - N_1 - N_2)}{(N_1 + N_2)^2 (N_1 + N_2 - 1)} \quad (13.2.3)$$

are the expected value and variance of R, respectively.

13.3 WILCOXON'S TEST FOR PAIRED DATA

The Wilcoxon test for paired samples is a test of the null hypothesis that two population medians are the same against the alternative that they are unequal. We assume that there are N pairs of observations, and that each pair is independent of every other pair. It is not necessary that the two observations in a pair be independent. For two populations, 1 and 2, consider the following definition.

Definition 13.3.1

a) The medians of the two populations will be called $Md(1)$ and $Md(2)$.
b) X_{1i} and X_{2i}, $(i = 1, 2, \ldots, N)$ will denote the observations from the two populations.
c) M will denote the number of pairs of observations.
d) N is the number of non-zero differences $D_i = X_{1i} - X_{2i}$, $(i = 1, 2, \ldots, N)$.

In order to test

$$H_0: Md(1) = Md(2),$$
$$H_A: Md(1) \neq Md(2),$$

we rank the absolute value of the differences $|D_i|$ as follows. Replace the smallest value of $|D_i|$ by 1, the next smallest by 2 and so on up to the largest, which is replaced by N. Tied differences are given by the average of the ranks that would have been given if no ties were present. Thus if two values tied for fourth place they would each be given the rank of 4.5 ($\frac{1}{2}$ of $4 + 5$). The "signs" of the original differences are then replaced and the sum of the positive ranks, called $T(+)$ is the test statistic. The critical region for $T(+)$ is given in Appendix Table 13. For example if $N = 10$, $\alpha = .05$ then the critical region from Appendix Table 13 is for $T(+) \leq 10$. If the alternative hypothesis is $Md(1) > Md(2)$, then reject H_0 if $T(+) \geq 45$. For the two-sided alternative reject H_0 if $T(+) \geq 47$ or $T(+) \leq 8$.

Example 13.3.1 (Pharmacology) A series of 12 pairs of Pharmaceutical products were analyzed for potency using two methods, (A and B). The differences d_i for each pair are given in Table 13.3.1. With $\alpha = .05$, test that the two methods differ.

TABLE 13.3.1
Differences of potency by two methods for 12 pairs of pharmaceutical observations

Pair	Difference	Absolute difference	Rank	Pair	Difference	Absolute difference	Rank
1	−.5	.5	8	7	+.2	.2	2.5
2	−.4	.4	6.5	8	+.3	.3	4.5
3	+.7	.7	10	9	−.4	.4	6.5
4	−.3	.3	4.5	10	−.8	.8	11
5	−.1	.1	1.0	11	−.2	.2	2.5
6	−.6	.6	9	12	+.9	.9	12

Solution

i) $T(+) = 10 + 2.5 + 4.5 + 12 = 29$.

ii) From Appendix Table 13, with $N = 12$, the critical region for $\alpha = .05$ is $T(+) \leq 13$ or $T(+) \geq 65$. Thus we cannot reject H_0. There is no evidence of a difference between the two methods of analysis.

Example 13.3.2 (Chemistry) In order to see if two laboratories are using the same techniques and giving the same results, 15 samples of chemical substances are divided in two. For each substance, laboratories A and B get a sample. If the results are as follows, test the hypothesis that the two laboritories differ, with a significance level of .05.

TABLE 13.3.2

(1) Chemical	(2) Lab. A	(3) Lab. B	(4) Difference	(5) Absolute difference	(6) Rank	(7) Ranks with signs restored
1	10.3	9.4	+.9	.9	3	+3
2	12.1	7.2	+4.9	4.9	12	+12
3	37.6	21.0	+16.6	16.6	15	+15
4	4.1	2.7	+1.4	1.4	4	+4
5	13.2	15.3	−2.1	2.1	6	−6
6	9.7	6.4	+3.3	3.3	9	+9
7	9.8	5.7	+4.1	4.1	10	+10
8	15.4	11.2	+4.3	4.3	11	+11
9	7.0	7.2	−.2	.2	1	−1
10	35.7	29.1	+6.6	6.6	14	+14
11	3.1	4.9	−1.8	1.8	5	−5
12	20.5	18.0	+2.5	2.5	7.5	+7.5
13	17.8	12.7	+5.1	5.1	13	+13
14	9.0	9.7	−.7	.7	2	−2
15	12.3	9.8	+2.5	2.5	7.5	+7.5

Solution

i) Columns 1, 2, and 3 of Table 13.3.2 show the chemical and the results of laboratories A and B. Column 4 shows the difference of laboratory A minus laboratory B, and the absolute differences are given in column 5. The ranks are assigned in column 6 without regard to sign and in the final column, the signs of the original differences are restored.

ii) $T(,+) = +3 + 12 + \cdots + 7.5 = 106$.

iii) From Appendix Table 13 we reject H_0 if $T(+) \geq 95$ or $T(+) \leq 25$. Thus we reject H_0. The two laboratories give different results.

The rationale for the use of $T(+)$ as the test statistic is as follows. If the two population medians are indeed equal, then the sum of the positive ranks and the sum of the negative ranks will be about equal, so that $T(+)$ will be approximately one half of $1 + 2 + \cdots + N$; that is $T(+)$ is approximately $N(N + 1)/2$. If $T(+)$ is very different from $N(N + 1)/2$, it will be considered significantly different, and this will take place if there are too many or too few positive ranks. Note as a check on the arithmetic, that $T(+) + T(-)$ (where $T(-)$ is the sum of the negative ranks) equals $N(N + 1)/2$.

For values of N larger than 25, we may use the fact that $T(+)$ is approximately normally distributed with

$$\text{Mean} = N(N + 1)/4$$

and

$$\text{Variance} = N(N + 1)(2N + 1)/24$$

so that

$$Z = \frac{T(+) - N(N + 1)/4}{\sqrt{N(N + 1)(2N + 1)/24}}$$

is approximately normally distributed with zero mean and unit variance.

Example 13.3.3 Consider the observations of Example 1.1.2 and test for a difference between the English and mathematics scores using a 5% significance level.

Solution

i) The differences (English scores minus mathematics scores) are as shown in column 1 of Table 13.3.3.
ii) In column (2) of Table 13.3.3 are the absolute differences, which are assigned their ranks in column (3). In column (4), the ranks with their algebriac signs restored are listed.
iii) $T(+) = 15 + 3 + 7 + \cdots + 12.5 = 290.5$.
iv) Assuming that $T(+)$ is approximately normal with

$$\text{Mean} = (29)(30)/4 = 217.5,$$

$$\text{Variance} = (29)(30)(59)/24 = 2138.75$$

then

$$z = \frac{290.5 - 217.5}{\sqrt{2138.75}} = \frac{73}{46.2} = 1.58.$$

v) Since $z_{.975} = 1.96$, the calculated z does not lie in the critical region; thus we cannot reject H_0. There is no proof of the difference between English and mathematics scores.

TABLE 13.3.3

(1) Differences	(2) Absolute differences	(3) Rank	(4) Rank with sign restored	(1) Differences	(2) Absolute differences	(3) Rank	(4) Rank with sign restored
8	8	15	+15	4	4	3	+3
4	4	3	+3	−13	13	24	−24
5	5	7	+7	6	6	10.5	+10.5
19	19	25.5	+25.5	11	11	21	+21
3	3	1	+1	−9	9	18	−18
7	7	12.5	+12.5	24	24	28	+28
−5	5	7	−7	9	9	18	+18
−12	12	22.5	−22.5	23	23	27	+27
−19	19	25.5	−22.5	−12	12	22.5	−22.5
5	5	7	+7	10	10	20	+20
8	8	15	+15	5	5	7	+7
−4	4	3	−3	9	9	18	+18
27	27	29	+29	0	0	*	*
−5	5	7	−7	7	7	12.5	+12.5
6	6	10.5	+10.5	−8	8	15	−15

13.4 WILCOXON RANK SUM TEST

If we wish to test the null hypothesis that two independent samples came from populations with the same distributions, then we may use a nonparametric procedure called the Wilcoxon Rank Sum test. This procedure is also known as the Wilcoxon test for independent samples, the rank sum test, and the Mann-Whitney test.

Suppose that we have samples of sizes N_1 and N_2. Combine the observations and arrange the entire group according to size. Give the smallest observation the rank 1 and the largest the rank $N_1 + N_2$. Ties are taken care of by giving them the average of the ranks that would have been assigned if no ties were present. The test statistic, called $T(1)$, is the sum of the ranks for the sample of size N_1.

Example 13.4.1 (Biology) The weight gains of five mice on a vitamin-rich diet and four mice on a vitamin-poor diet are as follows:

Vitamin-rich (VR)	27, 31, 45, 33, 40
Vitamin-poor (VP)	24, 30, 15, 21

Combining the two samples and ranking them gives

Sample	VP	VP	VP	VR	VP	VR	VR	VR	VR
Weight gain	15	21	24	27	30	31	33	40	45
Rank	1	2	3	4	5	6	7	8	9

$T(1) = 1 + 2 + 3 + 5 = 11$.

The critical values of $T(1)$ are given in Appendix Table 14. If we call A and B the two tabulated values ($A < B$) in that table, then we reject H_0 if $T(1) \leq A$ or $T(1) \geq B$.

Example 13.4.2 In Example 13.4.1, test the hypothesis that the weight gains of the vitamin-rich mice are greater than the vitamin-poor ones, with $\alpha = .05$.

Solution

i) H_0: weight of vitamin-rich mice = weight gain of vitamin-poor mice, H_A: weight gain of vitamin-rich mice > weight gain of vitamin-poor mice.

ii) From Example 13.4.1, $T(1) = 11$, $N_1 = 4$, $N_2 = 5$.

iii) From Appendix Table 14 with $N_1 = 4$, $N_2 = 5$, $\alpha = .05$ (one-sided), we reject H_0 if $T(1) \leq 12$. Thus we reject H_0 and conclude that mice fed vitamin-rich diets do gain more weight than those on vitamin-poor diets.

Note that if we had concentrated on the other sample, then we would have had $N_1 = 5$, $N_2 = 4$, $T(1) = 34$ and the same conclusion would have been reached. The justification for the use of $T(1)$ as the test statistic is that if the first sample has values which are all smaller than those of the second sample, then $T(1)$ will be small and the hypothesis of equality will be rejected. Similarly, if the values of the first sample are all larger than those of the second, $T(1)$ will be large.

For large values of N_1 and N_2, U is approximately normally distributed with

$$\text{Mean} = \frac{N_1 N_2}{2}$$

and

$$\text{Variance} = N_1 N_2 (N_1 + N_2 + 1)/12$$

and

$$Z = \frac{U - N_1 N_2/2}{\sqrt{N_1 N_2 (N_1 + N_2 + 1)/12}}$$

is approximately normal with mean zero and unit variance.

13.5 THE KRUSKAL-WALLIS TEST

The Kruskal-Wallis test, also called the H test, is a generalization of the Wilcoxon Rank Sum test for two samples to the case of K ($K > 2$) samples. The Kruskal-Wallis test is used to test the null hypothesis that K independent samples are from the same population. Thus it is a nonparametric counterpart of one-way analysis of variance. It is an extremely useful test when the researcher does not want to assume that the populations are normally distributed.

Let n_i ($i = 1, 2, \ldots, K$) be the size of the ith sample and $N = n_1 + n_2 + \cdots + n_K$ be the total number of observations. As in the Mann-Whitney test, each of the N observations is replaced by rank. Let R_i ($i = 1, 2, \ldots, K$) be the sum of ranks in the ith sample. Then the Kruskal-Wallis test is based on the statistic

$$H = \frac{12}{N(N+1)} \left[\frac{R_1^2}{n_1} + \frac{R_2^2}{n_2} + \cdots + \frac{R_K^2}{n_K} \right] - 3(N+1). \quad (13.5.1)$$

The statistic H defined above is approximately chi-square distributed with $K - 1$ degrees of freedom. If the value of H given by Equation (13.5.1) is greater than the chi-square value at a level of significance α and $K - 1$ degrees of freedom, then H_0 is rejected at a level of significance α.

Example 13.5.1 (Agriculture) Three varieties of wheat planted in 12 plots of land at random gave the following yield in bushels per acre.

Variety A	Variety B	Variety C
35	55	47
38	48	43
45	52	53
40	54	50

Apply the Kruskal-Wallis test to decide whether there is a significant difference between the three varieties. Use $\alpha = .05$.

Solution Calculations of the sum of ranks, R_i ($i = 1, 2, 3$), are shown in Table 13.5.1.

From Table 13.5.1 and Equation (13.5.1) we find

$$H = \frac{12}{N(N+1)} \left[\frac{R_1^2}{n_1} + \frac{R_2^2}{n_2} + \frac{R_3^2}{n_3} \right] - 3(N+1)$$

$$= \frac{12}{12(13)} \left[\frac{(11)^2}{4} + \frac{(39)^2}{4} + \frac{(28)^2}{4} \right] - 3(13)$$

$$= \frac{2426}{52} - 39 = 7.65. \tag{13.5.2}$$

TABLE 13.5.1

Variety A Observation	Rank	Variety B Observation	Rank	Variety C Observation	Rank
35	1	55	12	47	6
38	2	48	7	43	4
45	5	52	9	53	10
40	3	54	11	50	8
n_i 4		4		4	
R_i	11		39		28

Using a .05 level of significance and degrees of freedom $K - 1 = 3 - 1 = 2$, we find from Appendix Table 5, $\chi^2_{.95} = 5.99$. Since the computed value of $H = 7.65$ is greater than $\chi^2_{.95} = 5.99$, we reject H_0 and conclude that there is a difference between the three varieties of wheat.

13.6 FRIEDMAN'S TEST FOR K RELATED SAMPLES

If we extend the ideas of the Wilcoxon test for paired data to the case where each related set has K observations, then we can test the hypothesis that the K samples came from identical populations by means of the Friedman Test. We assume that there are N independent sets of data, each having K observations. The data may be arranged into a two-way table with N rows and K columns. Each row is then ranked from 1 for the smallest to K for the largest.

Example 13.6.1 (Psychology) Suppose that five students are each given a mathematics test, an English test, and a psychology test and their scores are as follows.

	Mathematics	English	Psychology
Student 1	71	53	66
Student 2	84	72	85
Student 3	57	94	47
Student 4	78	68	75
Student 5	65	80	70

In this example $K = 3$, $N = 5$. The ranked form of the data is:

300   Nonparametric statistics

	Mathematics	English	Psychology
Student 1	3	1	2
Student 2	2	1	3
Student 3	2	3	1
Student 4	3	1	2
Student 5	1	3	2
Total of ranks	11	9	10

If the K populations from which the samples came were indentical, then the ranks could be expected to occur in about equal numbers in each column. If on the other hand one population was larger (or smaller) than the others, its ranks could be expected to be higher (or lower).

Theorem 13.6.1 For N independent samples each with K observations, representing K populations, where the data are arranged in a two-way table of N rows and K columns, and ranked one row at a time from 1 to K, and where R_j is the sum of ranks in column j, then the null hypothesis of equality of populations is rejected if the statistic

$$S = \frac{12}{NK(K+1)} \sum_{j=1}^{K} R_j^2 - 3N(K+1)$$

is large. The upper significance levels of S are given in Appendix Table 15 for moderate values of K and N. For N and K outside this range, S has an approximate chi-square distribution with $K-1$ degrees of freedom.

Example 13.6.2 In Example 13.6.1 test the hypothesis that the mathematics, English, and psychology scores for the population from which the sample of 5 was drawn are in fact equal. Use $\alpha = .05$.

Solution

i) $R_1 = 11$, $R_2 = 9$, $R_3 = 10$, $N = 5$, $K = 3$.

ii) $S = \dfrac{12}{(5)(3)(4)} \{11^2 + 9^2 + 10^2\} - (3)(5)(4)$

$= 60.4 - 60 = .4$.

iii) From Appendix Table 15, S must be greater than 5.2 in order to reject H_0. Hence we cannot reject H_0. There is no evidence to support a difference in the population scores in mathematics, English, and psychology.

Example 13.6.3 (Chemistry) Six laboratories are asked to analyze five chemical substances in order to see if all laboratories are performing the analyses in the same manner. Use Friedman's test, with $\alpha = .05$, to see if any laboratories are significantly different from any others, if the ranked data are as follows.

	Lab. 1	Lab. 2	Lab. 3	Lab. 4	Lab. 5	Lab. 6
Chemical A	1	5	3	2	4	6
Chemical B	3	2	1	4	6	5
Chemical C	3	4	2	5	1	6
Chemical D	1	4	6	3	2	5
Chemical E	4	5	1	2	3	6
Total of Ranks	12	20	13	16	16	28

Solution

i) Here laboratories play the role of populations and chemicals play the role of observations. Thus $N = 5$, $K = 6$.

ii) $R_1 = 12$, $R_2 = 20$, $R_3 = 13$, $R_4 = 16$, $R_5 = 16$, $R_6 = 28$.

iii) $S = \dfrac{12}{(5)(6)(7)} \{12^2 + 20^2 + 13^2 + 16^2 + 16^2 + 28^2\} - (3)(5)(7)$

$= 114.8 - 105 = 9.8$.

iv) From Appendix Table 5, with $K - 1 = 5$ degrees of freedom $\chi^2_{5,.95} = 11.07$. Thus we cannot reject H_0. We cannot conclude that the laboratories are different.

13.7 RANK CORRELATION

There are many occasions when it is not possible to measure the variable under consideration quantitatively. For example, beauty, taste, and color cannot be measured quantitatively. But it is possible to rank such attributes in ascending or descending order. Researchers may like to know if there exists a relationship between two attributes. If a set of individuals are ranked according to two different attributes we may like to know the degree of association between the ranks of such attributes. For example, we may like to know the relationship between high school grades and college grades.

Definition 13.7.1 Let the ranks of N individuals according to attribute A be $x_1, x_2, \ldots, x_N$ and according to attribute B be $y_1, y_2, \ldots, y_N$, then correlation between the ranks of two attributes is defined as rank correlation.

Theorem 13.7.1 The coefficient of rank correlation, r_s, is given by

$$r_s = 1 - \frac{6 \sum d_i^2}{N(N^2 - 1)} \qquad (13.7.1)$$

where
N = number of pairs in the data,
$d_i = x_i - y_i \qquad i = 1, 2, \ldots, N$
= differences between the ranks of two attributes.

Proof. The coefficient of correlation, r, defined in Chapter 11, is given by

$$r = \frac{\sum_{i=1}^{N} (x_i - \bar{x})(y_i - \bar{y})}{\sqrt{\sum (x_i - \bar{x})^2} \sqrt{\sum (y_i - \bar{y})^2}}. \qquad (13.7.2)$$

Now if the x's and y's are given the ranks $1, 2, \ldots, N$ in some order, then it can be shown that

$$\sum x_i = \frac{N(N+1)}{2} \qquad (13.7.3)$$

and

$$\sum x_i^2 = \frac{N(N+1)(2N+1)}{6}. \qquad (13.7.4)$$

Therefore

$$\sum_{i=1}^{N}(x_i - \bar{x})^2 = \sum x_i^2 - \frac{(\sum x_i)^2}{N}$$

$$= \frac{N(N+1)(2N+1)}{6} - \frac{N^2(N+1)^2}{4N}$$

$$= \frac{N^3 - N}{12}. \qquad (13.7.5)$$

Similarly $\sum_{i=1}^{N} (y_i - \bar{y})^2 = (N^3 - N)/12.$ (13.7.6)

To find the value of $\sum_{i=1}^{N} (x_i - \bar{x})(y_i - \bar{y})$ we consider the difference between the ranks

$$d_i = x_i - y_i, \qquad i = 1, 2, \ldots, N. \qquad (13.7.7)$$

Since $\bar{x} = \bar{y}$, Equation (13.7.7) can be written as

$$\sum_{i=1}^{N} d_i^2 = \sum_{i=1}^{N} [(x_i - \bar{x}) - (y_i - \bar{y})]^2$$

or

$$\sum_{i=1}^{N} d_i^2 = \sum_{i=1}^{N} (x_i - \bar{x})^2 + \sum_{i=1}^{N} (y_i - \bar{y})^2 - 2\sum_{i=1}^{N} (x_i - \bar{x})(y_i - \bar{y}).$$

(13.7.8)

Since the observations are ranked, Equation (13.7.2) can be written as

$$r_s = \frac{\sum_{i=1}^{N}(x_i - \bar{x})(y_i - \bar{y})}{\sqrt{\sum_{i=1}^{N}(x_i - \bar{x})^2}\sqrt{\sum_{i=1}^{N}(y_i - \bar{y})^2}}.$$

Therefore

$$\sum_{i=1}^{N} d_i^2 = \sum_{i=1}^{N}(x_i - \bar{x})^2 + \sum_{i=1}^{N}(y_i - \bar{y})^2 - 2r_s\sqrt{\sum_{i=1}^{N}(x_i - \bar{x})^2 \sum_{i=1}^{N}(y_i - \bar{y})^2}.$$

Thus

$$r_s = \frac{\sum_{i=1}^{N}(x_i - \bar{x})^2 + \sum_{i=1}^{N}(y_i - \bar{y})^2 - \sum_{i=1}^{N} d_i^2}{2\sqrt{\sum_{i=1}^{N}(x_i - \bar{x})^2 \sum_{i=1}^{N}(y_i - \bar{y})^2}}.$$

(13.7.9)

Now using Equation (13.7.5) and Equation (13.7.6), Equation (13.7.9) can be written as

$$r_s = \frac{(N^3 - N)/12 + (N^3 - N)/12 - \sum_{i=1}^{N} d_i^2}{2\sqrt{[(N^3 - N)/12](N^3 - N12)/}}$$

$$= \frac{\dfrac{N^3 - N}{12} - \dfrac{\sum_{i=1}^{N} d_i^2}{2}}{(N^3 - N)/12}$$

$$= 1 - \frac{6\sum_{i=1}^{N} d_i^2}{N^3 - N}$$

(13.7.10)

Example 13.7.1 (Education) A random sample of ten college students is selected and their grades in mathematics and music are as follows.

Mathematics	75, 72, 80, 63, 45, 90, 65, 54, 85, 70
Music	93, 68, 75, 65, 53, 84, 71, 52, 81, 69

Calculate the coefficient of rank correlation.

Solution Calculation of coefficient of rank correlation is shown in Table 13.7.1.

TABLE 13.7.1

Grade in mathematics (x)	Grade in music (y)	Rank in mathematics (x)	Rank in music (y)	Difference (d)	(d^2)
75	93	4	1	3	9
72	68	5	7	−2	4
80	75	3	4	−1	1
63	65	8	8	0	0
45	53	10	9	1	1
90	84	1	2	−1	1
65	71	7	5	2	4
54	52	9	10	−1	1
85	81	2	3	−1	1
70	69	6	6	0	0
					22

Therefore

$$r_s = 1 - \frac{6 \sum_{i=1}^{10} d_i^2}{10^3 - 10}$$

$$= 1 - \frac{6(22)}{990} = 1 - \frac{132}{990}$$

$$= 1 - .1333 = .8667$$

indicating that there is a marked relationship between the grades in mathematics and grades in music.

Example 13.7.2 Calculate the coefficient of rank correlation for the data of Example 11.1.4 and compare it with the coefficient of correlation obtained in Example 11.2.1.

Solution Calculation of coefficient of rank correlation is shown in Table 13.7.2.

TABLE 13.7.2

Height of father (x)	Height of son (y)	Rank of father (x)	Rank of son (y)	d	d²
63	65	8	6	2	4
64	67	7	5	2	4
70	69	2	3	−1	1
72	70	1	2	−1	1
65	64	6	7	−1	1
67	68	4	4	0	0
68	71	3	1	2	4
66	63	5	8	−3	9
					24

Therefore

$$r_s = 1 - \frac{6 \sum_{i=1}^{N} d_i^2}{8^3 - 8}$$

$$= 1 - \frac{6(24)}{504} = 1 - \frac{144}{504}$$

$$= 1 - .2857 = .7143$$

which agrees well with the value $r = .6978$ obtained in Example 11.2.1.

13.8 SUMMARY

When the assumption concerning the normality of the population is not justified, one can use the techniques of nonparametric statistics for hypothesis testing. Those discussed are (i) the Runs test, to test for randomness; (ii) Wilcoxon's test for paired data, to compare the medians of two populations; (iii) the Wilcoxon Rank Sum test, to compare two populations on the basis of independent examples; (iv) the Kruskal-Wallis test, an extension of the Wilcoxon Rank Sum test from two to K populations; (v) Friedman's test, an extension of Wilcoxon's test for paired data from two to K populations; and (vi) Rank correlation, a measure of the relationship between two ranked variables.

EXERCISES

1. Refer to the sequence of defective items, D, and nondefective items, N, of Example 13.2.1. Use the method of runs to test whether the defectives are occurring at random or not. Use $\alpha = .05$.

2. Consider the diameter in inches of 11 bolts made by a machine

    1.3, 1.6, 1.5, 1.2, 1.4, 1.1, 1.7, 1.8, 1.9, 1.0, 1.1.

   Use the method of runs above and below the median to test whether the fluctuation from one bolt to another is random or whether a trend is present. Use $\alpha = .05$.

3. The following is a sequence of defective items, D, and nondefective items, N, produced by a machine:

    $NN\ D\ NNN\ DD\ NNNNN\ DDD\ NNNN.$

   Use the method of runs to test whether the defectives are occurring at random or not. Use $\alpha = .05$.

4. In order to measure the ability of rats to retain information, 15 matched pairs of rats are used in an experiment. The first rat in each pair (called rat A) is trained to pass through a maze, while the second rat in each pair (rat B) is left untrained. Two weeks after the end of training, the times for the rats to pass through the maze are as follows.

Pair no.	Rat A	Rat B	Pair no.	Rat A	Rat B
1	14	17	9	12	15
2	19	18	10	18	21
3	11	9	11	17	15
4	11	20	12	14	12
5	13	13	13	15	25
6	9	15	14	9	12
7	14	22	15	10	14
8	18	19			

   Use the Wilcoxon test to test the hypothesis that after two weeks, the trained rats retain none of their training.

5. Consider Chapter 9, Exercise 19. Use the Wilcoxon test to test the hypothesis that pulse rate after surgery is higher than pulse rate before surgery. Use $\alpha = .05$.

6. Take the data from Example 13.6.1 for the mathematics and English scores only (that is, ignore the psychology scores) and test the hypothesis that the two populations are equal using the Wilcoxon test.

7. Use the Wilcoxon Rank Sum test to test the hypothesis of Exercise 21 Chapter 9.

8. Use the Wilcoxon Rank Sum test to test the hypothesis of Exercise 22, Chapter 9.

9. In an attempt to assess the effects of socio-economic circumstances on the weight of newborn babies, an experimenter noted that 10 women from high-income families and 8 women from low-income families gave birth to babies with the following weights (in pounds).

High income	Low income
6.7	7.2
7.1	6.2
5.9	9.8
6.5	5.3
9.2	6.3
6.8	8.7
6.0	7.2
8.3	6.0
9.0	
6.4	

Test the hypothesis that socio-economic status has no effect on the birth of newborn babies.

10. Workers are randomly assigned to one of two machines (machines A and B) and told to perform a particular task. The times taken to perform the task are given below (in minutes).

Machine A	Machine B
17.3	18.3
15.7	14.8
12.8	12.7
14.2	19.2
20.9	21.3
24.0	18.7
16.7	16.4
9.2	15.0
	17.1

Can we say that there is any difference between the machines, in terms of the length of time needed to perform the task?

11. Two new drugs are put forward as comparable analgesics, and it is desired to test them. The variable measured is the number of hours before relief begins to take place. The following results are obtained.

Drug 1	Drug 2
.6	1.2
1.1	1.0
1.0	.9
.7	.9
.9	1.2
1.3	1.0
1.8	2.1

Use the Wilcoxon Rank Sum Test to test for a difference between the analgesics.

12. Ten batches of an industrial chemical are split into two sub-batches each, and one each is sent to laboratories A and B. Test that the two laboratories obtain the same results, if the following data (in %) are obtained.

Batch	1	2	3	4	5	6	7	8	9	10
Lab. A	12.3	13.2	9.7	12.4	14.0	10.9	13.7	8.1	15.6	12.5
Lab. B	14.7	15.6	8.3	11.2	16.8	11.3	14.0	12.6	11.0	11.8

13. Refer to the data of Exercise 4, Chapter 12. Use the Kruskal-Wallis H test to test the equality of the mean pollution level of the three rivers. Use $\alpha = .05$.

14. Refer to the data of Exercise 5, Chapter 12. Use the Kruskal-Wallis H test to test the hypothesis that four different types of feed are equally effective. Use $\alpha = .01$.

15. Refer to the data of Exercise 6, Chapter 12. Use the Kruskal-Wallis H test to test the hypothesis that the mean I.Q. of students in each district is the same. Use $\alpha = .05$.

16. Refer to the data of Exercise 7, Chapter 12. Use the Kruskal-Wallis H test to test whether the training has any effect. Use $\alpha = .01$.

17. Suppose in Exercise 12, that each batch had been split into three and the third sub-batch sent to laboratory C with the following results.

Batch	1	2	3	4	5	6	7	8	9	10
Lab. C	15.2	11.9	11.8	9.3	17.1	12.7	15.2	13.1	13.0	11.3

Use Friedman's test to see if all the laboratories obtain the same results.

18. Seven workers perform the same task on three different machines (in random order) since it is suspected that some machines are slower than others. The times (in minutes) taken to perform the tasks are given below.

Workers	Machine 1	Machine 2	Machine 3
1	24	22	29
2	28	30	35
3	26	21	27
4	21	25	31
5	27	22	21
6	25	28	26
7	20	24	30

Test the hypothesis that the machines are all the same in terms of time required to perform the task.

19. The rankings of 12 students in mathematics and statistics are as follows.

Mathematics	10	8	4	7	3	5	6	12	11	9	1	2
Statistics	8	6	3	5	1	2	11	10	12	7	4	9

Calculate the coefficient of rank correlation.

20. The following table gives ranks of 10 students in laboratory and lecture examinations of a physics course. Calculate the coefficient of rank correlation.

Laboratory	10	2	7	5	4	3	9	1	6	8
Lecture	9	1	6	10	2	5	4	3	8	7

21. Two judges in a beauty contest ranked eight competitors in the following order.

Judge A	8	3	5	1	6	2	4	7
Judge B	6	1	7	4	2	5	3	8

Do the judges have a common standard to the taste of beauty?

Appendix
TABLES

Appendix Table 1: Squares and Square Roots

N	N^2	$\sqrt{N}$	$\sqrt{10N}$	N	N^2	$\sqrt{N}$	$\sqrt{10N}$
1.0	1.00	1.000	3.162	5.5	30.25	2.345	7.416
1.1	1.21	1.049	3.317	5.6	31.36	2.366	7.483
1.2	1.44	1.095	3.464	5.7	32.49	2.387	7.550
1.3	1.69	1.140	3.606	5.8	33.64	2.408	7.616
1.4	1.96	1.183	3.742	5.9	34.81	2.429	7.681
1.5	2.25	1.225	3.873	6.0	36.00	2.449	7.746
1.6	2.56	1.265	4.000	6.1	37.21	2.470	7.810
1.7	2.89	1.304	4.123	6.2	38.44	2.490	7.874
1.8	3.24	1.342	4.243	6.3	39.69	2.510	7.937
1.9	3.61	1.378	4.359	6.4	40.96	2.530	8.000
2.0	4.00	1.414	4.472	6.5	42.25	2.550	8.062
2.1	4.41	1.449	4.583	6.6	43.56	2.569	8.124
2.2	4.84	1.483	4.690	6.7	44.89	2.588	8.185
2.3	5.29	1.517	4.796	6.8	46.24	2.608	8.246
2.4	5.76	1.549	4.899	6.9	47.61	2.627	8.307
2.5	6.25	1.581	5.000	7.0	49.00	2.646	8.367
2.6	6.76	1.612	5.099	7 1	50.41	2.665	8.426
2.7	7.29	1.643	5.196	7.2	51.84	2.683	8.485
2.8	7.84	1.673	5.292	7.3	53.29	2.702	8.544
2.9	8.41	1.703	5.385	7.4	54.76	2.720	8.602
3.0	9.00	1.732	5.477	7.5	56.25	2.739	8.660
3.1	9.61	1.761	5.568	7.6	57.76	2.757	8.718
3.2	10.24	1.789	5.657	7.7	59.29	2.775	8.775
3.3	10.89	1.817	5.745	7.8	60.84	2.793	8.832
3.4	11.56	1.844	5.831	7.9	62.41	2.811	8.888
3.5	12.25	1.871	5.916	8.0	64.00	2.828	8.944
3.6	12.96	1.897	6.000	8.1	65.61	2.846	9.000
3.7	13.69	1.924	6.083	8.2	67.24	2.864	9.055
3.8	14.44	1.949	6.164	8.3	68.89	2.881	9.110
3.9	15.21	1.975	6.245	8.4	70.56	2.898	9.165
4.0	16.00	2.000	6.325	8.5	72.25	2.915	9.220
4.1	16.81	2.025	6.403	8.6	73.96	2.933	9.274
4.2	17.64	2.049	6.481	8.7	75.69	2.950	9.327
4.3	18.49	2.074	6.557	8.8	77.44	2.966	9.381
4.4	19.36	2.098	6.633	8.9	79.21	2.983	9.434
4.5	20.25	2.121	6.708	9.0	81.00	3.000	9.487
4.6	21.16	2.145	6.782	9.1	82.81	3.017	9.539
4.7	22.09	2.168	6.856	9.2	84.64	3.033	9.592
4.8	23.04	2.191	6.928	9.3	86.49	3.050	9.644
4.9	24.01	2.214	7.000	9.4	88.36	3.066	9.695
5.0	25.00	2.236	7.071	9.5	90.25	3.082	9.747
5.1	26.01	2.258	7.141	9.6	92.16	3.098	9.798
5.2	27.04	2.280	7.211	9.7	94.09	3.114	9.849
5.8	28.09	2.302	7.280	9.8	96.04	3.130	9.899
5.4	29.16	2.324	7.348	9.9	98.01	3.146	9.950

Appendix Table 2: Areas Under the Normal Probability Curve*

A denotes the area between the line of symmetry (i.e. $z = 0$) and the given z-value.

z	A	z	A	z	A	z	A
0.00	0.0000	0.47	0.1808	0.94	0.3264	1.41	0.4207
.01	.0040	.48	.1844	.95	.3289	1.42	.4222
.02	.0080	.49	.1879	.96	.3315	1.43	.4236
.03	.0120	.50	.1915	.97	.3340	1.44	.4251
.04	.0160	.51	.1950	.98	.3365	1.45	.4265
.05	.0199	.52	.1985	.99	.3389	1.46	.4279
.06	.0239	.53	.2019	1.00	.3413	1.47	.4292
.07	.0279	.54	.2054	1.01	.3438	1.48	.4306
.08	.0319	.55	.2088	1.02	.3461	1.49	.4319
.09	.0359	.56	.2123	1.03	.3485	1.50	.4332
.10	.0398	.57	.2157	1.04	.3508	1.51	.4345
.11	.0438	.58	.2190	1.05	.3531	1.52	.4357
.12	.0478	.59	.2224	1.06	.3554	1.53	.4370
.13	.0517	.60	.2258	1.07	.3577	1.54	.4382
.14	.0557	.61	.2291	1.08	.3599	1.55	.4394
.15	.0596	.62	.2324	1.09	.3621	1.56	.4406
.16	.0636	.63	.2357	1.10	.3643	1.57	.4418
.17	.0675	.64	.2389	1.11	.3665	1.58	.4430
.18	.0714	.65	.2422	1.12	.3686	1.59	.4441
.19	.0754	.66	.2454	1.13	.3708	1.60	.4452
.20	.0793	.67	.2486	1.14	.3729	1.61	.4463
.21	.0832	.68	.2518	1.15	.3749	1.62	.4474
.22	.0871	.69	.2549	1.16	.3770	1.63	.4485
.23	.0910	.70	.2580	1.17	.3790	1.64	.4495
.24	.0948	.71	.2612	1.18	.3810	1.65	.4505
.25	.0987	.72	.2642	1.19	.3830	1.66	.4515
.26	.1026	.73	.2673	1.20	.3849	1.67	.4525
.27	.1064	.74	.2704	1.21	.3869	1.68	.4535
.28	.1103	.75	.2734	1.22	.3888	1.69	.4545
.29	.1141	.76	.2764	1.23	.3907	1.70	.4554
.30	.1179	.77	.2794	1.24	.3925	1.71	.4564
.31	.1217	.78	.2823	1.25	.3944	1.72	.4573
.32	.1255	.79	.2852	1.26	.3962	1.73	.4582
.33	.1293	.80	.2881	1.27	.3980	1.74	.4591
.34	.1331	.81	.2910	1.28	.3997	1.75	.4599
.35	.1368	.82	.2939	1.29	.4015	1.76	.4608
.36	.1406	.83	.2967	1.30	.4032	1.77	.4616
.37	.1443	.84	.2996	1.31	.4049	1.78	.4625
.38	.1480	.85	.3023	1.32	.4066	1.79	.4633
.39	.1517	.86	.3051	1.33	.4082	1.80	.4641
.40	.1554	.87	.3079	1.34	.4099	1.81	.4649
.41	.1591	.88	.3106	1.35	.4115	1.82	.4656
.42	.1628	.89	.3133	1.36	.4131	1.83	.4664
.43	.1664	.90	.3159	1.37	.4147	1.84	.4671
.44	.1700	.91	.3186	1.38	.4162	1.85	.4678
.45	.1736	.92	.3212	1.39	.4177	1.86	.4686
.46	.1772	.93	.3238	1.40	.4192	1.87	.4693

Appendix Table 2 (continued)

z	A	z	A	z	A	z	A
1.88	0.4700	2.41	0.4920	2.94	0.4984	3.47	0.4997
1.89	.4706	2.42	.4922	2.95	.4984	3.48	.4998
1.90	.4713	2.43	.4925	2.96	.4985	3.49	.4998
1.91	.4719	2.44	.4927	2.97	.4985	3.50	.4998
1.92	.4726	2.45	.4929	2.98	.4986	3.51	.4998
1.93	.4732	2.46	.4931	2.99	.4986	3.52	.4998
1.94	.4738	2.47	.4932	3.00	.4987	3.53	.4998
1.95	.4744	2.48	.4934	3.01	.4987	3.54	.4998
1.96	.4750	2.49	.4936	3.02	.4987	3.55	.4998
1.97	.4756	2.50	.4938	3.03	.4988	3.56	.4998
1.98	.4762	2.51	.4940	3.04	.4988	3.57	.4998
1.99	.4767	2.52	.4941	3.05	.4989	3.58	.4998
2.00	.4773	2.53	.4943	3.06	.4989	3.59	.4998
2.01	.4778	2.54	.4945	3.07	.4989	3.60	.4998
2.02	.4783	2.55	.4946	3.08	.4990	3.61	.4999
2.03	.4788	2.56	.4948	3.09	.4990	3.62	.4999
2.04	.4793	2.57	.4949	3.10	.4990	3.63	.4999
2.05	.4798	2.58	.4951	3.11	.4991	3.64	.4999
2.06	.4803	2.59	.4952	3.12	.4991	3.65	.4999
2.07	.4808	2.60	.4953	3.13	.4991	3.66	.4999
2.08	.4812	2.61	.4955	3.14	.4992	3.67	.4999
2.09	.4817	2.62	.4956	3.15	.4992	3.68	.4999
2.10	.4821	2.63	.4957	3.16	.4992	3.69	.4999
2.11	.4826	2.64	.4959	3.17	.4992	3.70	.4999
2.12	.4830	2.65	.4960	3.18	.4993	3.71	.4999
2.13	.4834	2.66	.4961	3.19	.4993	3.72	.4999
2.14	.4838	2.67	.4962	3.20	.4993	3.73	.4999
2.15	.4842	2.68	.4963	3.21	.4993	3.74	.4999
2.16	.4846	2.69	.4964	3.22	.4994	3.75	.4999
2.17	.4850	2.70	.4965	3.23	.4994	3.76	.4999
2.18	.4854	2.71	.4966	3.24	.4994	3.77	.4999
2.19	.4857	2.72	.4967	3.25	.4994	3.78	.4999
2.20	.4861	2.73	.4968	3.26	.4994	3.79	.4999
2.21	.4865	2.74	.4969	3.27	.4995	3.80	.4999
2.22	.4868	2.75	.4970	3.28	.4995	3.81	.4999
2.23	.4871	2.76	.4971	3.29	.4995	3.82	.4999
2.24	.4875	2.77	.4972	3.30	.4995	3.83	.4999
2.25	.4878	2.78	.4973	3.31	.4995	3.84	.4999
2.26	.4881	2.79	.4974	3.32	.4996	3.85	.4999
2.27	.4884	2.80	.4974	3.33	.4996	3.86	.4999
2.28	.4887	2.81	.4975	3.34	.4996	3.87	.5000
2.29	.4890	2.82	.4976	3.35	.4996	3.88	.5000
2.30	.4893	2.83	.4977	3.36	.4996	3.89	.5000
2.31	.4896	2.84	.4977	3.37	.4996		
2.32	.4898	2.85	.4978	3.38	.4996		
2.33	.4901	2.86	.4979	3.39	.4997		
2.34	.4904	2.87	.4980	3.40	.4997		
2.35	.4906	2.88	.4980	3.41	.4997		
2.36	.4909	2.89	.4981	3.42	.4997		
2.37	.4911	2.90	.4981	3.43	.4997		
2.38	.4910	2.91	.4982	3.44	.4997		
2.39	.4916	2.92	.4983	3.45	.4997		
2.40	.4918	2.93	.4983	3.46	.4997		

* Reproduced with permission from *Introduction to Probability and Statistics*, Fourth edition, by Henry L. Alder and Edward B. Roessler. W. H. Freeman and Company. Copyright © 1968.

Appendix Table 3: Random Numbers*

93108	77033	68325	10160	38667	62441	87023	94372	06164	30700
28271	08589	83279	48838	60935	70541	53814	95588	05832	80235
21841	35545	11148	34775	17308	88034	97765	35959	52843	44895
22025	79554	19698	25255	50283	94037	57463	92925	12042	91414
09210	20779	02994	02258	86978	85092	54052	18354	20914	28460
90552	71129	03621	20517	16908	06668	29916	51537	93658	29525
01130	06995	20258	10351	99248	51660	38861	49668	74742	47181
22604	56719	21784	68788	38358	59827	19270	99287	81193	43366
06690	01800	34272	65497	94891	14537	91358	21587	95765	72605
59809	69982	71809	64984	48709	43991	24987	69246	86400	29559
56475	02726	58511	95405	70293	84971	06676	44075	32338	31980
02730	34870	83209	03138	07715	31557	55242	61308	26507	06186
74482	33990	13509	92588	10462	76546	46097	01825	20153	36271
19793	22487	94238	81054	95488	23617	15539	94335	73822	93481
19020	27856	60526	24144	98021	60564	46373	86928	52135	74919
69565	60635	65709	77887	42766	86698	14004	94577	27936	47220
69274	23208	61035	84263	15034	28717	76146	22021	23779	98562
83658	14204	09445	41081	49630	34215	89806	40930	97194	21747
78612	51102	66826	40430	54072	62164	68977	95583	11765	81072
14980	74158	78216	38985	60838	82836	42777	85321	90463	11813
63172	28010	29405	91554	75195	51183	65805	87525	35952	83204
71167	37984	52737	06869	38122	95322	41356	19391	96787	64410
78530	56410	19195	34434	83712	50397	80920	15464	81350	18673
98324	03774	07573	67864	06497	20758	83454	22756	83959	96347
55793	30055	08373	32652	02654	75980	02095	87545	88815	80086
05674	34471	61967	91266	38814	44728	32455	17057	08339	93997
15643	22245	07592	22078	73628	60902	41561	54608	41023	98345
66750	19609	70358	03622	64898	82220	69304	46235	97332	64539
42320	74314	50222	82339	51564	42885	50482	98501	02245	88990
73752	73818	15470	04914	24936	65514	56633	72030	30856	85183
97546	02188	46373	21486	28221	08155	23486	66134	88799	49496
32569	52162	38444	42004	78011	16909	94194	79732	47114	23919
36048	93973	82596	28739	86985	58144	65007	08786	14826	04896
40455	36702	38965	56042	80023	28169	04174	65533	52718	55255
33597	47071	55618	51796	71027	46690	08002	45066	02870	60012
22828	96380	35883	15910	17211	42358	14056	55438	98148	35384
00631	95925	19324	31497	88118	06283	84596	72091	53987	01477
75722	36478	07634	63114	27164	15467	03983	09141	60562	65725
80577	01771	61510	17099	28731	41426	18853	41523	14914	76661
10524	20900	65463	83680	05005	11611	64426	59065	06758	02892
93815	69446	75253	51915	97839	75427	90685	60352	96288	34248
81867	97119	93446	20862	46591	97677	42704	13718	44975	67145
64649	07689	16711	12169	15238	74106	60655	56289	74166	78561
55768	09210	52439	33355	57884	36791	00853	49969	74814	09270
38080	49460	48137	61589	42742	92035	21766	19435	92579	27683
22360	16332	05343	34613	24013	98831	17157	44089	07366	66196
40521	09057	00239	51284	71556	22605	41293	54854	39736	05113
19292	69862	59951	49644	53486	28244	20714	56030	39292	45166
79504	40078	06838	05509	68581	39400	85615	52314	83202	40313
64138	27983	84048	42631	58658	62243	82572	45211	37060	15017

* Abstracted with permission from *A Million Random Digits with 100,000 Normal Deviates* by The Rand Corporation, The Free Press of Glencoe, New York (1955).

Appendix Table 4: Percentage Points of the t-Distribution*

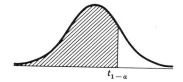

t_{1-a}

d.f.	$t_{.60}$	$t_{.70}$	$t_{.80}$	$t_{.90}$	$t_{.95}$	$t_{.975}$	$t_{.99}$	$t_{.995}$	$t_{.9995}$
1	0.3250	0.7270	1.376	3.078	6.3138	12.706	31.821	63.657	636.619
2	.2885	.6172	1.061	1.886	2.9200	4.3027	6.965	9.9248	31.598
3	.2766	.5840	.978	1.638	2.3534	3.1825	4.541	5.8409	12.924
4	.2707	.5692	.941	1.533	2.1318	2.7764	3.747	4.6041	8.610
5	.2672	.5598	.920	1.476	2.0150	2.5706	3.365	4.0321	6.869
6	.2648	.5536	.906	1.440	1.9432	2.4469	3.143	3.7074	5.959
7	.2632	.5493	.896	1.415	1.8946	2.3646	2.998	3.4995	5.408
8	.2619	.5461	.889	1.397	1.8595	2.3060	2.896	3.3554	5.041
9	.2610	.5436	.883	1.383	1.8331	2.2622	2.821	3.2498	4.781
10	.2602	.5416	.879	1.372	1.8125	2.2281	2.764	3.1693	4.587
11	.2596	.5400	.876	1.363	1.7939	2.2010	2.718	3.1058	4.437
12	.2590	.5387	.873	1.356	1.7823	2.1788	2.681	3.0545	4.318
13	.2586	.5375	.870	1.350	1.7709	2.1604	2.650	3.0123	4.221
14	.2582	.5366	.868	1.345	1.7613	2.1448	2.624	2.9768	4.140
15	.2579	.5358	.866	1.341	1.7530	2.1315	2.602	2.9467	4.073
16	.2576	.5351	.865	1.337	1.7459	2.1199	2.583	2.9208	4.015
17	.2574	.5344	.863	1.333	1.7396	2.1098	2.567	2.8982	3.965
18	.2571	.5338	.862	1.330	1.7341	2.1009	2.552	2.8784	3.922
19	.2569	.5333	.861	1.328	1.7291	2.0930	2.539	2.8609	3.883
20	.2567	.5329	.860	1.325	1.7247	2.0860	2.528	2.8453	3.850
21	.2566	.5325	.859	1.323	1.7207	2.0796	2.518	2.8314	3.819
22	.2564	.5321	.858	1.321	1.7171	2.0739	2.508	2.8188	3.792
23	.2563	.5318	.858	1.319	1.7139	2.0687	2.500	2.9073	3.767
24	.2562	.5315	.857	1.318	1.7109	2.0639	2.492	2.7969	3.745
25	.2561	.5312	.856	1.316	1.7081	2.0595	2.485	2.7874	3.725
26	.2560	.5309	.856	1.315	1.7056	2.0555	2.479	2.7787	3.707
27	.2559	.5307	.855	1.314	1.7033	2.0518	2.473	2.7707	3.690
28	.2558	.5304	.855	1.313	1.7011	2.0484	2.467	2.7633	3.674
29	.2557	.5302	.854	1.311	1.6991	2.0452	2.462	2.7564	3.659
30	.2556	.5300	.854	1.310	1.6973	2.0423	2.457	2.7500	3.616
35	.2553	.5292	.8521	1.3062	1.6896	2.0301	2.438	2.7239	3.5919
40	.2550	.5286	.8507	1.3031	1.6839	2.0211	2.423	2.7045	3.5511
45	.2549	.5281	.8497	1.3007	1.6794	2.0141	2.412	2.6896	3.5207
50	.2547	.5278	.8489	1.2987	1.6759	2.0086	2.403	2.6778	3.4955
60	.2545	.5272	.8477	1.2959	1.6707	2.0003	2.390	2.6603	3.4606
70	.2543	.5268	.8468	1.2938	1.6669	1.9945	2.381	2.6480	3.4355
80	.2542	.5265	.8462	1.2922	1.6641	1.9901	2.374	2.6388	3.4169
90	.2541	.5263	.8457	1.2910	1.6620	1.9867	2.368	2.6316	3.4022
100	.2540	.5261	.8452	1.2901	1.6602	1.9840	2.364	2.6260	3.3909
120	2539	.5258	.8446	1.2887	1.6577	1.9799	2.358	2.6175	3.3736
140	.2538	.5256	.8442	1.2876	1.6558	1.9771	2.353	2.6114	3.3615
160	.2538	.5255	.8439	1.2869	1 6545	1.9749	2.350	2.6070	3.3527
180	.2537	.5253	.8436	1.2863	1.6534	1.9733	2.347	2.6035	3.3456
200	.2537	.5252	.8434	1.2858	1.6525	1.9719	2.345	2.6006	3.3400
∞	.2533	.5244	.8416	1.2816	1.6449	1.9600	2.326	2.5758	3.2905

* Reproduced from *Documenta Geigy Scientific Tables*, 7th edition, by permission of CIBA-GEIGY Limited, Basle, Switzerland.

Appendix Table 5: Percentage Points of the χ^2-Distribution*

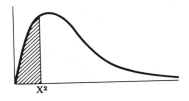

d.f.	$\chi^2_{.005}$	$\chi^2_{.010}$	$\chi^2_{.025}$	$\chi^2_{.050}$	$\chi^2_{.010}$	$\chi^2_{.250}$	$\chi^2_{.500}$
1	392704.10⁻¹⁰	157088.10⁻⁹	982069.10⁻⁹	393214.10⁻⁸	0·0157908	0·1015308	0·454936
2	0·0100251	0·0201007	0·0506356	0·102587	0·210721	0·575364	1·38629
3	0·0717218	0·114832	0·215795	0·351846	0·584374	1·212534	2·36597
4	0·206989	0·297109	0·484419	0·710723	1·063623	1·92256	3·35669
5	0·411742	0·554298	0·831212	1·145476	1·61031	2·67460	4·35146
6	0·675727	0·872090	1·23734	1·63538	2·20413	3·45460	5·34812
7	0·989256	1·239043	1·68987	2·16735	2·83311	4·25485	6·34581
8	1·34441	1·64650	2·17973	2·73264	3·48954	5·07064	7·34412
9	1·73493	2·08790	2·70039	3·32511	4·16816	5·89883	8·34283
10	2·15586	2·55821	3·24697	3·94030	4·86518	6·73720	9·34182
11	2·60322	3·05348	3·81575	4·57481	5·57778	7·58414	10·3410
12	3·07382	3·57057	4·40379	5·22603	6·30380	8·43842	11·3403
13	3·56503	4·10692	5·00875	5·89186	7·04150	9·29907	12·3398
14	4·07467	4·66043	5·62873	6·57063	7·78953	10·1653	13·3393
15	4·60092	5·22935	6·26214	7·26094	8·54676	11·0365	14·3389
16	5·14221	5·81221	6·90766	7·96165	9·31224	11·9122	15·3385
17	5·69722	6·40776	7·56419	8·67176	10·0852	12·7919	16·3382
18	6·26480	7·01491	8·23075	9·39046	10·8649	13·6753	17·3379
19	6·84397	7·63273	8·90652	10·1170	11·6509	14·5620	18·3377
20	7·43384	8·26040	9·59078	10·8508	12·4426	15·4518	19·3374
21	8·03365	8·89720	10·28293	11·5913	13·2396	16·3444	20·3372
22	8·64272	9·54249	10·9823	12·3380	14·0415	17·2396	21·3370
23	9·26043	10·19567	11·6886	13·0905	14·8480	18·1373	22·3369
24	9·88623	10·8564	12·4012	13·8484	15·6587	19·0373	23·3367
25	10·5197	11·5240	13·1197	14·6114	16·4734	19·9393	24·3366
26	11·1602	12·1981	13·8439	15·3792	17·2919	20·8434	25·3365
27	11·8076	12·8785	14·5734	16·1514	18·1139	21·7494	26·3363
28	12·4613	13·5647	15·3079	16·9279	18·9392	22·6572	27·3362
29	13·1211	14·2565	16·0471	17·7084	19·7677	23·5666	28·3361
30	13·7867	14·9535	16·7908	18·4927	20·5992	24·4776	29·3360
40	20·7065	22·1643	24·4330	26·5093	29·0505	33·6603	39·3353
50	27·9907	29·7067	32·3574	34·7643	37·6886	42·9421	49·3349
60	35·5345	37·4849	40·4817	43·1880	46·4589	52·2938	59·3347
70	43·2752	45·4417	48·7576	51·7393	55·3289	61·6983	69·3345
80	51·1719	53·5401	57·1532	60·3915	64·2778	71·1445	79·3343
90	59·1963	61·7541	65·6466	69·1260	73·2911	80·6247	89·3342
100	67·3276	70·0649	74·2219	77·9295	82·3581	90·1332	99·3341
X	−2·5758	−2·3263	−1·9600	−1·6449	−1·2816	−0·6745	0·0000

* Abridged with permission from *Biometrika Tables for Statisticians*, Vol. I, Edited by E. S. Pearson and H. O. Hartley, Cambridge University Press (1962).

Appendix Table 5 (continued)

d.f.	$\chi^2_{.750}$	$\chi^2_{.900}$	$\chi^2_{.950}$	$\chi^2_{.975}$	$\chi^2_{.990}$	$\chi^2_{.995}$	$\chi^2_{.999}$
1	1·32330	2·70554	3·84146	5·02389	6·63490	7·87944	10·828
2	2·77259	4·60517	5·99146	7·37776	9·21034	10·5966	13·816
3	4·10834	6·25139	7·81473	9·34840	11·3449	12·8382	16·266
4	5·38527	7·77944	9·48773	11·1433	13·2767	14·8603	18·467
5	6·62568	9·23636	11·0705	12·8325	15·0863	16·7496	20·515
6	7·84080	10·6446	12·5916	14·4494	16·8119	18·5476	22·458
7	9·03715	12·0170	14·0671	16·0128	18·4753	20·2777	24·322
8	10·2189	13·3616	15·5073	17·5345	20·0902	21·9550	26·125
9	11·3888	14·6837	16·9190	19·0228	21·6660	23·5894	27·877
10	12·5489	15·9872	18·3070	20·4832	23·2093	25·1882	29·588
11	13·7007	17·2750	19·6751	21·9200	24·7250	26·7568	31·264
12	14·8454	18·5493	21·0261	23·3367	26·2170	28·2995	32·909
13	15·9839	19·8119	22·3620	24·7356	27·6882	29·8195	34·528
14	17·1169	21·0641	23·6848	26·1189	29·1412	31·3194	36·123
15	18·2451	22·3071	24·9958	27·4884	30·5779	32·8013	37·697
16	19·3689	23·5418	26·2962	28·8454	31·9999	34·2672	39·252
17	20·4887	24·7690	27·5871	30·1910	33·4087	35·7185	40·790
18	21·6049	25·9894	28·8693	31·5264	34·8053	37·1565	42·312
19	22·7178	27·2036	30·1435	32·8523	36·1909	38·5823	43·820
20	23·8277	28·4120	31·4104	34·1696	37·5662	39·9968	45·315
21	24·9348	29·6151	32·6706	35·4789	38·9322	41·4011	46·797
22	26·0393	30·8133	33·9244	36·7807	40·2894	42·7957	48·268
23	27·1413	32·0069	35·1725	38·0756	41·6384	44·1813	49·728
24	28·2412	33·1962	36·4150	39·3641	42·9798	45·5585	51·179
25	29·3389	34·3816	37·6525	40·6465	44·3141	46·9279	52·618
26	30·4346	35·5632	38·8851	41·9232	45·6417	48·2899	54·052
27	31·5284	36·7412	40·1133	43·1945	46·9629	49·6449	55·476
28	32·6205	37·9159	41·3371	44·4608	48·2782	50·9934	56·892
29	33·7109	39·0875	42·5570	45·7223	49·5879	52·3356	58·301
30	34·7997	40·2560	43·7730	46·9792	50·8922	53·6720	59·703
40	45·6160	51·8051	55·7585	59·3417	63·6907	66·7660	73·402
50	56·3336	63·1671	67·5048	71·4202	76·1539	79·4900	86·661
60	66·9815	74·3970	79·0819	83·2977	88·3794	91·9517	99·607
70	77·5767	85·5270	90·5312	95·0232	100·425	104·215	112·317
80	88·1303	96·5782	101·879	106·629	112·329	116·321	124·839
90	98·6499	107·565	113·145	118·136	124·116	128·299	137·208
100	109·141	118·498	124·342	129·561	135·807	140·169	149·449
X	+0·6745	+1·2816	+1·6449	+1·9600	+2·3263	+2·5758	+3·0902

Appendix Table 6a: 95% Significance Points of the F-Distribution*

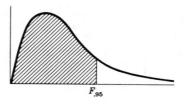

ν_1 \ ν_2	1	2	3	4	5	6	7	8	9
1	161.4	199.5	215.7	224.6	230.2	234.0	236.8	238.9	240.5
2	18.51	19.00	19.16	19.25	19.30	19.33	19.35	19.37	19.38
3	10.13	9.55	9.28	9.12	9.01	8.94	8.89	8.85	8.81
4	7.71	6.94	6.59	6.39	6.26	6.16	6.09	6.04	6.00
5	6.61	5.79	5.41	5.19	5.05	4.95	4.88	4.82	4.77
6	5.99	5.14	4.76	4.53	4.39	4.28	4.21	4.15	4.10
7	5.59	4.74	4.35	4.12	3.97	3.87	3.79	3.73	3.68
8	5.32	4.46	4.07	3.84	3.69	3.58	3.50	3.44	3.39
9	5.12	4.26	3.86	3.63	3.48	3.37	3.29	3.23	3.18
10	4.96	4.10	3.71	3.48	3.33	3.22	3.14	3.07	3.02
11	4.84	3.98	3.59	3.36	3.20	3.09	3.01	2.95	2.90
12	4.75	3.89	3.49	3.26	3.11	3.00	2.91	2.85	2.80
13	4.67	3.81	3.41	3.18	3.03	2.92	2.83	2.77	2.71
14	4.60	3.74	3.34	3.11	2.96	2.85	2.76	2.70	2.65
15	4.54	3.68	3.29	3.06	2.90	2.79	2.71	2.64	2.59
16	4.49	3.63	3.24	3.01	2.85	2.74	2.66	2.59	2.54
17	4.45	3.59	3.20	2.96	2.81	2.70	2.61	2.55	2.49
18	4.41	3.55	3.16	2.93	2.77	2.66	2.58	2.51	2.46
19	4.38	3.52	3.13	2.90	2.74	2.63	2.54	2.48	2.42
20	4.35	3.49	3.10	2.87	2.71	2.60	2.51	2.45	2.39
21	4.32	3.47	3.07	2.84	2.68	2.57	2.49	2.42	2.37
22	4.30	3.44	3.05	2.82	2.66	2.55	2.46	2.40	2.34
23	4.28	3.42	3.03	2.80	2.64	2.53	2.44	2.37	2.32
24	4.26	3.40	3.01	2.78	2.62	2.51	2.42	2.36	2.30
25	4.24	3.39	2.99	2.76	2.60	2.49	2.40	2.34	2.28
26	4.23	3.37	2.98	2.74	2.59	2.47	2.39	2.32	2.27
27	4.21	3.35	2.96	2.73	2.57	2.46	2.37	2.31	2.25
28	4.20	3.34	2.95	2.71	2.56	2.45	2.36	2.29	2.24
29	4.18	3.33	2.93	2.70	2.55	2.43	2.35	2.28	2.22
30	4.17	3.32	2.92	2.69	2.53	2.42	2.33	2.27	2.21
40	4.08	3.23	2.84	2.61	2.45	2.34	2.25	2.18	2.12
60	4.00	3.15	2.76	2.53	2.37	2.25	2.17	2.10	2.04
120	3.92	3.07	2.68	2.45	2.29	2.17	2.09	2.02	1.96
∞	3.84	3.00	2.60	2.37	2.21	2.10	2.01	1.94	1.88

* Abridged with permission from *Biometrika Tables for Statisticians*, Vol. I, Edited by E. S. Pearson and H. O. Hartley, Cambridge University Press (1962).

Appendix Table 6a (continued)

ν_1 / ν_2	10	12	15	20	24	30	40	60	120	∞
1	241·9	243·9	245·9	248·0	249·1	250·1	251·1	252·2	253·3	254·3
2	19·40	19·41	19·43	19·45	19·45	19·46	19·47	19·48	19·49	19·50
3	8·79	8·74	8·70	8·66	8·64	8·62	8·59	8·57	8·55	8·53
4	5·96	5·91	5·86	5·80	5·77	5·75	5·72	5·69	5·66	5·63
5	4·74	4·68	4·62	4·56	4·53	4·50	4·46	4·43	4·40	4·36
6	4·06	4·00	3·94	3·87	3·84	3·81	3·77	3·74	3·70	3·67
7	3·64	3·57	3·51	3·44	3·41	3·38	3·34	3·30	3·27	3·23
8	3·35	3·28	3·22	3·15	3·12	3·08	3·04	3·01	2·97	2·93
9	3·14	3·07	3·01	2·94	2·90	2·86	2·83	2·79	2·75	2·71
10	2·98	2·91	2·85	2·77	2·74	2·70	2·66	2·62	2·58	2·54
11	2·85	2·79	2·72	2·65	2·61	2·57	2·53	2·49	2·45	2·40
12	2·75	2·69	2·62	2·54	2·51	2·47	2·43	2·38	2·34	2·30
13	2·67	2·60	2·53	2·46	2·42	2·38	2·34	2·30	2·25	2·21
14	2·60	2·53	2·46	2·39	2·35	2·31	2·27	2·22	2·18	2·13
15	2·54	2·48	2·40	2·33	2·29	2·25	2·20	2·16	2·11	2·07
16	2·49	2·42	2·35	2·28	2·24	2·19	2·15	2·11	2·06	2·01
17	2·45	2·38	2·31	2·23	2·19	2·15	2·10	2·06	2·01	1·96
18	2·41	2·34	2·27	2·19	2·15	2·11	2·06	2·02	1·97	1·92
19	2·38	2·31	2·23	2·16	2·11	2·07	2·03	1·98	1·93	1·88
20	2·35	2·28	2·20	2·12	2·08	2·04	1·99	1·95	1·90	1·84
21	2·32	2·25	2·18	2·10	2·05	2·01	1·96	1·92	1·87	1·81
22	2·30	2·23	2·15	2·07	2·03	1·98	1·94	1·89	1·84	1·78
23	2·27	2·20	2·13	2·05	2·01	1·96	1·91	1·86	1·81	1·76
24	2·25	2·18	2·11	2·03	1·98	1·94	1·89	1·84	1·79	1·73
25	2·24	2·16	2·09	2·01	1·96	1·92	1·87	1·82	1·77	1·71
26	2·22	2·15	2·07	1·99	1·95	1·90	1·85	1·80	1·75	1·69
27	2·20	2·13	2·06	1·97	1·93	1·88	1·84	1·79	1·73	1·67
28	2·19	2·12	2·04	1·96	1·91	1·87	1·82	1·77	1·71	1·65
29	2·18	2·10	2·03	1·94	1·90	1·85	1·81	1·75	1·70	1·64
30	2·16	2·09	2·01	1·93	1·89	1·84	1·79	1·74	1·68	1·62
40	2·08	2·00	1·92	1·84	1·79	1·74	1·69	1·64	1·58	1·51
60	1·99	1·92	1·84	1·75	1·70	1·65	1·59	1·53	1·47	1·39
120	1·91	1·83	1·75	1·66	1·61	1·55	1·50	1·43	1·35	1·25
∞	1·83	1·75	1·67	1·57	1·52	1·46	1·39	1·32	1·22	1·00

Appendix Table 6b: 99% Significant Points of the F-Distribution*

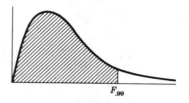

$F_{.99}$

v_2 \ v_1	1	2	3	4	5	6	7	8	9
1	4052	4999.5	5403	5625	5764	5859	5928	5981	6022
2	98.50	99.00	99.17	99.25	99.30	99.33	99.36	99.37	99.39
3	34.12	30.82	29.46	28.71	28.24	27.91	27.67	27.49	27.35
4	21.20	18.00	16.69	15.98	15.52	15.21	14.98	14.80	14.66
5	16.26	13.27	12.06	11.39	10.97	10.67	10.46	10.29	10.16
6	13.75	10.92	9.78	9.15	8.75	8.47	8.26	8.10	7.98
7	12.25	9.55	8.45	7.85	7.46	7.19	6.99	6.84	6.72
8	11.26	8.65	7.59	7.01	6.63	6.37	6.18	6.03	5.91
9	10.56	8.02	6.99	6.42	6.06	5.80	5.61	5.47	5.35
10	10.04	7.56	6.55	5.99	5.64	5.39	5.20	5.06	4.94
11	9.65	7.21	6.22	5.67	5.32	5.07	4.89	4.74	4.63
12	9.33	6.93	5.95	5.41	5.06	4.82	4.64	4.50	4.39
13	9.07	6.70	5.74	5.21	4.86	4.62	4.44	4.30	4.19
14	8.86	6.51	5.56	5.04	4.69	4.46	4.28	4.14	4.03
15	8.68	6.36	5.42	4.89	4.56	4.32	4.14	4.00	3.89
16	8.53	6.23	5.29	4.77	4.44	4.20	4.03	3.89	3.78
17	8.40	6.11	5.18	4.67	4.34	4.10	3.93	3.79	3.68
18	8.29	6.01	5.09	4.58	4.25	4.01	3.84	3.71	3.60
19	8.18	5.93	5.01	4.50	4.17	3.94	3.77	3.63	3.52
20	8.10	5.85	4.94	4.43	4.10	3.87	3.70	3.56	3.46
21	8.02	5.78	4.87	4.37	4.04	3.81	3.64	3.51	3.40
22	7.95	5.72	4.82	4.31	3.99	3.76	3.59	3.45	3.35
23	7.88	5.66	4.76	4.26	3.94	3.71	3.54	3.41	3.30
24	7.82	5.61	4.72	4.22	3.90	3.67	3.50	3.36	3.26
25	7.77	5.57	4.68	4.18	3.85	3.63	3.46	3.32	3.22
26	7.72	5.53	4.64	4.14	3.82	3.59	3.42	3.29	3.18
27	7.68	5.49	4.60	4.11	3.78	3.56	3.39	3.26	3.15
28	7.64	5.45	4.57	4.07	3.75	3.53	3.36	3.23	3.12
29	7.60	5.42	4.54	4.04	3.73	3.50	3.33	3.20	3.09
30	7.56	5.39	4.51	4.02	3.70	3.47	3.30	3.17	3.07
40	7.31	5.18	4.31	3.83	3.51	3.29	3.12	2.99	2.89
60	7.08	4.98	4.13	3.65	3.34	3.12	2.95	2.82	2.72
120	6.85	4.79	3.95	3.48	3.17	2.96	2.79	2.66	2.56
∞	6.63	4.61	3.78	3.32	3.02	2.80	2.64	2.51	2.41

* Abridged with permission from *Biometrika Tables for Statisticians*, Vol. I, Edited by E. S. Pearson and H. O. Hartley, Cambridge University Press (1962).

Appendix Table 6b (continued)

ν_2 \ ν_1	10	12	15	20	24	30	40	60	120	∞
1	6056	6106	6157	6209	6235	6261	6287	6313	6339	6366
2	99.40	99.42	99.43	99.45	99.46	99.47	99.47	99.48	99.49	99.50
3	27.23	27.05	26.87	26.69	26.60	26.50	26.41	26.32	26.22	26.13
4	14.55	14.37	14.20	14.02	13.93	13.84	13.75	13.65	13.56	13.46
5	10.05	9.89	9.72	9.55	9.47	9.38	9.29	9.20	9.11	9.02
6	7.87	7.72	7.56	7.40	7.31	7.23	7.14	7.06	6.97	6.88
7	6.62	6.47	6.31	6.16	6.07	5.99	5.91	5.82	5.74	5.65
8	5.81	5.67	5.52	5.36	5.28	5.20	5.12	5.03	4.95	4.86
9	5.26	5.11	4.96	4.81	4.73	4.65	4.57	4.48	4.40	4.31
10	4.85	4.71	4.56	4.41	4.33	4.25	4.17	4.08	4.00	3.91
11	4.54	4.40	4.25	4.10	4.02	3.94	3.86	3.78	3.69	3.60
12	4.30	4.16	4.01	3.86	3.78	3.70	3.62	3.54	3.45	3.36
13	4.10	3.96	3.82	3.66	3.59	3.51	3.43	3.34	3.25	3.17
14	3.94	3.80	3.66	3.51	3.43	3.35	3.27	3.18	3.09	3.00
15	3.80	3.67	3.52	3.37	3.29	3.21	3.13	3.05	2.96	2.87
16	3.69	3.55	3.41	3.26	3.18	3.10	3.02	2.93	2.84	2.75
17	3.59	3.46	3.31	3.16	3.08	3.00	2.92	2.83	2.75	2.65
18	3.51	3.37	3.23	3.08	3.00	2.92	2.84	2.75	2.66	2.57
19	3.43	3.30	3.15	3.00	2.92	2.84	2.76	2.67	2.58	2.49
20	3.37	3.23	3.09	2.94	2.86	2.78	2.69	2.61	2.52	2.42
21	3.31	3.17	3.03	2.88	2.80	2.72	2.64	2.55	2.46	2.36
22	3.26	3.12	2.98	2.83	2.75	2.67	2.58	2.50	2.40	2.31
23	3.21	3.07	2.93	2.78	2.70	2.62	2.54	2.45	2.35	2.26
24	3.17	3.03	2.89	2.74	2.66	2.58	2.49	2.40	2.31	2.21
25	3.13	2.99	2.85	2.70	2.62	2.54	2.45	2.36	2.27	2.17
26	3.09	2.96	2.81	2.66	2.58	2.50	2.42	2.33	2.23	2.13
27	3.06	2.93	2.78	2.63	2.55	2.47	2.38	2.29	2.20	2.10
28	3.03	2.90	2.75	2.60	2.52	2.44	2.35	2.26	2.17	2.06
29	3.00	2.87	2.73	2.57	2.49	2.41	2.33	2.23	2.14	2.03
30	2.98	2.84	2.70	2.55	2.47	2.39	2.30	2.21	2.11	2.01
40	2.80	2.66	2.52	2.37	2.29	2.20	2.11	2.02	1.92	1.80
60	2.63	2.50	2.35	2.20	2.12	2.03	1.94	1.84	1.73	1.60
120	2.47	2.34	2.19	2.03	1.95	1.86	1.76	1.66	1.53	1.38
∞	2.32	2.18	2.04	1.88	1.79	1.70	1.59	1.47	1.32	1.00

Appendix Table 7a: Binomial Confidence Intervals with $1 - \alpha = .95$*

The numbers printed along the curves indicate the sample size n. If for a given value of the abscissa c/n, p_A and p_B are the ordinates read from (or interpolated between) the appropriate lower and upper curves, then
$$Pr\{p_A \leqslant p \leqslant p_B\} \leqslant 1 - 2\alpha.$$

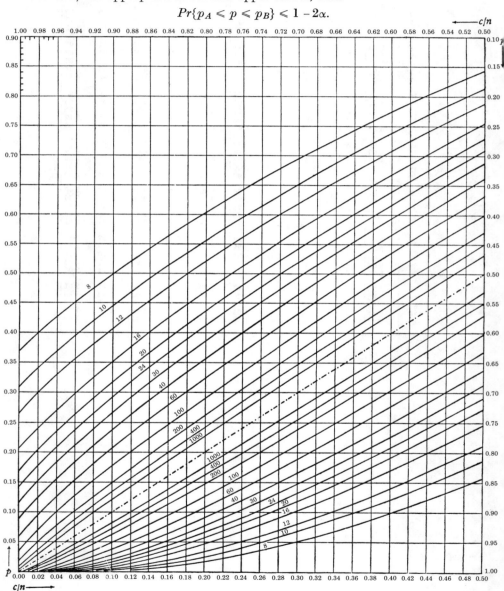

* Reproduced with permission from *Biometrika Tables for Statisticians*, Vol. I, Edited by E. S. Pearson and H. O. Hartley, Cambridge University Press (1962).

Appendix Table 7b: Binomial Confidence Intervals with $1 - \alpha = .99$*

The numbers printed along the curves indicate the sample size n.

Note: the process of reading from the curves can be simplified with the help of the right-angled corner of a loose sheet of paper or thin card, along the edges of which are marked off the scales shown in the top left-hand corner of each chart.

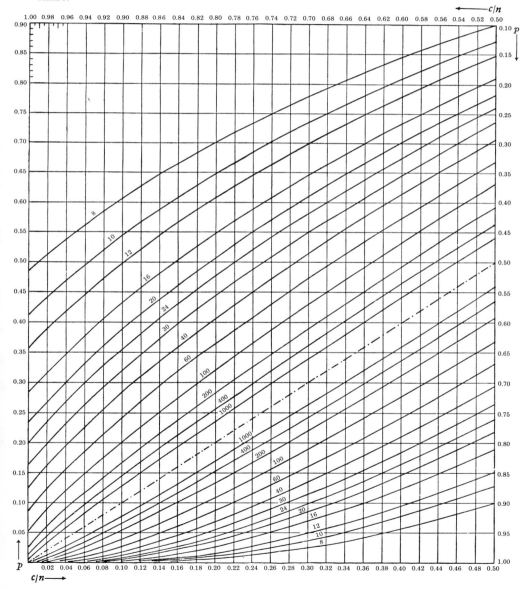

* Reproduced with permission from *Biometrika Tables for Statisticians*, Vol. I, Edited by E. S. Pearson and H. O. Hartley, Cambridge University Press (1962).

Appendix Table 8: Poisson Confidence Intervals*

If c is the observed frequency or count and m_A, m_B are the lower and upper confidence limits for its expectation, m, then

$$Pr\{m_A \leq m \leq m_B\} \leq 1 - 2\alpha$$

$1-2\alpha$	0.998		0.99		0.98		0.95		0.90		$1-2\alpha$
α	0.001		0.005		0.01		0.025		0.05		α
c	Lower	Upper	Lower	Upper	Lower	Upper	Lower	Upper	Lower	Upper	c
0	0.00000	6.91	0.00000	5.30	0.0000	4.61	0.0000	3.69	0.0000	3.00	0
1	·00100	9.23	·00501	7.43	·0101	6.64	·0253	5.57	·0513	4.74	1
2	·0454	11.23	·103	9.27	·149	8.41	·242	7.22	·355	6.30	2
3	·191	13.06	·338	10.98	·436	10.05	·619	8.77	·818	7.75	3
4	·429	14.79	·672	12.59	·823	11.60	1.09	10.24	1.37	9.15	4
5	0.739	16.45	1.08	14.15	1.28	13.11	1.62	11.67	1.97	10.51	5
6	1.11	18.06	1.54	15.66	1.79	14.57	2.20	13.06	2.61	11.84	6
7	1.52	19.63	2.04	17.13	2.33	16.00	2.81	14.42	3.29	13.15	7
8	1.97	21.16	2.57	18.58	2.91	17.40	3.45	15.76	3.98	14.43	8
9	2.45	22.66	3.13	20.00	3.51	18.78	4.12	17.08	4.70	15.71	9
10	2.96	24.13	3.72	21.40	4.13	20.14	4.80	18.39	5.43	16.96	10
11	3.49	25.59	4.32	22.78	4.77	21.49	5.49	19.68	6.17	18.21	11
12	4.04	27.03	4.94	24.14	5.43	22.82	6.20	20.96	6.92	19.44	12
13	4.61	28.45	5.58	25.50	6.10	24.14	6.92	22.23	7.69	20.67	13
14	5.20	29.85	6.23	26.84	6.78	25.45	7.65	23.49	8.46	21.89	14
15	5.79	31.24	6.89	28.16	7.48	26.74	8.40	24.74	9.25	23.10	15
16	6.41	32.62	7.57	29.48	8.18	28.03	9.15	25.98	10.04	24.30	16
17	7.03	33.99	8.25	30.79	8.89	29.31	9.90	27.22	10.83	25.50	17
18	7.66	35.35	8.94	32.09	9.62	30.58	10.67	28.45	11.63	26.69	18
19	8.31	36.70	9.64	33.38	10.35	31.85	11.44	29.67	12.44	27.88	19
20	8.96	38.04	10.35	34.67	11.08	33.10	12.22	30.89	13.25	29.06	20
21	9.62	39.38	11.07	35.95	11.82	34.36	13.00	32.10	14.07	30.24	21
22	10.29	40.70	11.79	37.22	12.57	35.60	13.79	33.31	14.89	31.42	22
23	10.96	42.02	12.52	38.48	13.33	36.84	14.58	34.51	15.72	32.59	23
24	11.65	43.33	13.25	39.74	14.09	38.08	15.38	35.71	16.55	33.75	24
25	12.34	44.64	14.00	41.00	14.85	39.31	16.18	36.90	17.38	34.92	25
26	13.03	45.94	14.74	42.25	15.62	40.53	16.98	38.10	18.22	36.08	26
27	13.73	47.23	15.49	43.50	16.40	41.76	17.79	39.28	19.06	37.23	27
28	14.44	48.52	16.24	44.74	17.17	42.98	18.61	40.47	19.90	38.39	28
29	15.15	49.80	17.00	45.98	17.96	44.19	19.42	41.65	20.75	39.54	29
30	15.87	51.08	17.77	47.21	18.74	45.40	20.24	42.83	21.59	40.69	30
35	19.52	57.42	21.64	53.32	22.72	51.41	24.38	48.68	25.87	46.40	35
40	23.26	63.66	25.59	59.36	26.77	57.35	28.58	54.47	30.20	52.07	40
45	27.08	69.83	29.60	65.34	30.88	63.23	32.82	60.21	34.56	57.69	45
50	30.96	75.94	33.66	71.27	35.03	69.07	37.11	65.92	38.96	63.29	50

* Reproduced with permission from *Biometrika Tables for Statisticians*, Vol. I, Edited by E. S. Pearson and H. O. Hartley, Cambridge University Press (1962).

Appendix Table 9: Values of Fisher's Z Transformation*: $Z = \frac{1}{2}\log_e(1+r)/(1-r)$

r	Z	r	Z	r	Z
.00	.000				
.01	.010	.36	.377	.71	.887
.02	.020	.37	.388	.72	.908
.03	.030	.38	.400	.73	.929
.04	.040	.39	.412	.74	.950
.05	.050	.40	.424	.75	.973
.06	.060	.41	.436	.76	.996
.07	.070	.42	.448	.77	1.020
.08	.080	.43	.460	.78	1.045
.09	.090	.44	.472	.79	1.071
.10	.100	.45	.485	.80	1.099
.11	.110	.46	.497	.81	1.127
.12	.121	.47	.510	.82	1.157
.13	.131	.48	.523	.83	1.188
.14	.141	.49	.536	.84	1.221
.15	.151	.50	.549	.85	1.256
.16	.161	.51	.563	.86	1.293
.17	.172	.52	.576	.87	1.333
.18	.182	.53	.590	.88	1.376
.19	.192	.54	.604	.89	1.422
.20	.203	.55	.618	.90	1.472
.21	.213	.56	.633	.91	1.528
.22	.224	.57	.648	.92	1.589
.23	.234	.58	.662	.93	1.658
.24	.245	.59	.678	.94	1.738
.25	.255	.60	.693	.95	1.832
.26	.266	.61	.709	.96	1.946
.27	.277	.62	.725	.97	2.092
.28	.288	.63	.741	.98	2.298
.29	.299	.64	.758	.99	2.647
.30	.310	.65	.775		
.31	.321	.66	.793		
.32	.332	.67	.811		
.33	.343	.68	.829		
.34	.354	.69	.848		
.35	.365	.70	.867		

* Abridged from Table VII of Fisher and Yates ; *Statistical Tables for Biological, Agricultural, and Medical Research* ; published by Oliver and Boyd Limited, Edinburgh, by permission of the authors and publishers.

Appendix Table 10: Duncan's Critical Difference Multipliers for α = .05*

d.f. \ p	2	3	4	5	6	7	8	9	10	12	14	16
1	17.97											
2	6.09	6.09										
3	4.50	4.52	4.52									
4	3.93	4.01	4.03	4.03								
5	3.64	3.75	3.80	3.81	3.81							
6	3.46	3.59	3.65	3.68	3.69	3.70						
7	3.34	3.48	3.55	3.59	3.61	3.62	3.63					
8	3.26	3.40	3.48	3.52	3.55	3.57	3.57	3.58				
9	3.20	3.34	3.42	3.47	3.50	3.52	3.54	3.54	3.55			
10	3.15	3.29	3.38	3.43	3.47	3.49	3.51	3.52	3.52			
11	3.11	3.26	3.34	3.40	3.44	3.46	3.48	3.49	3.50	3.51		
12	3.08	3.23	3.31	3.37	3.41	3.44	3.46	3.47	3.48	3.50		
13	3.06	3.20	3.29	3.35	3.39	3.42	3.44	3.46	3.47	3.48	3.49	
14	3.03	3.18	3.27	3.33	3.37	3.40	3.43	3.44	3.46	3.47	3.48	
15	3.01	3.16	3.25	3.31	3.36	3.39	3.41	3.43	3.45	3.47	3.48	3.48
16	3.00	3.14	3.23	3.30	3.34	3.38	3.40	3.42	3.44	3.46	3.47	3.48
17	2.98	3.13	3.22	3.28	3.33	3.37	3.39	3.41	3.43	3.45	3.46	3.47
18	2.97	3.12	3.21	3.27	3.32	3.36	3.38	3.40	3.42	3.44	3.46	3.47
19	2.96	3.11	3.20	3.26	3.31	3.35	3.38	3.40	3.41	3.44	3.46	3.47
20	2.95	3.10	3.19	3.25	3.30	3.34	3.37	3.39	3.41	3.43	3.45	3.46
24	2.92	3.07	3.16	3.23	3.28	3.31	3.35	3.37	3.39	3.42	3.44	3.45
30	2.89	3.03	3.13	3.20	3.25	3.29	3.32	3.35	3.37	3.40	3.43	3.45
40	2.86	3.01	3.10	3.17	3.22	3.27	3.30	3.33	3.35	3.39	3.42	3.44
60	2.83	2.98	3.07	3.14	3.20	3.24	3.28	3.31	3.33	3.37	3.40	3.43
120	2.80	2.95	3.04	3.12	3.17	3.22	3.25	3.29	3.31	3.36	3.39	3.42
∞	2.77	2.92	3.02	3.09	3.15	3.19	3.23	3.27	3.29	3.34	3.38	3.41

* Reproduced from H. L. Harter, of the Aerospace Research Laboratories, Wright-Patterson Air Force Base. "Critical values for Duncan's new multiple range test." It contains some corrected values to replace those given by D. B. Duncan in "Multiple Range and Multiple F Tests," *Biometrics*, Vol. 11, 1955. Reproduced with the permission of the author and the editor of *Biometrics*.

Appendix Table 11: Transformation of Percentage to Arcsin $\sqrt{\text{Percentage}}$*

The numbers in this table are the angles (in degrees) corresponding to given percentages under the transformation arcsin $\sqrt{\text{percentage}}$.

%	0	1	2	3	4	5	6	7	8	9
.0	0	.57	.81	.99	1.15	1.28	1.40	1.52	1.62	1.72
.1	1.81	1.90	1.99	2.07	2.14	2.22	2.29	2.36	2.43	2.50
.2	2.56	2.63	2.69	2.75	2.81	2.87	2.92	2.98	3.03	3.09
.3	3.14	3.19	3.24	3.29	3.34	3.39	3.44	3.49	3.53	3.58
.4	3.63	3.67	3.72	3.76	3.80	3.85	3.89	3.93	3.97	4.01
.5	4.05	4.09	4.13	4.17	4.21	4.25	4.29	4.33	4.37	4.40
.6	4.44	4.48	4.52	4.55	4.59	4.62	4.66	4.69	4.73	4.76
.7	4.80	4.83	4.87	4.90	4.93	4.97	5.00	5.03	5.07	5.10
.8	5.13	5.16	5.20	5.23	5.26	5.29	5.32	5.35	5.38	5.41
.9	5.44	5.47	5.50	5.53	5.56	5.59	5.62	5.65	5.68	5.71
1	5.74	6.02	6.29	6.55	6.80	7.04	7.27	7.49	7.71	7.92
2	8.13	8.33	8.53	8.72	8.91	9.10	9.28	9.46	9.63	9.81
3	9.98	10.14	10.31	10.47	10.63	10.78	10.94	11.09	11.24	11.39
4	11.54	11.68	11.83	11.97	12.11	12.25	12.39	12.52	12.66	12.79
5	12.92	13.05	13.18	13.31	13.44	13.56	13.69	13.81	13.94	14.06
6	14.18	14.30	14.42	14.54	14.65	14.77	14.89	15.00	15.12	15.23
7	15.34	15.45	15.56	15.68	15.79	15.89	16.00	16.11	16.22	16.32
8	16.43	16.54	16.64	16.74	16.85	16.95	17.05	17.16	17.26	17.36
9	17.46	17.56	17.66	17.76	17.85	17.95	18.05	18.15	18.24	18.34
10	18.44	18.53	18.63	18.72	18.81	18.91	19.00	19.09	19.19	19.28
11	19.37	19.46	19.55	19.64	19.73	19.82	19.91	20.00	20.09	20.18
12	20.27	20.36	20.44	20.53	20.62	20.70	20.79	20.88	20.96	21.05
13	21.13	21.22	21.30	21.39	21.47	21.56	21.64	21.72	21.81	21.89
14	21.97	22.06	22.14	22.22	22.30	22.38	22.46	22.55	22.63	22.71
15	22.79	22.87	22.95	23.03	23.11	23.19	23.26	23.34	23.42	23.50
16	23.58	23.66	23.73	23.81	23.89	23.97	24.04	24.12	24.20	24.27
17	24.35	24.43	24.50	24.58	24.65	24.73	24.80	24.88	24.95	25.03
18	25.10	25.18	25.25	25.33	25.40	25.48	25.55	25.62	25.70	25.77
19	25.84	25.92	25.99	26.06	26.13	26.21	26.28	26.35	26.42	26.49
20	26.56	26.64	26.71	26.78	26.85	26.92	26.99	27.06	27.13	27.20
21	27.28	27.35	27.42	27.49	27.56	27.63	27.69	27.76	27.83	27.90
22	27.97	28.04	28.11	28.18	28.25	28.32	28.38	28.45	28.52	28.59
23	28.66	28.73	28.79	28.86	28.93	29.00	29.06	29.13	29.20	29.27
24	29.33	29.40	29.47	29.53	29.60	29.67	29.73	29.80	29.87	29.93
25	30.00	30.07	30.13	30.20	30.26	30.33	30.40	30.46	30.53	30.59
26	30.66	30.72	30.79	30.85	30.92	30.98	31.05	31.11	31.18	31.24
27	31.31	31.37	31.44	31.50	31.56	31.63	31.69	31.76	31.82	31.88
28	31.95	32.01	32.08	32.14	32.20	32.27	32.33	32.39	32.46	32.52
29	32.58	32.65	32.71	32.77	32.83	32.90	32.96	33.02	33.09	33.15
30	33.21	33.27	33.34	33.40	33.46	33.52	33.58	33.65	33.71	33.77

Appendix Table 11 (continued)

%	0	1	2	3	4	5	6	7	8	9
31	33.83	33.89	33.96	34.02	34.08	34.14	34.20	34.27	34.33	34.39
32	34.45	34.51	34.57	34.63	34.70	34.76	34.82	34.88	34.94	35.00
33	35.06	35.12	35.18	35.24	35.30	35.37	35.43	35.49	35.55	35.61
34	35.67	35.73	35.79	35.85	35.91	35.97	36.03	36.09	36.15	36.21
35	36.27	36.33	36.39	36.45	36.51	36.57	36.63	36.69	36.75	36.81
36	36.87	36.93	36.99	37.05	37.11	37.17	37.23	37.29	37.35	37.41
37	37.47	37.52	37.58	37.64	37.70	37.76	37.82	37.88	37.94	38.00
38	38.06	38.12	38.17	38.23	38.29	38.35	38.41	38.47	38.53	38.59
39	38.65	38.70	38.76	38.82	38.88	38.94	39.00	39.06	39.11	39.17
40	39.23	39.29	39.35	39.41	39.47	39.52	39.58	39.64	39.70	39.76
41	39.82	39.87	39.93	39.99	40.05	40.11	40.16	40.22	40.28	40.34
42	40.40	40.46	40.51	40.57	40.63	40.69	40.74	40.80	40.86	40.92
43	40.98	41.03	41.09	41.15	41.21	41.27	41.32	41.38	41.44	41.50
44	41.55	41.61	41.67	41.73	41.78	41.84	41.90	41.96	42.02	42.07
45	42.13	42.19	42.25	42.30	42.36	42.42	42.48	42.53	42.59	42.65
46	42.71	42.76	42.82	42.88	42.94	42.99	43.05	43.11	43.17	43.22
47	43.28	43.34	43.39	43.45	43.51	43.57	43.62	43.68	43.74	43.80
48	43.85	43.91	43.97	44.03	44.08	44.14	44.20	44.25	44.31	44.37
49	44.43	44.48	44.54	44.60	44.66	44.71	44.77	44.83	44.89	44.94
50	45.00	45.06	45.11	45.17	45.23	45.29	45.34	45.40	45.46	45.52
51	45.57	45.63	45.69	45.75	45.80	45.86	45.92	45.97	46.03	46.09
52	46.15	46.20	46.26	46.32	46.38	46.43	46.49	46.55	46.61	46.66
53	46.72	46.78	46.83	46.89	46.95	47.01	47.06	47.12	47.18	47.24
54	47.29	47.35	47.41	47.47	47.52	47.58	47.64	47.70	47.75	47.81
55	47.87	47.93	47.98	48.04	48.10	48.16	48.22	48.27	48.33	48.39
56	48.45	48.50	48.56	48.62	48.68	48.73	48.79	48.85	48.91	48.97
57	49.02	49.08	49.14	49.20	49.26	49.31	49.37	49.43	49.49	49.54
58	49.60	49.66	49.72	49.78	49.84	49.89	49.95	50.01	50.07	50.13
59	50.18	50.24	50.30	50.36	50.42	50.48	50.53	50.59	50.65	50.71
60	50.77	50.83	50.89	50.94	51.00	51.06	51.12	51.18	51.24	51.30
61	51.35	51.41	51.47	51.53	51.59	51.65	51.71	51.77	51.83	51.88
62	51.94	52.00	52.06	52.12	52.18	52.24	52.30	52.36	52.42	52.48
63	52.53	52.59	52.65	52.71	52.77	52.83	52.89	52.95	53.01	53.07
64	53.13	53.19	53.25	53.31	53.37	53.43	53.49	53.55	53.61	53.67
65	53.73	53.79	53.85	53.91	53.97	54.03	54.09	54.15	54.21	54.27
66	54.33	54.39	54.45	54.51	54.57	54.63	54.70	54.76	54.82	54.88
67	54.94	55.00	55.06	55.12	55.18	55.24	55.30	55.37	55.43	55.49
68	55.55	55.61	55.67	55.73	55.80	55.86	55.92	55.98	56.04	56.11
69	56.17	56.23	56.29	56.35	56.42	56.48	56.54	56.60	56.66	56.73
70	56.79	56.85	56.91	56.98	57.04	57.10	57.17	57.23	57.29	57.35
71	57.42	57.48	57.54	57.61	57.67	57.73	57.80	57.86	57.92	57.99
72	58.05	58.12	58.18	58.24	58.31	58.37	58.44	58.50	58.56	58.63
73	58.69	58.76	58.82	58.89	58.95	59.02	59.08	59.15	59.21	59.28
74	59.34	59.41	59.47	59.54	59.60	59.67	59.74	59.80	59.87	59.93
75	60.00	60.07	60.13	60.20	60.27	60.33	60.40	60.47	60.53	60.60

Appendix Table 11 (continued)

%	0	1	2	3	4	5	6	7	8	9
76	60.67	60.73	60.80	60.87	60.94	61.00	61.07	61.14	61.21	61.27
77	61.34	61.41	61.48	61.55	61.62	61.68	61.75	61.82	61.89	61.96
78	62.03	62.10	62.17	62.24	62.31	62.37	62.44	62.51	62.58	62.65
79	62.72	62.80	62.87	62.94	63.01	63.08	63.15	63.22	63.29	63 36
80	63.44	63.51	63.58	63.65	63.72	63.79	63.87	63.94	64.01	64.08
81	64.16	64.23	64.30	64.38	64.45	64.52	64.60	64.67	64.75	64.82
82	64.90	64.97	65.05	65.12	65.20	65.27	65.35	65.42	65.50	65.57
83	65.65	65.73	65.80	65.88	65.96	66.03	66.11	66.19	66.27	66.34
84	66.42	66.50	66.58	66.66	66.74	66.81	66.89	66.97	67.05	67.13
85	67.21	67.29	67.37	67.45	67.54	67.62	67.70	67.78	67.86	67.94
86	68.03	68.11	68.19	68.28	68.36	68.44	68.53	68.61	68.70	68.78
87	68.87	68.95	69.04	69.12	69.21	69.30	69.38	69.47	69.56	69.64
88	69.73	69.82	69.91	70.00	70.09	70.18	70.27	70.36	70.45	70.54
89	70.63	70.72	70.81	70.91	71.00	71.09	71.19	71.28	71.37	71.47
90	71.56	71.66	71.76	71.85	71.95	72.05	72.15	72.24	72.34	72.44
91	72.54	72.64	72.74	72.84	72.95	73.05	73.15	73.26	73.36	73.46
92	73.57	73.68	73.78	73.89	74.00	74.11	74.21	74.32	74.44	74.55
93	74.66	74.77	74.88	75.00	75.11	75.23	75.35	75.46	75.58	75.70
94	75.82	75.94	76.06	76.19	76.31	76.44	76.56	76.69	76.82	76.95
95	77.08	77.21	77.34	77.48	77.61	77.75	77.89	78.03	78.17	78.32
96	78.46	78.61	78.76	78.91	79.06	79.22	79.37	79.53	79.69	79.86
97	80.02	80.19	80.37	80.54	80.72	80.90	81.09	81.28	81.47	81.67
98	81.87	82.08	82.29	82.51	82.73	82.96	83.20	83.45	83.71	83.98
99.0	84.26	84.29	84.32	84.35	84.38	84.41	84.44	84.47	84.50	84.53
99.1	84.56	84.59	84.62	84.65	84.68	84.71	84.74	84.77	84.80	84.84
99.2	84.87	84.90	84.93	84.97	85.00	85.03	85.07	85.10	85.13	85.17
99.3	85.20	85.24	85.27	85.31	85.34	85.38	85.41	85.45	85.48	85.52
99.4	85.56	85.60	85.63	85.67	85.71	85.75	85.79	85.83	85.87	85.91
99.5	85.95	85.99	86.03	86.07	86.11	86.15	86.20	86.24	86.28	86.33
99.6	86.37	86.42	86.47	86.51	86.56	86.61	86.66	86.71	86.76	86.81
99.7	86.86	86.91	86.97	87.02	87.08	87.13	87.19	87.25	87.31	87.37
99.8	87.44	87.50	87.57	87.64	87.71	87.78	87.86	87.93	88.01	88.10
99.9	88.19	88.28	88.38	88.48	88.60	88.72	88.85	89.01	89.19	89.43
100.0	90.00	—	—	—	—	—	—	—	—	—

* From *Plant Protection* (Leningrad), Vol. 12 (1937). Reproduced with permission of the author, C. I. Bliss.

Appendix Table 12: Critical Values for a 5% Two-sided Runs Test*

	5	6	7	8	9	10	11	12	13	14	15	16	17	18	19	20
2								2/6	2/6	2/6	2/6	2/6	2/6	2/6	2/6	2/6
3		2/8	2/8	2/8	2/8	2/8	2/8	2/8	2/8	2/8	3/8	3/8	3/8	3/8	3/8	3/8
4	2/9	2/9	2/10	3/10	3/10	3/10	3/10	3/10	3/10	3/10	3/10	4/10	4/10	4/10	4/10	4/10
5	2/10	3/10	3/11	3/11	3/12	3/12	4/12	4/12	4/12	4/12	4/12	4/12	4/12	5/12	5/12	5/12
6		3/11	3/12	3/12	4/13	4/13	4/13	4/13	5/14	5/14	5/14	5/14	5/14	5/14	6/14	6/14
7			3/13	4/13	4/14	5/14	5/14	5/14	5/15	5/15	6/15	6/16	6/16	6/16	6/16	6/16
8				4/14	5/14	5/15	5/15	6/16	6/16	6/16	6/16	6/17	7/17	7/17	7/17	7/17
9					5/15	5/16	6/16	6/16	6/17	7/17	7/18	7/18	7/18	8/18	8/18	8/18
10						6/16	6/17	7/17	7/18	7/18	7/18	8/19	8/19	8/19	8/20	9/20
11							7/17	7/18	7/19	8/19	8/19	8/20	9/20	9/20	9/21	9/21
12								7/19	8/19	8/20	8/20	9/21	9/21	9/21	10/22	10/22
13									8/20	9/20	9/21	9/21	10/22	10/22	10/23	10/23
14										9/21	9/22	10/22	10/23	10/23	11/23	11/24
15											10/22	10/23	11/23	11/24	11/24	12/25
16												11/23	11/24	11/25	12/25	12/25
17													11/25	12/25	12/26	13/26
18														12/26	13/26	13/27
19															13/27	13/27
20																14/28

* From C. Eisenhart and F. Swed, "Tables for Testing Randomness of Grouping in a Sequence of Alternatives," *Annals of Mathematical Statistics*, Vol. 14 (1943), p. 66. Reprinted by permission of the authors and the publisher.

Appendix Table 13: Critical Values for the Wilcoxon Test for Paired Data*

n	$2\alpha \leq 0.10$	$2\alpha \leq 0.05$	$2\alpha \leq 0.02$	$2\alpha \leq 0.01$
5	0– 15	–	–	–
6	2– 19	–	–	–
7	3– 25	0– 21	–	–
8	5– 31	2– 26	0– 28	–
9	8– 37	3– 33	1– 35	0– 36
		5– 40	3– 42	1– 44
10	10– 45	8– 47	5– 50	3– 52
11	13– 53	10– 56	7– 59	5– 61
12	17– 61	13– 65	9– 69	7– 71
13	21– 70	17– 74	12– 79	9– 82
14	25– 80	21– 84	15– 90	12– 93
15	30– 90	25– 95	19–101	15–105
16	35–101	29–107	23–113	19–117
17	41–112	34–119	28–125	23–130
18	47–124	40–131	32–139	27–144
19	53–137	46–144	37–153	32–158
20	60–150	52–158	43–167	37–173
21	67–164	58–173	49–182	42–189
22	75–178	66–187	55–198	48–205
23	83–193	73–203	62–214	54–222
24	91–209	81–219	69–231	61–239
25	100–225	89–236	76–249	68–257

* Reproduced from *Documenta Geigy Scientific Tables*, 7th edition, by permission of CIBA-GEIGY Limited, Basle, Switzerland.

Appendix Table 14: Critical Values for the Wilcoxon Rank Sum Test*

For one-sided tests $\alpha = .05$, for two-sided tests $\alpha = .10$

N_2 \ N_1	4	5	6	7	8	9	10	11	12	13	14	15	16	17	18	19	20	21	22	23	24	25
	T_l – T_r	T_l – T_r	T_l – T_r	T_l – T_r	T_l – T_r	T_l – T_r	T_l – T_r	T_l – T_r	T_l – T_r	T_l – T_r	T_l – T_r	T_l – T_r	T_l – T_r	T_l – T_r	T_l – T_r	T_l – T_r	T_l – T_r	T_l – T_r	T_l – T_r	T_l – T_r	T_l – T_r	T_l – T_r
4	**11–25**	17–33	24–42	32–52	41–63	51–75	62–88	74–102	87–117	101–133	116–150	132–168	150–186	168–206	187–227	207–249	228–272	250–296	273–321	297–347	322–374	348–402
5	12–28	**19–36**	26–46	34–57	44–68	54–81	66–94	78–109	91–125	106–141	121–159	138–177	155–197	173–218	193–239	213–262	235–285	257–310	281–335	305–362	330–390	357–418
6	13–31	20–40	**28–50**	36–62	46–74	57–87	69–101	82–116	95–133	110–150	126–168	143–187	161–207	179–229	199–251	220–274	242–298	265–323	289–349	313–377	339–405	366–434
7	14–34	21–44	29–55	**39–66**	49–79	60–93	72–108	85–124	99–141	115–158	131–177	148–197	166–218	186–239	206–262	227–286	249–311	272–337	297–363	322–391	348–420	375–450
8	15–37	23–47	31–59	41–71	**51–85**	63–99	75–115	89–131	104–148	119–167	136–186	153–207	172–228	192–250	212–274	234–298	257–323	280–350	305–377	330–406	357–435	385–465
9	16–40	24–51	33–63	43–76	54–90	**66–105**	79–121	93–138	108–156	124–178	141–195	159–216	178–238	198–261	219–285	241–310	264–336	288–363	313–391	339–420	366–450	394–481
10	17–43	26–54	35–67	45–81	56–96	69–111	**82–128**	97–145	112–164	128–184	146–204	164–226	184–248	204–272	226–297	248–322	272–348	296–376	321–405	348–434	375–465	403–497
11	18–46	27–58	37–71	47–86	59–101	72–117	86–134	**100–153**	116–172	133–192	151–213	170–235	190–258	210–283	232–308	255–334	279–361	304–389	329–419	356–449	384–480	413–512
12	19–49	28–62	38–76	49–91	62–106	75–123	89–141	104–160	**120–180**	138–200	156–222	175–245	196–268	217–293	239–319	262–346	286–374	312–402	338–432	365–463	393–495	423–527
13	20–52	30–65	40–80	52–95	64–112	78–129	92–148	108–167	125–187	**142–209**	161–231	181–254	201–279	223–304	245–331	269–358	294–386	320–415	346–446	374–477	403–509	433–542
14	21–55	31–69	42–84	54–100	67–117	81–135	96–154	112–174	129–195	147–217	**166–240**	186–264	207–289	229–315	252–342	276–370	301–399	328–428	355–459	383–491	412–524	442–558
15	22–58	33–72	44–88	56–105	69–123	84–141	99–161	116–181	133–203	152–225	171–249	**191–274**	213–299	235–326	259–353	284–381	309–411	336–441	363–473	392–505	422–538	452–573
16	24–60	34–76	46–92	58–110	72–128	87–147	103–167	120–188	138–210	156–234	176–258	197–283	**219–309**	242–336	266–364	291–393	317–423	344–454	372–486	401–519	431–553	462–588
17	25–63	35–80	47–97	61–114	75–133	90–153	106–174	123–196	142–218	161–242	181–267	202–293	225–319	**248–347**	273–375	298–405	325–435	352–467	380–500	410–533	440–568	472–603
18	26–66	37–83	49–101	63–119	77–139	93–159	110–180	127–203	146–226	165–251	186–276	208–302	231–329	255–357	**280–386**	305–417	332–448	360–480	389–513	419–547	450–582	482–618
19	27–69	38–87	51–105	65–124	80–144	96–165	113–187	131–210	150–234	170–259	191–285	214–311	237–339	261–368	286–398	**313–428**	340–460	368–493	398–526	428–561	459–597	492–633
20	28–72	40–90	53–109	67–129	83–149	99–171	117–193	135–217	154–242	175–267	196–294	219–321	243–349	268–378	293–409	320–440	**348–472**	376–506	406–540	437–575	469–611	501–649
21	29–75	41–94	55–113	69–134	85–155	102–177	120–200	139–224	159–249	180–275	202–302	225–330	249–359	274–389	300–420	327–452	355–485	**385–518**	415–553	446–589	478–626	511–664
22	30–78	43–97	57–117	72–138	88–160	105–183	123–207	142–232	163–257	184–284	207–311	230–340	255–369	280–400	307–431	335–463	363–497	393–531	**423–567**	455–603	488–640	521–679
23	31–81	44–101	58–122	74–143	90–166	108–189	127–213	146–239	167–265	189–292	212–320	236–349	261–379	287–410	314–442	342–475	371–509	401–544	432–580	**464–617**	497–655	531–694
24	32–84	45–105	60–126	76–148	93–171	111–195	130–220	150–246	171–273	194–300	217–329	242–358	267–389	293–421	321–453	349–487	379–521	409–557	441–593	473–631	**507–669**	541–709
25	33–87	47–108	62–130	78–153	96–176	114–201	133–227	154–253	176–280	199–308	222–338	247–368	273–399	300–431	328–464	357–498	386–534	417–570	449–607	482–645	516–684	**551–724**
26	34–90	48–112	64–134	81–157	98–182	117–207	137–233	158–260	180–288	203–317	227–347	253–377	279–409	306–442	335–475	364–510	394–546	426–582	458–620	491–659	526–698	561–739
27	35–93	50–115	66–138	83–162	101–187	120–213	140–240	162–267	184–296	208–325	233–355	258–387	285–419	313–452	342–486	371–522	402–558	434–595	467–633	500–673	535–713	571–754
28	36–96	51–119	67–143	85–167	103–193	123–219	144–246	166–274	189–303	213–333	238–364	264–396	291–429	319–463	348–498	379–533	410–570	442–608	475–647	510–686	545–727	581–769
29	37–99	53–122	69–147	87–172	106–198	126–225	147–253	170–281	193–311	218–341	243–373	270–405	297–439	326–473	355–509	386–545	418–582	450–621	484–660	519–700	554–742	591–784
30	38–102	54–126	71–151	89–177	109–203	129–231	151–259	174–288	197–319	222–350	248–382	275–415	303–449	332–484	362–520	393–557	425–595	459–633	493–673	528–714	564–756	601–799
31	39–105	55–130	73–155	92–181	111–209	132–237	154–266	178–295	202–326	227–358	253–391	281–424	309–459	339–494	369–531	401–568	433–607	467–646	501–687	537–728	574–770	611–814
32	40–108	57–133	75–159	94–186	114–214	135–243	158–272	181–303	206–334	232–366	259–399	286–434	315–469	345–505	376–542	408–580	441–619	475–659	510–700	546–742	583–785	621–829
33	41–111	58–137	77–163	96–191	117–219	138–249	161–279	185–310	210–342	237–374	264–408	292–443	321–479	352–515	383–553	415–592	449–631	483–672	519–713	555–756	593–799	631–844
34	42–114	60–140	78–168	98–196	119–225	141–255	165–285	189–317	215–349	241–383	269–417	298–452	327–489	358–526	390–564	423–603	457–643	492–684	527–727	564–770	602–814	641–859
35	43–117	61–144	80–172	100–201	122–230	144–261	168–292	193–324	219–357	246–391	274–426	303–462	333–499	365–536	397–575	430–615	464–656	500–697	536–740	574–783	612–828	651–874
36	44–120	62–148	82–176	102–206	124–236	148–266	172–298	197–331	223–365	251–399	279–435	309–471	340–508	371–547	404–586	438–626	472–668	508–710	545–753	583–797	621–843	661–889
37	45–123	64–151	84–180	105–210	127–241	151–272	175–305	201–338	228–372	256–407	285–443	315–480	346–518	378–557	411–597	445–638	480–680	516–723	554–766	592–811	631–857	671–904
38	46–126	65–155	85–185	107–215	130–246	154–278	179–311	205–345	232–380	260–416	290–452	320–490	352–528	384–568	418–608	452–650	488–692	525–735	562–780	601–825	641–871	681–919
39	47–129	67–158	87–189	109–220	132–252	157–284	182–318	209–352	236–388	265–424	295–461	326–499	358–538	391–578	425–619	460–661	496–704	533–748	571–793	610–839	650–886	691–934
40	48–132	68–162	89–193	111–225	135–257	160–290	186–324	213–359	241–395	270–432	300–470	331–509	364–548	397–589	432–630	467–673	504–716	541–761	580–806	619–853	660–900	702–948
41	49–135	69–166	91–197	114–229	138–262	163–296	189–331	217–366	245–403	275–440	305–479	337–518	370–558	404–599	439–641	474–685	511–729	549–774	588–820	628–867	669–915	712–963
42	50–138	71–169	93–201	116–234	140–268	166–302	193–337	221–373	249–411	279–449	311–487	343–527	376–568	410–610	446–652	482–696	519–741	558–786	597–833	638–880	679–929	722–978
43	51–141	72–173	95–205	118–239	143–273	169–308	196–344	224–381	254–418	284–457	316–496	348–537	382–578	417–620	452–664	489–708	527–753	566–799	606–846	647–894	689–943	732–993
44	52–144	74–176	96–210	120–244	146–278	172–314	200–350	228–388	258–426	289–465	321–505	354–546	388–588	423–631	459–675	497–719	535–765	574–812	615–859	656–908	698–958	742–1008
45	53–147	75–180	98–214	123–248	148–284	175–320	203–357	232–395	262–434	294–473	326–514	360–555	394–598	430–641	466–686	504–731	543–777	583–824	623–873	665–922	708–972	752–1023
46	55–149	77–183	100–218	125–253	151–289	178–326	207–363	236–402	267–441	299–481	331–523	365–565	400–608	436–652	473–697	511–743	551–789	591–837	632–886	674–936	718–986	762–1038
47	56–152	78–187	102–222	127–258	154–294	181–332	210–370	240–409	271–449	303–490	337–531	371–574	406–618	443–662	480–708	519–754	558–802	599–850	641–899	683–950	727–1001	772–1053
48	57–155	79–191	104–226	129–263	156–300	184–338	214–376	244–416	275–457	308–498	342–540	377–583	412–628	449–673	487–719	526–766	566–814	607–863	649–913	693–963	737–1015	782–1068
49	58–158	81–194	106–230	132–267	159–305	187–344	217–383	248–423	280–464	313–506	347–549	382–593	418–638	456–683	494–730	534–777	574–826	616–875	658–926	702–977	746–1030	792–1083
50	59–161	82–198	107–235	134–272	162–310	190–350	221–387	252–430	284–472	318–514	352–558	388–602	425–647	462–694	501–741	541–789	582–838	624–888	667–939	711–991	756–1044	802–1098

* Reproduced from *Documenta Geigy Scientific Tables*, 7th edition, by permission of CIBA-GEIGY Limited, Basle, Switzerland.

Appendix table 14 (continued)

For one-sided tests $\alpha = .025$, for two-sided tests $\alpha = .05$

(Table of critical values for the Wilcoxon rank-sum test. Columns are indexed by $N_1 = 4, 5, 6, \ldots, 25$, each giving lower and upper critical values T_r, T_l. Rows are indexed by $N_2 = 4, 5, \ldots, 50$.)

For one-sided tests $\alpha = .01$, for two-sided tests $\alpha = .02$

N_2 \ N_1	4	5	6	7	8	9	10	11	12	13	14	15	16	17	18	19	20	21	22	23	24	25
	T_l T_r	T_l T_r	T_l T_r	T_l T_r	T_l T_r	T_l T_r	T_l T_r	T_l T_r	T_l T_r	T_l T_r	T_l T_r	T_l T_r	T_l T_r	T_l T_r	T_l T_r	T_l T_r	T_l T_r	T_l T_r	T_l T_r	T_l T_r	T_l T_r	T_l T_r
4	—	15-35	22-44	29-55	38-66	48-78	58-92	70-106	83-121	96-138	111-155	127-173	143-193	161-213	180-234	199-257	220-280	242-304	264-330	288-356	313-383	338-412
5	10-30	**16-39**	23-49	31-60	40-72	50-85	61-99	73-114	86-130	100-147	115-165	131-184	148-204	166-225	185-247	205-270	226-294	248-319	271-345	295-372	320-400	346-429
6	11-33	17-43	**24-54**	32-66	42-78	52-92	63-107	75-123	89-139	103-157	118-176	135-195	152-216	171-237	190-260	210-284	232-308	254-334	277-361	302-388	327-417	354-446
7	11-37	18-47	25-59	**34-71**	43-85	54-99	66-114	78-131	92-148	107-166	122-186	139-206	157-227	176-249	195-273	216-297	238-322	261-348	284-376	309-404	335-433	361-464
8	12-40	19-51	27-63	35-77	**45-91**	56-106	68-122	81-139	95-157	111-175	127-195	144-216	162-238	181-261	201-285	222-310	244-336	267-363	291-391	316-420	342-450	370-480
9	13-43	20-55	28-68	37-82	47-97	**59-112**	71-129	84-147	99-165	114-185	131-205	148-227	167-249	186-273	207-297	228-323	250-350	274-377	298-406	324-435	350-466	378-497
10	13-47	21-59	29-73	39-87	49-103	61-119	**74-136**	88-154	102-174	118-194	135-215	153-237	172-260	191-285	212-310	234-336	257-363	281-391	306-420	331-451	358-482	386-514
11	14-50	22-63	30-78	40-93	51-109	63-126	77-143	**91-162**	106-182	122-203	139-225	157-248	177-271	197-296	218-322	240-349	263-377	288-405	313-435	339-466	366-498	394-531
12	15-53	23-67	32-82	42-98	53-115	66-132	79-151	94-170	**109-191**	126-212	143-235	162-258	182-282	202-308	224-334	246-362	270-390	295-419	320-450	347-481	374-514	403-547
13	15-57	24-71	33-87	44-103	56-120	68-139	82-158	97-178	113-199	**130-221**	148-244	167-268	187-293	208-319	230-346	253-374	276-404	301-434	327-465	354-497	382-530	411-564
14	16-60	25-75	34-92	45-109	58-126	71-145	85-165	100-186	116-208	134-230	**152-254**	171-279	192-304	213-331	235-359	259-387	283-417	308-448	335-479	362-512	391-545	420-580
15	17-63	26-79	36-96	47-114	60-132	73-152	88-172	103-194	120-216	138-239	156-264	**176-289**	197-315	218-343	241-371	265-400	290-430	315-462	342-494	370-527	399-561	429-596
16	17-67	27-83	37-101	49-119	62-138	76-158	91-179	107-201	124-224	142-248	161-273	181-299	**202-326**	224-354	247-383	271-413	296-444	323-475	350-508	378-542	407-577	437-613
17	18-70	28-87	39-105	51-124	64-144	78-165	93-187	110-209	127-233	146-257	165-283	185-310	207-337	**229-366**	253-395	278-425	303-457	330-489	357-523	386-557	416-592	446-629
18	19-73	29-91	40-110	52-130	66-150	81-171	96-194	113-217	131-241	150-266	169-293	190-320	212-348	235-377	**259-407**	284-438	310-470	337-503	365-537	394-572	424-608	455-645
19	19-77	30-95	41-115	54-135	68-156	83-178	99-201	116-225	134-250	154-275	174-302	195-330	217-359	241-388	265-419	**290-451**	317-483	344-517	373-551	402-587	432-624	464-661
20	20-80	31-99	43-119	56-140	70-162	85-185	102-208	119-233	138-258	157-285	178-312	200-340	222-370	246-400	271-431	297-463	**323-497**	351-531	380-566	410-602	441-639	473-677
21	21-83	32-103	44-124	58-145	72-168	88-191	105-215	123-240	142-266	161-294	182-322	204-351	228-380	252-411	277-443	303-476	330-510	**359-544**	388-580	418-617	449-655	482-693
22	21-87	33-107	45-129	59-151	74-174	90-198	108-222	126-248	145-275	165-303	187-331	209-361	233-391	257-423	283-455	310-488	337-523	366-558	**395-595**	426-632	458-670	490-710
23	22-90	34-111	47-133	61-156	76-180	93-204	110-230	129-256	149-283	169-312	191-341	214-371	238-402	263-434	289-467	316-501	344-536	373-572	403-609	**434-647**	466-686	499-726
24	23-93	35-115	48-138	63-161	78-186	95-211	113-237	132-264	152-292	173-321	196-350	219-381	243-413	269-445	295-479	322-514	351-549	380-586	411-623	442-662	**475-701**	508-742
25	23-97	36-119	50-142	64-167	81-191	98-217	116-244	135-272	156-300	177-330	200-360	224-391	248-424	274-457	301-491	329-526	358-562	388-599	418-638	450-677	483-717	**517-758**
26	24-100	37-123	51-147	66-172	83-197	100-224	119-251	138-280	159-309	181-339	204-370	229-401	254-434	280-468	307-503	335-539	365-575	395-613	426-652	458-692	492-732	526-774
27	25-103	38-127	52-152	68-177	85-203	103-230	121-259	142-287	163-317	185-348	209-379	233-412	259-445	285-480	313-515	342-551	371-589	402-627	434-666	466-707	500-748	535-790
28	26-106	39-131	54-156	70-182	87-209	105-237	124-266	145-295	167-325	189-357	213-389	238-422	264-456	291-491	319-527	348-564	378-602	409-641	441-681	475-721	509-763	544-806
29	26-110	40-135	55-161	71-188	89-215	108-243	127-273	148-303	170-334	193-366	218-398	243-432	269-467	297-502	325-539	355-576	385-615	417-654	449-695	483-736	517-779	553-822
30	27-113	41-139	56-166	73-193	91-221	110-250	130-280	151-311	174-342	198-374	222-408	248-442	275-477	302-514	331-551	361-589	392-628	424-668	457-709	491-751	526-794	562-838
31	28-116	42-143	58-170	75-198	93-227	112-257	133-287	155-318	178-350	202-383	227-417	253-452	280-488	308-525	337-563	368-601	399-641	431-682	465-723	499-766	534-810	571-854
32	28-120	43-147	59-175	77-203	95-233	115-263	136-294	158-326	181-359	206-392	231-427	258-462	285-499	314-536	343-575	374-614	406-654	439-695	472-738	507-781	543-825	580-870
33	29-123	44-151	61-179	78-209	97-239	117-270	139-301	161-334	185-367	210-401	235-437	262-473	290-510	319-548	349-587	381-626	413-667	446-709	480-752	515-796	552-840	589-886
34	30-126	45-155	62-184	79-215	99-245	120-276	141-309	164-342	189-375	214-410	240-446	267-483	296-520	325-559	356-598	387-639	420-680	453-723	488-766	523-811	560-856	598-902
35	30-130	46-159	63-189	81-220	101-251	122-283	144-316	168-349	192-384	218-419	244-456	272-493	301-531	331-570	362-610	394-651	427-693	461-736	496-780	532-825	569-871	607-918
36	31-133	47-163	65-193	83-225	103-257	125-289	147-323	171-357	196-392	222-428	249-465	277-503	306-542	336-582	368-622	400-664	433-707	468-750	503-795	540-840	577-887	616-934
37	32-136	48-167	66-198	84-231	105-263	127-296	150-330	174-365	199-401	226-437	253-475	282-513	311-553	342-593	374-634	407-676	440-720	475-764	511-809	548-855	586-902	625-950
38	32-140	49-171	67-203	86-236	107-269	129-303	153-337	177-373	203-409	230-446	258-484	287-523	317-563	348-604	380-646	413-689	447-733	483-777	519-823	556-870	595-917	634-966
39	33-143	50-175	69-207	88-241	109-275	132-309	156-344	181-380	207-417	234-455	262-494	292-533	322-574	353-616	386-658	420-701	454-746	490-791	527-837	564-885	603-933	643-982
40	34-146	51-179	70-212	90-246	111-281	134-316	159-351	184-388	210-426	238-464	267-503	296-544	327-585	359-627	392-670	426-714	461-759	497-805	534-852	573-899	612-948	652-998
41	34-150	52-183	72-216	91-252	113-287	137-322	161-359	187-396	214-434	242-473	271-513	301-554	333-595	365-638	398-682	433-726	468-772	505-818	542-866	581-914	620-964	661-1014
42	35-153	53-187	73-221	93-257	116-292	139-329	164-366	190-404	218-442	246-482	276-522	306-564	338-606	371-649	404-694	439-739	475-785	512-832	550-880	589-929	629-979	670-1030
43	35-157	54-191	74-226	95-262	118-298	142-335	167-373	194-411	221-451	250-491	280-532	311-574	343-617	376-661	410-706	446-751	482-798	519-846	558-894	597-944	638-994	679-1046
44	36-160	55-195	76-230	97-267	120-304	144-342	170-380	197-419	225-459	254-500	284-542	316-584	348-628	382-672	416-718	452-764	489-811	527-859	565-909	605-959	646-1010	688-1062
45	37-163	56-199	77-235	98-273	122-310	147-348	173-387	200-427	229-467	258-509	289-551	321-594	354-638	388-683	423-729	459-776	496-824	534-873	573-923	614-973	655-1025	697-1078
46	37-167	57-203	78-240	100-278	124-316	149-355	176-394	203-435	232-476	262-518	293-561	326-604	359-649	393-695	429-741	465-789	503-837	541-887	581-937	622-988	663-1041	706-1094
47	38-170	58-207	80-244	102-283	126-322	152-361	179-401	207-442	236-484	266-527	298-570	330-615	364-660	399-706	435-753	472-801	510-850	549-900	589-951	630-1003	672-1056	715-1110
48	39-173	59-211	81-249	103-289	128-328	154-368	181-409	210-450	240-492	270-536	302-580	335-625	369-671	405-717	441-765	478-814	517-863	556-914	597-965	638-1018	681-1071	724-1126
49	39-177	60-215	82-254	105-294	130-334	157-374	184-416	213-458	243-501	274-545	307-589	340-635	375-681	410-729	447-777	485-826	524-876	563-928	604-980	646-1033	689-1087	733-1142
50	40-180	61-219	84-258	107-299	132-340	159-381	187-423	216-466	247-509	279-553	311-599	345-645	380-692	416-740	453-789	491-839	531-889	571-941	612-994	655-1047	698-1102	743-1157

Appendix Table 15a: Table of Probabilities Associated for Obtaining Values as Large as Observed Values with Friedman's Test. $k = 3$*

χ_r^2	p	χ_r^2	p	χ_r^2	p	χ_r^2	p
\multicolumn{2}{c	}{$N = 2$}	\multicolumn{2}{c	}{$N = 3$}	\multicolumn{2}{c	}{$N = 4$}	\multicolumn{2}{c}{$N = 5$}

$N = 2$		$N = 3$		$N = 4$		$N = 5$	
χ_r^2	p	χ_r^2	p	χ_r^2	p	χ_r^2	p
0	1.000	.000	1.000	.0	1.000	.0	1.000
1	.833	.667	.944	.5	.931	.4	.954
3	.500	2.000	.528	1.5	.653	1.2	.691
4	.167	2.667	.361	2.0	.431	1.6	.522
		4.667	.194	3.5	.273	2.8	.367
		6.000	.028	4.5	.125	3.6	.182
				6.0	.069	4.8	.124
				6.5	.042	5.2	.093
				8.0	.0046	6.4	.039
						7.6	.024
						8.4	.0085
						10.0	.00077

$N = 6$		$N = 7$		$N = 8$		$N = 9$	
χ_r^2	p	χ_r^2	p	χ_r^2	p	χ_r^2	p
.00	1.000	.000	1.000	.00	1.000	.000	1.000
.33	.956	.286	.964	.25	.967	.222	.971
1.00	.740	.857	.768	.75	.794	.667	.814
1.33	.570	1.143	.620	1.00	.654	.889	.865
2.33	.430	2.000	.486	1.75	.531	1.556	.569
3.00	.252	2.571	.305	2.25	.355	2.000	.398
4.00	.184	3.429	.237	3.00	.285	2.667	.328
4.33	.142	3.714	.192	3.25	.236	2.889	.278
5.33	.072	4.571	.112	4.00	.149	3.556	.187
6.33	.052	5.429	.085	4.75	.120	4.222	.154
7.00	.029	6.000	.052	5.25	.079	4.667	.107
8.33	.012	7.143	.027	6.25	.047	5.556	.069
9.00	.0081	7.714	.021	6.75	.038	6.000	.057
9.33	.0055	8.000	.016	7.00	.038	6.222	.048
10.33	.0017	8.857	.0084	7.75	.018	6.889	.031
12.00	.00013	10.286	.0036	9.00	.0099	8.000	.019
		10.571	.0027	9.25	.0080	8.222	.016
		11.143	.0012	9.75	.0048	8.667	.010
		12.286	.00032	10.75	.0024	9.556	.0060
		14.000	.000021	12.00	.0011	10.667	.0035
				12.25	.00086	10.889	.0029
				13.00	.00026	11.556	.0013
				14.25	.000061	12.667	.00066
				16.00	.0000036	13.556	.00035
						14.000	.00020
						14.222	.000097
						14.889	.000054
						16.222	.000011
						18.000	.0000006

* Adapted from Friedman, M. "The Use of Ranks to Avoid the Assumption of Normality Implicit in the Analysis of Variance," *J. Amer. Statist. Ass.*, **32**, 688–689 (1937), with permission of the author and publisher.

Appendix Table 15b: Table of Probabilities Associated for Obtaining Values as Large as the Observed Values with Friedman's Test. $k = 4$*

$N = 2$		$N = 3$		$N = 4$			
χ_r^2	p	χ_r^2	p	χ_r^2	p	χ_r^2	p
.0	1.000	.2	1.000	.0	1.000	5.7	.141
.6	.958	.6	.958	.3	.992	6.0	.105
1.2	.834	1.0	.910	.6	.928	6.3	.094
1.8	.792	1.8	.727	.9	.900	6.6	.077
2.4	.625	2.2	.608	1.2	.800	6.9	.068
3.0	.542	2.6	.524	1.5	.754	7.2	.054
3.6	.458	3.4	.446	1.8	.677	7.5	.052
4.2	.375	3.8	.342	2.1	.649	7.8	.036
4.8	.208	4.2	.300	2.4	.524	8.1	.033
5.4	.167	5.0	.207	2.7	.508	8.4	.019
6.0	.042	5.4	.175	3.0	.432	8.7	.014
		5.8	.148	3.3	.389	9.3	.012
		6.6	.075	3.6	.355	9.6	.0069
		7.0	.054	3.9	.324	9.9	.0062
		7.4	.033	4.5	.242	10.2	.0027
		8.2	.017	4.8	.200	10.8	.0016
		9.0	.0017	5.1	.190	11.1	.00094
				5.4	.158	12.0	.000072

* Adapted from Friedman, M. "The Use of Ranks to Avoid the Assumption of Normality Implicit in the Analysis of Variance," *J. Amer. Statist. Ass.*, **32**, 688-689 (1937), with permission of the author and the publisher.

ANSWERS TO EXERCISES

ANSWERS TO EXERCISES

Chapter 2

1. a) $\{1, 2, 3, \ldots\}$, $\{x \mid x \text{ is a positive integer}\}$
   b) $\{2, 4, 6, \ldots\}$, $\{2x \mid x \text{ is a positive integer}\}$
   c) $\{H, T\}$, $\{x \mid x \text{ is } H \text{ or } T\}$
   d) $\{1, 2, 3, 4, 5, 6\}$, $\{x \mid x = 1, 2, 3, 4, 5\ 6\}$

2. a) $\{1, 3\}$; b) $\{1, 2, 3, 5\}$; c) $\{2\}$;
   d) $\{1, 2, 3, 4, 6, 8\}$; e) $\{4, 5, 6, 7, 8\}$; f) $\{4, 6, 7, 8\}$;
   g) $\{4, 6, 7, 8\}$; h) $\varnothing$; i) $\{1, 2, 3, 4, 5, 6, 8\}$

3. a) $\{a\}, \{b\}, \{c\}, \{d\}$;
   b) $\{a, b\}, \{a, c\}, \{a, d\}, \{b, c\}, \{b, d\}, \{c, d\}$;
   c) $\{a, b, c\}, \{a, b, d\}, \{a, c, d\}, \{b, c, d\}$

4. a) $\{t \mid t \geq 50\}$; b) $\{t \mid 50 \leq t \leq 100\}$; c) $\{t \mid t > 50\}$;
   d) $\{t \mid t > 75\}$; e) $\{t \mid t \geq 50\}$; f) $\{t \mid t \leq 50\}$

8. a) $\binom{15}{4}$; b) $\binom{15}{1}\binom{14}{1}$

9. a) 10; b) 455; c) 1;
   d) 1; e) 500; f) 50

10. a) 1; b) 1; c) n;
    d) $n(n-1)/2$; e) $n!/x!(n-x)!$; f) $n!/(x-1)!(n-x+1)!$

12. $\binom{52}{13}$ 13. 6^6 14. 6^7

15. a) 5; b) 20

16. $26^2\ 10^4$ 17. $5!$ 18. $10!/5!$

19. 1080

20. a) $\frac{1}{2}$; b) $\frac{2}{3}$; c) $\frac{1}{2}$;
    d) $\frac{1}{3}$

21. $\frac{3}{5}$

22. a) $\frac{3}{5}$; b) $\frac{1}{5}$

342 Answers to exercises

23. $\frac{5}{136}$

24. a) $\frac{4}{21}$; b) $\frac{2}{7}$; c) $\frac{1}{7}$;
    d) $\frac{1}{7}$; e) $\frac{1}{42}$

25. a) $\frac{1}{9}$; b) $\frac{1}{36}$; c) $\frac{5}{36}$

26. a) $\frac{4}{52}$; b) $\frac{4}{52}$; c) $\frac{8}{52}$;
    d) $\frac{12}{52}$

27. $\frac{1}{20}$ 28. $6.7!/2.6^7$ 29. $\binom{7}{2}\binom{4}{1}\binom{3}{1}/\binom{14}{4}$

30. $6/6^3$ 31. $\frac{1}{3}$

32. a) .9; b) .4; c) .5;
    d) .8; e) .1; f) .3

33. a) .3; b) $\frac{35}{40}$; c) $\frac{3}{40}$

34. a) $\frac{1}{2}$; b) $\frac{1}{2}$; c) $\frac{1}{2}$;
    d) $\frac{1}{4}$

35. .06, .56

36. a) $\frac{11}{12}$; b) $\frac{9}{24}$

37. a) $\frac{5}{9}$; b) $\frac{16}{27}$; c) $\frac{6}{9}$;
    d) $\frac{1}{3}$

38. $(.9)^{50}$

39. a) $\binom{6}{4}\binom{5}{3}/\binom{11}{7}$; b) $\dfrac{\binom{6}{4}\binom{5}{3}+\binom{6}{5}\binom{5}{2}+\binom{6}{6}\binom{5}{1}}{\binom{11}{7}}$

40. $1/\binom{20}{4}$ 41. $(6)(364)(363)/365^3$

42. a) $\frac{7}{20}$; b) $\frac{1}{2}$; c) $\frac{17}{20}$

43. $\frac{1}{40}$ 44. $\frac{279}{442}$ 45. $\frac{14}{17}$

46. $\frac{5}{12}$

47. a) $\frac{1}{3}$; b) $\frac{1}{2}$

48. .5, 1 49. $\frac{11}{14}$ 50. $\frac{4}{7}$

51. $2\binom{5}{2}\binom{21}{11}/\binom{26}{13}$ 52. $\frac{1}{7}$ 53. $\binom{6}{3}/\binom{8}{4}$

54. $\frac{7}{15}$

Chapter 3

2. a) discrete; b) continuous; c) discrete;
    d) continuous; e) continuous

3. a) $S = \{WW, RR, BB, WR, WB, RB, BW, BR, RW\}$;
    b) 9
    c) $\{\frac{16}{144}, \frac{25}{144}, \frac{9}{144}, \frac{20}{144}, \frac{12}{144}, \frac{15}{144}, \frac{12}{144}, \frac{15}{144}, \frac{20}{144}\}$;
    d) $\{WW, WR, RW, RR\}$; 4; $\{\frac{16}{81}, \frac{20}{81}, \frac{20}{81}, \frac{25}{81}\}$

4. $S = \{HHHH, HHHT, HHTH, HTHH, THHH, HHTT, THHT, HTHT,$
$THTH, TTHH, THHT, HTTT, THTT, TTHT, TTTH, TTTT\}$

$X = x$	0	1	2	3	4
$P(X = x)$	$\frac{1}{16}$	$\frac{4}{16}$	$\frac{6}{16}$	$\frac{4}{16}$	$\frac{1}{16}$

6. $\frac{1}{32}$

7. a) $\frac{1}{18}$; b) $\frac{43}{9}$; c) $\frac{95}{81}$

8. a) $c = \frac{1}{12}$; b) (i) $\frac{1}{2}$, (ii) $\frac{23}{24}$, (iii) $\frac{1}{6}$;
   c) $F(0) = \frac{1}{12}$, $F(1) = \frac{3}{12}$, $F(2) = \frac{6}{12}$, $F(3) = \frac{10}{12}$, $F(4) = \frac{23}{24}$, $F(5) = 1$

9. a) $\frac{\binom{13}{x}\binom{39}{4-x}}{\binom{52}{4}}$, $x = 0, 1, 2, 3, 4$;

   b) $\frac{\binom{4}{x}\binom{48}{4-x}}{\binom{52}{4}}$, $x = 0, 1, 2, 3, 4$

11. $f(x) = (\frac{1}{2})^x$, $x = 1, 2, \ldots$ 12. $f(x) = (\frac{3}{4})^{x-1}(\frac{1}{4})$, $x = 1, 2, \ldots$

13. $f(x) = \frac{\binom{4}{x}\binom{8}{3-x}}{\binom{12}{3}}$, $x = 0, 1, 2, 3$

14. $x = 0, 1, 2, 3, 4, 5$; 36

15. $f(x) = \frac{1}{10}$, $x = 1, 2, \ldots, 10$

16. $f(x) = \binom{5}{x}(\frac{1}{2})^5$, $x = 0, 1, 2, 3, 4, 5$

17. a) $f(x) = \frac{\binom{3}{x}\binom{5}{2-x}}{\binom{8}{2}}$, $x = 0, 1, 2$; b) $F(x) = \sum_{u=0}^{x} \frac{\binom{3}{u}\binom{5}{2-u}}{\binom{8}{2}}$

18. a) $\frac{1}{2}$; b) $\frac{7}{8}$; c) $\frac{3}{8}$; d) 0

19. a) $\frac{5}{16}$; b) $\frac{1}{16}$; c) .2775; d) 1

20. a) $\frac{1}{2}$; b) $\frac{1}{8}$; c) $\frac{3}{4}$

21. a) $f_1 x_1^2 + f_2 x_2^2 + \ldots + f_{10} x_{10}^2$; b) $x_6^2 + x_7^2 + x_8^2 + x_9^2$;
    c) $(y_1(y_1 - 2) + y_2(y_2 - 2) + y_3(y_3 - 2) + y_4(y_4 - 2))^2$

22. a) $\sum_{i=1}^{5} x_i$; b) $\sum_{i=1}^{N} f_i x_i$; c) $\sum_{i=1}^{5} x_i^2$;
    d) $\sum_{i=1}^{4} (x_i - 2)^2$; e) $\sum_{i=1}^{10} f_i(x_i - 4)^2$

23. a) 8; b) 0; c) 26

24. 2   25. 1   26. 1

27. | $X = x$ | 2 | 3 | 4 | 5 | 6 | 7 | 8 | 9 | 10 | 11 | 12 |
|---|---|---|---|---|---|---|---|---|---|---|---|
| $f(x)$ | $\frac{1}{36}$ | $\frac{2}{36}$ | $\frac{3}{36}$ | $\frac{4}{36}$ | $\frac{5}{36}$ | $\frac{6}{36}$ | $\frac{5}{36}$ | $\frac{4}{36}$ | $\frac{3}{36}$ | $\frac{2}{36}$ | $\frac{1}{36}$ |

$E(X) = 7$

28. a) $\frac{17}{11}$; b) 2   29. $44

30. $E(Y) = 7$; $\text{Var}(Y) = 18$ **31.** 136

33. a) $\frac{1}{32}$; b) $\frac{2x^2 + 1}{32}$, $x = 0, 1, 2, 3$;

c) $\frac{2y^2 + 7}{16}$, $y = 0, 1$

34. a) $f(x) = x/6$, $x = 1, 2, 3$ b) $f(y) = y/6$, $y = 1, 2, 3$

35. a) $f(x) = (\frac{2}{3})^x(\frac{1}{3})^{1-x}$, $x = 0, 1$ b) $f(y) = (\frac{2}{3})^y(\frac{1}{3})^{1-y}$, $y = 0, 1$

36. $f(x, y) = \dfrac{\binom{13}{x}\binom{13}{y}\binom{26}{5-x-y}}{\binom{52}{6}}$, $x + y = 0, 1, 2, 3, 4, 5$

37. a) $\{HHHH, HHHT, HHTH, HTHH, THHH, HHTT, HTHT, THHT,$
$TTHH, THTH, HTTH, HTTT, TTTH, THTT, TTHT, TTTT\}$

b)

$Y=y$ \ $X=x$	0	1	2	3	4
0	$\frac{1}{16}$	$\frac{3}{16}$	$\frac{3}{16}$	$\frac{1}{16}$	0
1	0	$\frac{1}{16}$	$\frac{3}{16}$	$\frac{3}{16}$	$\frac{1}{16}$

38.

$X = x$	0	1	2	3	4
$f(x)$	$\frac{1}{16}$	$\frac{4}{16}$	$\frac{6}{16}$	$\frac{4}{16}$	$\frac{1}{16}$

;

$Y = y$	0	1
$f(y)$	$\frac{1}{2}$	$\frac{1}{2}$

39. a) 70%   b) 60%

40. 6   **41.** 51

42. a)

$X = x$	0	1	2	3
$f(x)$	.76	.12	.07	.05

;

b)

$Y = y$	0	1	2	3
$f(y)$	.90	.05	.03	.02

;

c) .41, .17, .58, .11;

d) No; $E(XY) \neq E(X)E(Y)$

43. 1   **44.** a) $\frac{12}{17}$;   b) $\frac{768}{2873}$

45. $\frac{6}{11}$ **46.** a) .6819;   b) .3211

47. a) 8;   b) 4

Chapter 4

1. $(\frac{1}{6})^x (\frac{5}{6})^{1-x}$, $x = 0, 1$ **4.** a) $\frac{20}{1296}$;   b) $\frac{150}{1296}$

5. a) 1;   b) 4;   c) 6

6. .7660   **7.** $p = \frac{1}{5}$, $q = \frac{4}{5}$, $n = 35$

8. a) .729;   b) .243;   c) .027;   d) .001

10. 40   **11.** a) .3669;   b) $n \geq 5$

12. $1 - (.9)^{10}$

13. a) $1 - (.9)^8$;   b) $(.9)^8$;   c) 5

14. $\frac{4!}{6^4}$ 15. $\frac{6!}{6^6}$ 16. $.192$

17. $.0972$ 18. $.0576$

19. a) $.1036$;   b) $.0386$

20. a) $.1029$;   b) $.343$

21. $.0127$ 22. $.0132$ 23. $\frac{16}{243}$

24. 36 25. $.0864$ 26. $.0528$

27. $.0272$ 28. $\frac{255}{256}$ 29. $.0000003$

30. $.2223$ 31. $(.8)^{10}$ 32. $\frac{462}{924}$

33. $\frac{\binom{6}{3}\binom{6}{2}}{\binom{12}{5}}$ 34. $.7038$ 35. $.0965$

36. a) $\frac{1}{70}$;   b) $\frac{53}{70}$

37. a) $\frac{\binom{4}{2}\binom{48}{2}}{\binom{52}{4}}$;   b) $1 - \frac{\binom{48}{4}}{\binom{52}{4}}$

38. $\sum_{x=0}^{3} \frac{\binom{24}{x}\binom{126}{15-x}}{\binom{150}{15}}$ 39. $.0758$ 40. $.1255$

41. 5 42. a) $.0498$;   b) $.8008$

43. a) $.0183$;   b) $.4334$

44. a) $.0842$;   b) $.1353$;   c) $\frac{e^{-s}s^k}{k!}$

45. a) $.0025$;   b) $.9975$

Chapter 5

1. a) $.2580$;   b) $.1915$;   c) $.7119$;
   d) $.2638$;   e) $.8643$;   f) $.1151$

2. a) 1.30;   b) $.70$;   c) $-.70$;   d) $.70$

3. a) $.0250$;   b) $.9974$;   c) $.1587$;
   d) $.6876$;   e) $.0026$

4. a) $.0228$;   b) $.4773$

5. a) 12.90;   b) -6.90

6. a) 53;   b) 3;   c) 192

7. 86.50 8. $.9546$ 9. 61.24

10. a) 53;   b) 61.0;   c) $64.68 - 85.32$

11. $\mu = 49.93$, $\sigma = 11.69$ 12. $\mu = 73.2$, $\sigma = 7.14$ 13. 31.72%

14. $\sigma = 6.06$ 15. 68 years   16. 2.14%

346 Answers to exercises

17. 366   18. 881   19. 86.44%
20. $\mu = 3$, $\sigma = 5$
21. a) 1.96;   b) 2.57;   c) 1.65
22. a) .5;   b) .84;   c) .0013
23. a) .0062;   b) .4938;   c) .1056
24. $\mu = 140$, $\sigma = 1$ 25. .68   26. .9838
27. a) .9099;   b) .9713
28. a) .2563;   b) .0034;   c) .2578
29. .0019   30. .0002
31. a) .0346;   b) .4364;   c) .0805
32. .0202

Chapter 6

4. a) continuous;   b) discrete;   c) continuous;
     d) discrete;   e) discrete;   f) continuous;
     g) discrete

5. a) continuous;   b) discrete;   c) continuous;
     d) discrete;   e) continuous;   f) discrete;
     g) continuous;   h) continuous;   i) continuous;
     j) discrete

Chapter 7

7. a) 6.5;   b) 6.25;   c) 6.054;
     d) 5.84;   e) no mode   f) 4;
     g) 2.9166;   h) 1.705;   i) 1.25
14. 78
15. a) 12.22;   b) 12.29;   c) 12.95
16. $Q_1 = 57.50$; $Q_3 = 74.54$ 17. Mean = 68.57; s.d. = 3.83
18. 83.50   19. 58.38
20. Mean = 123.33; standard deviation = 2.98
21. Mean = 52.76; standard deviation = 8.53
22. 75.4   23. 3.162   24. 2, 8
25. 13.48   27. 119.54   28. 5.76
29. 53.95

Chapter 8

4. Normal, 7, .2   5. $\overline{X}$ 8. σ^2/n

9. Both are unbiased; $\dfrac{\sigma^2}{n} = \text{Var}(\overline{X}) < \text{Var}\left(\dfrac{X_1 + X_2}{2}\right) = \dfrac{\sigma^2}{2}$

10. No
11. .1056, .0001, .00, .00
12. .9430
13. .0143
14. $N\mu, N\sigma^2$
15. .0008
16. 11
17. 10, 17.5
18. $(N-1)\sigma^2/N$
19. $(-1.41, 4.13); (-2.29, 5.01)$
20. (.57, 2.15); (.04, 2.68)
21. (56.58, 63.42)
22. (.1652, .3348); (.17, .34) from table
23. (.286, .414)
24. (4.80, 18.39)
25. (.103, 9.27)
26. (.0208, .2792)
27. a) (.4931, .5069); b) (.4909, .5091)
28. a) (148.25, 151.75); b) (147.27, 152.73)
29. a) (7.62, 12.38); b) (6.85, 13.15)
30. (.0003, .0393)
31. a) (.5069, .6931); b) (.4891, .7109)
32. (a) (.0789, .1877)
33. (.0540, .2660)
34. a) (1445.87, 5332.88); b) (1231.14, 6940.38)
35. a) 544; b) 1331
36. a) (.1648, .4333); b) (.1396, .5527)
37. (.152, 1.341)
38. (95.593, 161.682)
39. (.355, 6.30)
40. 165

Chapter 9

1. No
2. a) .8612; b) .1388
3. Reject the hypothesis if $z > 1.65$ or $z < -1.65$
4. Yes, $z = -2$
5. Yes, $z = 3$
6. New diet is better, $t = 5.07$
7. .018, 4, .217
8. a) $\alpha = .1330, \beta = .4114$; b) $\alpha = .264, \beta = .6778$
9. Course is of value, $z = 2.776$
10. .936, .984, .997, .999, 1.000
11. Yes, $z = 4.381$
12. No, $t = 1.6556$
13. Accept claim, $t = -.62$
14. The program is not effective, $t = 1.19$
15. $\sigma^2 > 10, \chi^2 = 25.15$
16. Accept $H_0 : \sigma^2 = 10, \chi^2 = 10.38$
17. Yes, $\chi^2 = 11.11$
18. Yes, $\chi^2 = 81.89$
19. Yes, $z = 1.79$
20. No increase, $t = 2.13$
21. No difference, $t = -.071$
22. Means equal, $t = -2.58$
23. Variances equal, $F = .70$
24. Variances differ, $F = .089$
25. Yes, $z = -2.98$
26. Reject the hypothesis, $t = 33.33$
27. No difference, $t = 1.70$

348   Answers to exercises

28. a) Variances equal, $F = 1.83$; b) Means equal, $t = 1.18$
29. Variances unequal, $F = .084$
30. No decrease, $\chi^2 = 4.66$
31. There is a difference, $z = 3.86$
32. $\mu \neq 1550$, $z = -2.250$
33. Suspicion confirmed, $z = 3.45$
34. No difference, $z = 1.78$
35. Reject claim, $z = 4.083$
36. Campaign has failed, $z = 5.24$
37. .0008, .9999

Chapter 10
1. Coin biased, $\chi^2 = 9$
2. No, $\chi^2 = 1.696$
3. No, $\chi^2 = 21.33$
4. $f(x) = \binom{6}{x}(.59)^{6-x}(.49)^x$, x = 0, 1, 2, 3, 4, 5, 6; the fit is good
5. $e^{-.8}(.8)^x/x!$; the fit is good
6. Die not biased, $\chi^2 = 9.87$
7. Accept claim, $\chi^2 = 3.84$
8. Reject hypothesis, $\chi^2 = 27.31$
9. $\chi^2 = 53.8$; air pollution affects lung cancer
10. Reject hypothesis, $\chi^2 = 9.21$
11. The serum has an effect, $\chi^2 = 6.653$
12. Reject hypothesis, $\chi^2 = 25.63$
13. Reject hypothesis, $\chi^2 = 10.367$
16. Yes, $\chi^2 = 8.261$
17. Serum has an effect, $\chi^2 = 4.811$
18. No, $\chi^2 = 1.96$
19. Yes, $\chi^2 = 8.163$
20. No, $\chi^2 = 2.587$

Chapter 11
1. a) $\hat{y} = 54.39 + .525x$; b) 83.26
2. a) $\hat{y} = 79.72 + 1.19x$; b) 146.36
3. a) $\hat{y} = 31.80 + 2.28x$;
   b) $t = -.81$, accept $\alpha = -25$;
   c) $t = -1.19$, accept $\beta = 2.5$
4. $\hat{y} = -1.83 + .16x$; $t = 16.30$, reject $\beta = 0$
5. b) $\hat{y} = 4414.67 + 48.40x$; c) 6834.65;
   d) yes; e) $t = -.161$, accept $\beta = 50$;
   f) $r = .87$; g) $.53 < \rho < .97$
6. $\hat{y} = -1.13 + 7.54x$
7. a) $\hat{x} = 12.20 + 6.08y$; b) $r = .98$; c) $.89 < \rho < .99$
8. a) $r = .94$; b) $\hat{y} = .074 + .03x$; $\hat{x} = .567 + 59.06y$
9. $t = 59.65$, reject $\beta = 0$
10. $t = 17.53$ reject $\alpha = 60$
11. $50.53 < \alpha < 108.91$; $.6248 < \beta < 1.7552$
12. $-4.36 < \alpha < 2.10$; $6.82 < \beta < 8.26$
13. $t = -8.77$, reject $\beta = 10$
14. $t = 1.49$, accept $\alpha = -3$
15. $t = 3.00$, yes
16. $.16 < \rho < .62$

17. $z = 6.82$, reject $\rho = .70$
19. $N = 24$
22. $r = .8$, $\bar{x} = 22.5$, $\bar{y} = -21.0$
23. $z = .943$, no

18. $N = 95$
20. $.21 < \rho < .65$

24. $z = -1.75$, no

Chapter 12

1.

Source of variation	d.f.	S.S.	M.S.	F
Variety	2	340.17	170.08	11.24
Error	9	136.5	15.17	
Total	11	476.67		

Significant difference between the varieties

2.

Source of variation	d.f.	S.S.	M.S.	F
Feeds	3	10764.5	3588.17	74.69
Error	12	576.5	48.07	
Total	15	11341.0		

Significant difference between the feeds

3. a)

Source of variation	d.f.	S.S.	M.S.	F
Groups	2	26.83	13.42	<1
Error	9	289.42	32.16	
Total	11	316.25		

b)

Group	Mean
1	16.00
2	16.25
3	12.67

c) No significant difference between the group means

4.

Source of variation	d.f.	S.S.	M.S.	F
River	2	3.54	1.77	4.18
Error	12	5.08	.42	
Total	14	8.62		

Mean pollution levels of the three rivers not equal

Answers to exercises

5. | Source of variation | d.f. | S.S. | M.S. | F |
    |---|---|---|---|---|
    | Feed | 3 | 2093.88 | 697.96 | 6.64 |
    | Error | 13 | 1366.35 | 105.10 | |
    | Total | 16 | 3460.23 | | |

    Significant difference between the weight gains of pigs on four different types of feed

6. | Source of variation | d.f. | S.S. | M.S. | F |
    |---|---|---|---|---|
    | District | 2 | 165.64 | 82.82 | <1 |
    | Error | 13 | 2762.8 | 212.52 | |
    | Total | 15 | 2928.44 | | |

    Mean I.Q. in each district is the same

7. | Source of variation | d.f. | S.S. | M.S. | F |
    |---|---|---|---|---|
    | Training | 2 | 4.80 | 2.40 | 9.8 |
    | Error | 21 | 5.14 | .24 | |
    | Total | 23 | 9.94 | | |

    Significant effect of training

9. | Source of variation | d.f. | S.S. | M.S. | F |
    |---|---|---|---|---|
    | Drugs | 2 | .348 | .174 | 4.81 |
    | Error | 10 | .362 | .0362 | |
    | Total | 12 | .71 | | |

    Yes, there is a difference

10. $B = .62$; variances equal

13. If these three rivers are of interest, treat the factor as fixed; if these three are a sample from a larger population, treat the factor as random

14. Fixed   15. Fixed   16. Log transformation

17. $B = .0034$: variances equal

18. | Source of variation | d.f. | S.S. | M.S. | F |
    |---|---|---|---|---|
    | District | 2 | 159.6 | 79.8 | <1 |
    | Error | 12 | 2758.0 | 229.8 | |
    | Total | 14 | 2917.6 | | |

19. Variety: A   C   B
    Mean: 39.5   48.25   52.25

20. River   C   A   B
    Mean: 1.42   1.88   2.60

21. $B = .287$; variances equal

22. $B = .56$; variances equal

23.

Source of variation	d.f.	S.S.	M.S.	F
Firm	2	36.4	18.2	3.309
Error	12	66.0	5.5	
Total	14	102.4		

No significant difference between firms

24.

Source of variation	d.f.	S.S.	M.S.	F
Types of apple	2	98.38	49.19	6.49
Error	18	136.28	7.57	
Total	20	234.66		

Significant difference between the mean yields of different types of apples

25.

Source of variation	d.f.	S.S.	M.S.	F
Variety	3	1497.8	499.26	19.22
Locations	4	694.7	173.67	6.68
Error	12	311.7	25.975	
Total	19	2504.2		

a) Significant difference between the varieties
b) Significant difference between the locations

26.

Source of variation	d.f.	S.S.	M.S.	F
Feeds	2	5379.16	2689.58	86.08
Litters	3	625.00	208.33	6.66
Error	6	187.50	31.25	
Total	11	6191.66		

a) Significant difference between feeds
b) Litters have some effect on the gain in weight

27.

Source of variation	d.f.	S.S.	M.S.	F
Machines	3	312.50	104.16	9.72
Employees	3	266.00	88.66	8.27
Errors	9	96.50	10.72	
Total	15	675.00		

a) Machines affect production
b) Employees produce at different rates

Chapter 13

1. $r = 5$; accept the null hypothesis that the defectives are occurring at random
2. $r = 6$; accept the null hypothesis that the fluctuation from one bolt to another is random
3. $r = 7$; accept the null hypothesis that the defectives are occurring at random
4. $T(+) = 17.5$ or 102.5
5. Accept H_0
6. Accept H_0
7. $T(1) = 100.5$ or 130.5; accept H_0
8. $T(1) = 91$ or 185; accept H_0
9. $T(1) = 73.5$ or 97.5; accept H_0
10. $T(1) = 67$ or 86; accept H_0
11. $T(1) = 48$ or 57; accept H_0
12. $T(+) = 22$ or 33; laboratories are the same
13. $H = 5.58$; the mean pollution level of the three rivers is equal
14. $H = 9.48$; accept the hypothesis that four different types of feed are equally effective
15. $H = .87$; mean I.Q. in each district is the same
16. $H = 11.76$; significant effect of training
17. $S = 1.4$; no difference between laboratories
18. $S = 2.3$; no difference between laboratories
19. $r_s = .5875$
20. $r_s = .5758$
21. $r_s = .4286$

INDEX

INDEX

A

Acceptance region (*see* Hypothesis testing), 195, 213
Alpha, α, error (*see* Error, type I), 195, 213
Alpha, α, level (*see* Level of significance)
Alternate hypothesis, one-sided, two-sided (*see* Hypothesis)
Analysis of variance (ANOVA), 264, 283
  notation in, 265–266, 274
  table, 270
  use
    in multiple comparisons, 272–273
    of F-statistics in, 270
    with two factors, 273–277
    with fixed and random effects, 277–280
Approximation, of binomial (*see* Normal)
Approximation, Welch (*see* Welch)
Area under a probability density function (*see* Probability)
Arithmetic mean (*see* Location)
Associative laws (*see* Sets)
Average, 55

B

Bartlett's test (*see* Hypothesis testing), 280–281
  use of χ^2 in, 281
Bayes' theorem, 31–34
Behrens-Fisher, 181
Bernoulli distribution (*see* Distributions)
Best fit, line of (*see* Regression), 239

Beta, β, error (*see* Error, type II), 195
Binomial distribution (*see* Distributions)

C

Center (*see* Location)
Central limit theorem, 165, 188
Characteristic of a population, 125
Chi-square, distribution, 175–178, 188
Chi-square statistic, use of:
  in Bartlett's test, 281
  in confidence intervals for normal population standard deviation and variance, 175–178, 188
  in test of location, 298, 300
  in testing
    categorical, qualitative or enumerative data, 219–231
    goodness of fit of, 223–226; binomial, 223–224; Poisson, 224–225; Normal, 225–226
    independence of criteria of classification, 227–228
    standard deviation and variance of normal, 206–207
Class frequency, 137, 157
Class interval, 136, 157
Class limits, 136
Class mark, 136, 157
Class width, 137
Commutative laws (*see* Sets)
Complement (*see* Sets)
Complement laws (*see* Sets)
Composite hypothesis (*see* Hypothesis)
Confidence, 3, 4
Confidence coefficient, 170

355

Confidence interval, 170
Confidence interval compared to hypothesis testing, 205, 213
Confidence level (*see* Confidence coefficient)
Contingency table, definitions, degrees of freedom of, 226–231
Continuity, Yates' correction for, 228
Continuous random variable (*see* Random variable)
Correlation, 253–256
Correlation coefficient, 254
  properties of, 255, 256
  significance of
    use of normal statistics in tests of, 257
    use of t-statistics in tests of, 257
Counting techniques
  addition principle, 17–18
  combinations, 21–23
  multiplication principle, 18
  permutations, 19–21
Critical region (*see* Hypothesis testing)
Cumulative frequency, 143–144, 157

D

Data, bivariate, 236
Data, organization and analysis of, 135–157
Degrees of freedom, 172–173, 175, 183, 221, 227, 246
De Moivre, 106
Destructive sampling (*see* Sampling)
Determination, coefficient of, 255
Discrete random variable (*see* Random variable)
Dispersion, measures of, 151–157
  comparison of, 157
  deciles, 157
  mean absolute deviation (m.a.d.), 155
  percentiles, 157
  properties of sample variance, 153
  quartiles, 155–157
  sample standard deviation, 152, 157
  sample variance, s^2, 151–155, 157
  special case, pooled sample variance, 154
Distributions, discrete probability, 76–101
  Bernoulli, 76–77
  binomial, 77–82
  geometric, 85–88
  hypergeometric, 91–94
  multinomial, 82–84
  negative binomial, 88–91
  Poisson, 94–98
  uniform, 98–100
Distributive laws (*see* Sets)
Duncan, 272
Duncan multiple range test, 272–273

E

Error, experimental, 7
Error, type I, 194–213
  probability of (α), 195
Error, type II, 194, 213
Estimation, 162–189
Estimation, interval, 168–188
  of difference of two binomial parameters (*see* Normal distribution), 187–188
  of difference of two normal means, paired data, 182–183, 188
  of difference of two normal means, variances known (*see* Normal distribution), 178–179, 188
  of difference of two normal means, variances unknown, equal (*see* t-distribution), 179–181, 188
  of difference of two normal means, variance unknown, unequal (*see* t-distribution), 181–182, 188
  of intercept of regression line (*see* t-distribution), 249–250
  of μ, the Poisson parameter, 188
  of p in binomial probability function (*see* Normal distribution), 185–187, 188
  of population mean, variance known (*see* Normal distribution), 168–172, 188
  of population mean, variance unknown (*see* t-distribution), 172–175, 188
  of population variance and standard deviation (*see* χ^2), 175–178, 188
  of ratio of two normal variances (*see* F-distribution), 183–185, 188
  of slope of regression line (*see* t-distribution, 251–252

Estimation, point, 166–168
    point estimator, 166
        criteria for choosing between unbiased point estimators, 168
        minimum variance unbiased point estimators, 168
        of the slope and intercept, of the population regression line, 239–240, 243
        unbiased, 167
Event, 16
Events, mutually exclusive, 17
Expectation of a product of random variables and independence, 61
Expected value, properties of, 58–63, 69
    of a finite sum of random variables, 61
    of a function of a random variable, 59
    of a product of independent random variables, 61
    special case, 60
    of a sum of two random variables, 60
Expected value, of a discrete random variable, 55
Expected value, of a random variable, 55–58, 59
Experiment, 14–17

F

F-distribution, 183–185, 188
    analysis of variance, F-test percentage points; degrees of freedom, 183, 184
F-statistic, use of
    in ANOVA, 270
    in confidence intervals for the ratio of two normal variances, 184–185
    in tests of the ratio of two normal variances, 210–211
Factor, fixed, random, 277
Factorial (*see* Counting), 20
Finite set (*see* Set)
Fisher, R. A. (*see* Behrens-Fisher), 1, 257
Fit, goodness of (*see* Hypothesis testing, χ^2)
Frequency, relative, 25
Frequency tables, 136–144, 157
    conditions on construction of, 138
    cumulative, 143–144, 157
    graphical representation of, 141–143
    polygon, 141–142, 157
    problems of, 137
Friedman's test (*see* Location, tests of)

G

Games of chance (*see* Chance)
Gaussian distribution (*see* Normal)
Geometric distribution (*see* Distributions, discrete probability)
Geometric mean (*see* Location)
Goodness of fit tests (*see* Hypothesis testing, χ^2)
Gosset, W. S., 172

H

Harmonic mean (*see* Location)
Histogram, 141–142, 157
Hypergeometric distribution (*see* Distributions, discrete probability)
Hypothesis, simple, 193
    alternate, 194, 213
    composite, 193
    null, 194, 213
Hypothesis testing, 193–213
    acceptance region, 195, 213
    α, level of significance, 195, 213
    β, type II error, 195
    comparison with confidence intervals, 205, 213
    critical region, 195, 213
    one-sided *vs* two-sided test, 196
    power, 195, 213
    test of
        the difference of two binomial parameters (*see* Normal), 212–213
        difference of two normal means, 207–208: variance known (*see* Normal), 207–208; variance unknown, equal (*see* t), 208; variance unknown, unequal (*see* t), 209; paired data, variance known (*see* Normal), 209; paired data, variance unknown (*see* t), 210

358 Index

equality of variances, 280–283
goodness of fit (*see* χ^2), 223–226
independence of criteria of classification (*see* χ^2), 226–231
the intercept of regression line (*see* t), 248–249
mean of normal population, variance known (*see* Normal), 198–203
mean of normal population, variance unknown (*see* t), 203–205
of categorical, qualitative, or enumerative data (*see* χ^2), 223–226
the parameter of a binomial distribution (*see* Normal), 211–212
the ratio of two normal variances (*see* F), 210–211
the slope of regression line (*see* t), 250
use of χ^2 in, 281
variance and standard deviation of normal population (*see* χ^2), 206–207
test statistic, 196
tests of, simple hypothesis, 194–213
    error, type I, 194–213; type II, 194, 213

I

Identity laws (*see* Sets)
Independence, 30–31, 34, 55, 61
    tests of (*see* Hypothesis testing, χ^2)
Inference, 8
Infinite set (*see* Sets)
Intersection (*see* Sets)
Interval, confidence, 170
Interval estimation (*see* Estimation)

J

Joint probability function (*see* Random variable), 53, 69
Joint probability table, 53

K

Keuls, 272
Kruskal-Wallis test (*see* Location, tests of)

L

Least squares, method of (*see* Regression), 238
Level of significance (*see* Hypothesis testing)
Line of best fit (*see* Regression), 230–240, 242, 243
Location, measures of (center), 144–151
  arithmetic mean, 144–147, 157
  comparison of, 150–151, 161 (#20)
  geometric mean, 149–150, 157
  grouped observations, 145
  harmonic mean, 150
  median, 147–149, 157
  mode, 149, 157
  of a linear function, 145
  overall mean of two samples using, 147
  sum of deviations of observations from, 146
  ungrouped observations, 144–145
Location, tests of, 292–300
  K independent samples, Kruskall-Wallis test, 289, 298–299
  K related samples, Friedman's test, 289, 299–301
  Mann-Whitney test, 296
  paired data, Wilcoxon's test for, 289, 292, 296
  two independent samples, Wilcoxon's rank sum test, 289, 296–297

M

Mann-Whitney test (*see* Location, tests of)
Mean, arithmetic, geometric, harmonic (*see* Location)
Mean, estimation of (*see* Estimation)
Mean, sample, 163–166, 188
  properties of, 163–166, 188
    expectation of, 163
    distribution of, 163–164
    distribution of difference of, 165–166
Mean, tests of (*see* Hypothesis testing)
Mean absolute deviation (m.a.d.) (*see* Dispersion)

Measures of
  dispersion (*see* Dispersion)
  location (center) (*see* Location)
Median (*see* Location)
Mendel, G. J., 5
Mode (*see* Location)
Mu, μ, population mean, 56
Multinomial distribution (*see* Distributions, discrete probability)
Multiple linear regression (*see* Regression), 252–253
Multiplication principle (*see* Counting techniques)
Mutually exclusive events (*see* Events)
Mutually exclusive sets (*see* Sets)

N

Negative binomial distribution (*see* Distributions, discrete probability)
Newman, 272
Newmans-Keuls multiple-range test, 272
Nonparametric statistics, 289–305
Normal distribution, the, 106–120
  distribution of
    difference of two independent normal random variables, 108
    a linear function of a normal random variable, 108
    a sum of independent normal random variables, 108
  effect of σ, the standard deviation on, 107
  probability as area under, 106
  properties, 107–108
Normal distribution, the standard, 108–116
  as an approximation to the binomial distribution, 116–119
  probability as area under, 109
  probability density function, 109
  to a standard normal variable, 109
  used to evaluate normal probabilities, 110–116
Normal equations (*see* Regression), 239–240, 243, 253
Normal random variable, 106
  probability density of, 106
Normal variable, standard (z), 108–116
  use in confidence intervals for
    the difference of two binomial parameters, 187
    difference of two normal means, 178–183
    p, in the binomial probability function, 186
    paired data, 182
    population mean, variance known, 168–172, 188
    population mean, variance unknown, 178–179, 188
  use in tests of
    the difference of two binomial parameters, 212
    difference of two normal means, variance known, 207
    p, in the binomial probability function, 212
    paired observations, 209
    population mean, variance known, 198–203
    significance of correlation coefficient, 257
Null hypothesis (*see* Hypothesis)
Null set (*see* Set)

O

Observation, 6
One-sided tests, 198, 213
Operations on sets (*see* Sets)

P

Parameter, 125
Particulate inheritance, 6
Pearson, K., 1
Permutation, 19–21
Point estimator (*see* Estimator)
Pooled sample variance (*see* Variance)
Population, 124
Power, of a test, 195, 213
Prediction equation (*see* Line of best fit, regression)
Probability, 1, 7, 8, 23–34
  frequency definition of, 25
  laws of, 26–28
    as area under a probability density function, 107
    complementary events, 27

union of n mutually exclusive events, 27
union of two events, 26
union of two mutually exclusive events, 27
properties of, 24
Probability, conditional, 28–30

Q

Quartile (*see* Dispersion, measures of)

R

Rand corporation, 131
Random number tables, 130–132
Random variable, 40–42, 69
  continuous, 41–42, 45–48, 69
    probability density function, 46
  discrete, 40, 69
    distribution of, 42–45, 69
    distribution function of, 43
  two-dimensional, 51–55
    joint probability function, 53, 69
    marginal probability function, 54, 69
Randomness, tests of, the runs test, 289, 290–292
Rank correlation, 301–305
Regression, linear, 238, 244
  error, 238
  least squares, 239
  normal equations (coefficients a and b), 239, 243
  prediction (regression) equation
    y on x, 238
    x on y, 243
Regression, multiple linear, 252–253
Relative frequency (*see* Frequency)
Run, 290
  length of, 290

S

Sample, 124–132
  randomly selected sample, 129–132
  selection of, 127–130
  stratified random sample, 132
Sampling, 8, 126–127
  destructive, 126–127
  reasons, for, 126–127
Sets, 9–10
  equality of, 10
  finite, infinite, 9, 34
  null, 10
  operations on, 11–14
  subsets, 10
  universal, 10
    associative laws, 14
    commutative laws, 13
    complement, 13
    complement laws, 13, 14
    De Morgan's laws, 14
    disjoint, 11, 14
    distributive laws, 14
    identity laws, 13
    intersection, 11
    mutually exclusive, 11
    union, 12, 14
Slope coefficient, β (*see* Regression)
Standard error of estimate, 247
Summation, 48–51
  properties of, 50–51
Sums of squares, in estimation of variance, 244–247
  partitioning of, 246
  ANOVA, 266, 267, 275
  properties of, 245

T

t, student's, distribution, degrees of freedom, 172
t-statistic
  use in confidence intervals, 173–175, 203–205
  use in correlation, 257
  use in regression, 249
Tally, 139
Test of hypothesis (*see* Hypothesis testing)
Two-sided tests of hypothesis (*see* Hypothesis testing)

U

Unbiasedness (*see* Estimation)
Uniform distribution (*see* Distributions, discrete probability)

Union (*see* Sets)
Universal set (*see* Sets)

V

Variable, 6
Variance, estimation of (*see* Estimation)
Variance, σ^2 (*see* Dispersion), 63–66, 69
    properties of, 67–69, 70
        a constant times a random variable, 67
        a finite sum of independent random variables, 69
        a random variable plus a constant, 67
        a sum of two independent random variables, 68
    standard deviation, 64
    in terms of expected values, 64

Variance, sample, properties of, 166, 188
Variance stabilizing transformations, 282
Variation, 7
    uncontrollable effects of, 1
Venn diagram, 11

W

Welch, 209
Welch approximation, 181
Wilcoxon's test for paired data (*see* Location, tests of)
Wilcoxon rank sum test (*see* Location, tests of)

Y

Yates (*see* Continuity)